Palgrave Studies in Water Governance:
Policy and Practice

Series Editors
Christian Bréthaut
Institute for Environmental Sciences
University of Geneva
Geneva, Switzerland

Thomas Bolognesi
University of Geneva
Geneva, Switzerland

Looking at the issues of water governance through the perspective of the social sciences, books in the Palgrave Series in Water Governance take a global perspective on one of the key challenges facing society today: the sustainable development of water resources and services for all. In stepping away from the traditional focus on engineering and geophysics, the series takes a more holistic approach to both consolidate and generate knowledge that can be applied to different geographic areas by academics, researchers, policy-makers, NGOs and the private sector. This series emphasises the link between science and policy through considering water as a socio-ecological system, water and the territoriality of action, and water in the context of conflicts.

More information about this series at
http://www.palgrave.com/gp/series/15054

Simon Porcher • Stéphane Saussier
Editors

Facing the Challenges of Water Governance

palgrave
macmillan

Editors
Simon Porcher
IAE Paris - Sorbonne Business School
University of Paris 1 Panthéon-Sorbonne
Paris, France

Stéphane Saussier
IAE Paris - Sorbonne Business School
University of Paris 1 Panthéon-Sorbonne
Paris, France

Palgrave Studies in Water Governance: Policy and Practice
ISBN 978-3-319-98514-5 ISBN 978-3-319-98515-2 (eBook)
https://doi.org/10.1007/978-3-319-98515-2

Library of Congress Control Number: 2018952460

Cover illustration: Trevor Pearson / Alamy Stock Photo

This Palgrave Macmillan imprint is published by the registered company Springer Nature Switzerland AG
The registered company address is: Gewerbestrasse 11, 6330 Cham, Switzerland

Contents

Notes on Contributors

Jaime Melo Baptista is principal researcher at the National Laboratory of Civil Engineering (LNEC) in Portugal. He is a project leader at the Lisbon International Centre for Water (LIS-Water). His expertise lies in water supply and wastewater services—in relation to public policies, governance, regulation, quality of service, and performance assessment. He has served as the head of the Hydraulics Department (1990–2000) and the Sanitary Hydraulics Division (1984–1989) of LNEC. He was a member of the board of directors of the International Water Association (IWA) from 2012 to 2016. He was the chairperson of the Portuguese Water and Waste Services Regulatory Authority (ERSAR) from 2003 to 2015, regulating 500 utilities.

Lorenzo Bardelli is director of the Water Systems Directorate of the Italian Regulatory Authority for Energy, Networks and Environment (ARERA) and Adjunct Professor of Regulatory Economics at the University of Milan. Before joining ARERA in 2012, Lorenzo was director of the Utilitatis Foundation, whose aim is the promotion of researches and studies in public utilities sectors, and a member of organizations focused on services of general interests and on the form of public interventions in the economy (CEEP, CEDEC, CIRIEC, Federutility, Federambiente, Confservizi, Fonservizi). His interests include design of allocative mechanisms for efficiency and equity improvement in public utilities.

Salvador Bertoméu-Sánchez is a PhD candidate at the European Center for Advanced Research in Economics and Statistics (ECARES), Université libre de

Bruxelles, Belgium. He graduated as an economist from the Solvay Brussels School of Economics and Management (ULB). His fields of research are public economics, especially public infrastructure services and their regulation, and political economy. He has worked as an external consultant for the World Bank and the Inter-American Development Bank (IDB) on projects related to the water and sanitation sectors and their regulation in developing countries. He has published scientific articles in *Transport Policy* and *Utilities Policy* academic journals.

Astrid Cullmann is a senior researcher at the German Institute for Economic Research (DIW Berlin) and a senior lecturer in the Department of Econometrics and Statistics, Faculty of Economics and Management at Technische Universität Berlin. Her current research focuses on public service provision in sectors of general interest, innovation economics, advanced structural production function models, and a variety of econometric techniques.

Antonio Estache is Professor of Economics at the Université libre de Bruxelles, Belgium, and a member of ECARES in Brussels. He also advises governments and international organizations on infrastructure regulation, privatization, environmental regulation, public sector management, and tax reforms. His work in these fields has been published extensively, including in the *Journal of Public Economics*, *Journal of Regulatory Economics*, *Journal of Development Economics*, *Journal of Economic Literature*, *World Development*, and *World Bank Economic Review*.

María A. García-Valiñas is an associate professor (tenured) at the University of Oviedo, Spain. She is a member of the *Oviedo Efficiency Group* (Department of Economics). Her main research interests are in the fields of public economics and water economics. She has collaborated with several public institutions, such as the Spanish Ministry of Environmental Issues and the River Basin Authorities, in order to implement the economic analysis required by the EU Water Framework Directive (WFD). Moreover, she has assessed water demand, pricing, and regulatory issues. Her work has appeared in peer-reviewed journals (more than 35 papers) and prestigious editorials (9 book chapters).

Olivia Jensen is a political economist specializing in water and international development. Since 2014, she has been a senior research fellow at the Institute of Water Policy, Lee Kuan Yew School of Public Policy, National University of Singapore. The focus of her research is urban water governance and regulation. She has experience in consultancy and media as well as in academia and has led projects for public and private clients including the World Bank and the

European Commission. She has lived and worked throughout Asia, including in China, Japan, and India.

Michael Klien is a researcher at the Austrian Institute of Economic Research (WIFO). His research centers on the organization and performance of public services, ranging from organizational studies to political economy. In addition to a number of scientific publications in peer-reviewed international journals, Michael has provided consultation in a range of related areas—to, for example, World Bank and OECD. He has worked extensively in the area of water services with a focus on utility performance, efficiency measurement, and utility aggregations.

Emanuele Lobina is a principal lecturer with the Public Services International Research Unit (PSIRU), Business Faculty, University of Greenwich, United Kingdom. Emanuele joined PSIRU in 1998 to further their research and policy work on water service reform and has written extensively on the topic. Taking the urban water sector in developed, transition, and developing countries as reference, his research revolves around the relative efficiency of public and private enterprise, and the policy process as networked governance. From September 2017 to August 2018, Emanuele was an ISRF (Independent Social Research Foundation) Political Economy Research Fellow (http://www.isrf.org/about/fellows-and-projects/emanuele-lobina).

A. Wren Montgomery is an assistant professor at the Ivey Business School, Western University, Canada, and a visiting research scholar at the University of Michigan's Erb Institute for Business Sustainability (United States). Montgomery's research focuses on emerging pressures on water resources and institutions including tensions over ownership, corporate water management, and access and affordability challenges. She also studies environmental entrepreneurship, cross-sectoral collaboration, and corporate greenwashing. Prior to academia, she worked as a government relations and management consultant as well as a policy analyst at senior levels of government.

Simon Porcher is Associate Professor of Management and vice-director of the Economics of Public-Private Partnerships Chair at Sorbonne Business School, IAE Paris, Université Paris 1 Panthéon-Sorbonne, France. He was a fellow, lecturer, and then visiting lecturer (2014–2017) at the London School of Economics; an invited professor at the Institute of Water Policy of the Lee Kuan Yew School of Public Policy, National University of Singapore (2018); and a

Robert Schuman Fellow (2013) at the European University Institute. He holds a PhD from the Sorbonne Business School and graduated from the Paris School of Economics and Sciences Po. During his PhD, he has specialized in the study of privatization of water public services in France and has then extended his research to other countries including the United Kingdom and Chile and has ongoing works on China, Germany, the Netherlands, and Portugal. His research has been published in the *Journal of Public Administration Research and Theory*, *Energy Economics*, and the *Journal of Cleaner Production*, among others. He is an associate editor of *M@n@gement*, the French top management journal.

Julia Rechlitz is a researcher in the Department of Energy, Transportation, Environment at DIW Berlin. Her research focuses on applied microeconomics and energy and water economics.

Maria Salvetti is an independent consultant who has worked for the World Bank at several occasions since 2014. She also collaborates with the Economics of Public-Private Partnerships research chair at the Sorbonne Business School since 2013. Within the framework of H2020 program, she presently coordinates, for the Sorbonne Business School and with Portuguese and English partners, the implementation of an international center of excellence on water based in Lisbon. As the former chief economist for the French National Agency for Water and Aquatic Environments (Onema), Salvetti produced the first report on water services performance which was presented during the World Water Forum in 2012. She has also worked on several environmental and socio-economic studies required by the European WFD. From 2008 to 2010, she worked for the largest water service in Europe (SEDIF) providing economic and financial advice in the preparation of the service management contract. She has also worked as a principal economist for Seine Normandy Water Agency (France) and the Environment Agency (England and Wales).

Stéphane Saussier is Professor of Economics and Management at the Sorbonne Business School, France, where he is running a research group working on the Economics of Public-Private Partnerships. He is also director of the water research area at the Florence School of Regulation, Italy. Specializing in issues of water governance and public-private partnerships (PPPs), he regularly contributes to reports for international institutions such as the OECD and the European Parliament. His work has been widely published (more than 60 articles and book chapters)—many in international journals.

Tomás Serebrisky is the principal economic advisor of the Infrastructure Department of the IDB. Prior to joining the IDB, Serebrisky was the chief economist of the Antitrust Commission in Argentina, a visiting professor at Universidad Torcuato Di Tella, and a senior economist in the World Bank. His areas of expertise are the economics of infrastructure investment, PPPs, economic regulation, and antitrust. His work has been published extensively in refereed journals. He is the author of the book *Airport Economics in Latin America: Benchmarking, Regulation and Pricing*. He holds a PhD in Economics from the University of Chicago and a BA in Economics from Universidad de San Andres.

Caroline Stiel is a researcher at DIW Berlin. Her research focuses on firm performance and innovation, with a particular emphasis on the role of firms' organizational structures. She has conducted several empirical studies on firm behavior in the German energy and water sector.

Takuya Urakami is a professor at the Faculty of Business Administration, Kindai University, Osaka, Japan. His main areas of research interest are performance measurement in water and sewerage utilities as well as optimal industry structure of these sectors. He has served as council committee member of the central and local governments in the field of water supply and sewerage.

Paola Valbonesi is a Professor of Economics at the University of Padova and a research professor at the Higher School of Economics, National Research University, Russian Federation. She obtained her Laurea in Political Science at the University of Padova and MA and PhD in Economics at the European University Institute (EUI), Florence. Her main research interests focus on the design of efficient rules and economic incentives, primarily where government intervention plays a role. She has investigated the design of regulated sectors (energy, water, tlc, motorway, university), procurement contracts, and auction formats; and the effects of state aids to business and to foster innovation.

Dan Zhao is a PhD candidate in Business Economics at the Ross School of Business, University of Michigan, Ann Arbor, United States. Zhao's research on water focuses on topics related to the quality of drinking water in the United States. In particular, he has studied how an amalgam of factors including different paradigms of interaction between government regulators and water systems, governance forms of water systems, and pressure from local stakeholders work together to shape the behavior of a water system and its regulator. He also studies the drivers of corporate social responsibility.

List of Figures

List of Tables

List of Boxes

1

Introduction

Facing the Challenges of Water Governance: Moving from the Public Versus Private Debate to Other (At Least) Equally Important Governance Issues

Simon Porcher and Stéphane Saussier

Good water management is essential for human development, people's well-being, social inclusion, and environmental sustainability. The Sustainable Development Goals of the United Nations set forward an ambitious agenda of providing universal access to good-quality water supply and sanitation services within a financially constrained environment. However, the various peculiarities of each country regarding water governance make it difficult to identify and implement the best practices and benchmarks. Responses should be adapted to territorial specificities, in a case-by-case methodology, but at the same time, information should flow more easily and research on water governance topics should make progress and potentially shed light on inefficiencies, failures, and poor

S. Porcher • S. Saussier (✉)
IAE Paris - Sorbonne Business School, University of Paris 1 Panthéon-Sorbonne, Paris, France
e-mail: Simon.Porcher@univ-paris1.fr; saussier@univ-paris1.fr

S. Porcher, S. Saussier (eds.), *Facing the Challenges of Water Governance*,
Palgrave Studies in Water Governance: Policy and Practice,
https://doi.org/10.1007/978-3-319-98515-2_1

practices that may help to identify what are the future challenges associated with water governance and potential avenues to overcome them.

When it comes to water governance issues, the private versus public management question stays at the core of the public debate, for ideological reasons but also for practical ones; when there is a need for investments, the temptation to look for private finance is natural. It pushes to the background other issues which are at least equally important, such as the proper scale to operate water services, stakeholders' engagement, or the need for independent regulatory agencies.

In this introduction, we will first focus on the public versus private debate (Sect. 1) reminding what are the main trade-offs at stakes and what we learn from empirical studies (Sect. 2). We will then argue that empirical studies are not conclusive and review other issues which are at least equally important to face water challenges (Sect. 3).

1 The Big Question: Public Versus Private Management of Water Services

Many countries face the double challenge of growing demand and aging physical assets. This is not specific to developing countries. As noted by Lyon et al. (2018), aging infrastructure is causing water rates to soar in the United States and is driving poorer Americans to pay significant amounts of their income for water and wastewater services, and often to default. It also fosters poor quality and water contamination. Such underinvestment issues are also present in European countries like France or the United Kingdom (see Chap. 7) as well as in Japan (see Chap. 13) and many other parts in the world.

The need for infrastructure investment worldwide in the coming decades is significant and, in many countries, far beyond the government's capacity. Global infrastructure investment needs are estimated to be approximately US$50 trillion for roads, water, electricity, telecommunications, and rail in OECD countries between 2005 and 2030 (OECD 2012). Global financing needs for water infrastructure range from US$6.7 trillion by 2030 to US$22.6 trillion by 2050 (OECD 2015a).

As a consequence, there is an important role for private sector participation in funding the development of these essential services. Public financial constraints have been translated by a changing role of the government itself. Moving from its own production to delegation and externalization, the public sector has shifted its focus from addressing internal bureaucracy to managing relations with external partners, and the public sector currently favors private participation through public-private partnerships (PPPs) to seek more efficient uses of increasingly limited resources.

However, at the same time, we can observe many failed PPPs (Estache 2006), especially in the water sector (Guasch et al. 2014). These failures attest the difficult challenges that face policy makers. Infrastructure investment involves contracts that are complex and that operate under the double imperative of ensuring financial sustainability and meeting user needs and social objectives, and this type of investment is often also very exposed to public opinion and political scrutiny (Spiller 2008).

To understand better this back-and-forth motion and why this love/hate relationship is often encountered in regard to public-private relationships—the history of partnerships between the public and private sectors to provide public services goes as far back as the history of the public sector itself—let us start by defining what PPPs are and their expected advantages and drawbacks.

PPPs for Water Services

There is no single well-defined "type" of PPP but rather various types that differ depending on whether the contract is global (i.e., is bundling investment needs and the public service provision) or simple, whether payment is made upon delivery or deferred, and whether the operator is remunerated mostly based on the service operating results or, on the contrary, on its ability to meet the performance objectives that are described in the contract (Box 1.1). The "landscape" of PPPs is thus a complex one that comprises various subcategories that range from traditional public procurement contracts, user-pay PPPs (concessions), and public-budget pay

Box 1.1 PPPs in Water Services

The most common PPP type in the water sector is the concession contract in which the operator manages the service, invests in the network, and obtains financial compensation through consumer receipts. In such contractual agreements, public authorities transfer some risks (especially part of the demand risk) that are supported by private companies in exchange for greater decision rights and claims on revenues. If infrastructures pre-exist, investments are essentially maintenance costs (a "light investment concession"). If infrastructures do not pre-exist, concession contracts are usually very long-term agreements that increase their complexity.

Other contractual agreements are possible, such as a management contract in which the operator is paid a fixed fee in exchange for the obligation to perform ancillary services: for example, the operation and maintenance of water and sewage facilities, the provision of technical assistance, and the collection of charges on behalf of the public authorities. In these types of contracts, a small part of the operator's revenues may depend on performance. In management contracts, operators do not assume the risks of the cost of operation and maintenance or of financing improvements. The risk of the operator is to be able to achieve and maintain the service standards.

The following table summarizes the different forms of private sector participation in water supply and provides examples.

Table 1.1 The different forms of private sector participation

Form	Ownership	Financing	Operations	Examples
"Light" concession	Public	Public	Private	Cote d'Ivoire, France
Concession	Public	Private	Private	Used in Buenos Aires-Argentina, Jakarta
Service contract	Public	Public	Public then private	Mexico City, Santiago-Chile, Madras
Management contract	Public	Public	Private	Cartagena-Colombia, Johannesburg, Mali
Build-operate-transfer	Private then public	Private	Private	Izmit-Turkey, Mendoza-Argentina
Sale of full divestiture	Private	Private	Private	England and Wales

PPPs (availability contracts, which are most of the private finance initiatives or PFIs). Worse still, there are many possible variants within each group of PPPs. For instance, some concession contracts provide for risk-sharing mechanisms that can take the form of profit sharing above some threshold or revenue compensation in the case of underperformance. This type of risk-sharing mechanism, for example, is in place in the French city of Dijon's water management contract (Saussier and de Brux 2018). This mechanism enables risk mitigation while still benefiting from private sector efficiency but at a lower cost. As noted in Saussier and de Brux (2018), such evolutions of user-pay PPP contracts indicate the parties' acknowledgment that a significant counter-performance most often is due to exogenous factors for which the operator cannot reasonably be held responsible alone. This evolution brings user-pay PPP contracts (concession contracts) closer to availability-based contracts (public-budget pay PPPs), which creates a continuum of public-private contracts (Estache and Saussier 2014).

Although it is difficult to analyze PPPs as discrete and alternative forms of public service organization, they all constitute a somewhat partial outsourcing of activities that contribute to the realization of a public service. However, it is essential to emphasize that economic theory sees PPPs as having advantages and drawbacks that are associated not only with the outsourcing issue but also with the public-private nature of the relationship that makes these contracts so particular and much more difficult to manage than private-private relationships.

The Promise of PPPs for Water Services

A search for expertise—The first reason that is identified by the economic literature to justify outsourcing is a lack of in-house expertise. For the same reason that private companies might outsource part of their activities, public authorities see outsourcing as an alternative to the integration of a partner or the development of skills, two processes that can take a long time and generate costly irreversibility. Outsourcing is an opportunity for public authorities to focus on their "core business", namely, the supervision of public services rather than their provision. The provision of public services can be realized at a lower cost by private operators, which are experts that benefit from economies of scale, experience, and scope. This deficit in the expertise of the public party compared with that of private operators depends on the size of the public body, as well as the complexity of the services in question.

Economies of scale—The reason that is most often advanced to justify outsourcing is the search for economies of scale. One can easily understand that the in-house provision of public services (i.e., direct public management—which is the alternative to externalization, whatever its form) does not enable the same economies of scale as PPPs. Operators that are present in several markets can realize economies of scale, which is not the case for public authorities, as they only operate in a single market, unless the optimal output level—beyond which the average costs increase—is low or the public body is sufficiently large to be able to realize economies of scale itself. Therefore, these advantages of contracting out are stronger when the value of the investments to be established is substantial (especially in the case of the construction/renovation of infrastructure). The existence of such economies of scale can also explain municipalities' wish to manage their public services within organizations for intercommunal cooperation and also why some countries push for more centralization and less fragmentation of their water services (see the case of Italy in Chap. 3, Eastern European countries in Chap. 10, and Japan in Chap. 13). Finally, because private operators usually operate at a very large scale to benefit from economies of scale, the sectors that relate to the provision of public services often suffer from high degrees of concentration. The existence of a competitive price is thus not assured.

Several studies have attempted to estimate the optimal size of a water service depending on scale economies. For example, Mizutani and Urakami (2001) found that in Japan, a water supply organization of optimal size would supply a population of approximately 766,000 people, whereas Marques and De Witte (2011) estimated that the optimal scale of the water utilities in Portugal is located between 160,000 and 180,000 inhabitants. In both cases, this size is far greater than the size of many French and Spanish water services where PPPs are widely used.

Economies of experience—Some activities allow companies to benefit from economies that result from the accumulated experience of their employees and the organizational routines that they have established by coping with and overcoming the problems that they have often encountered throughout their history. This experience effect allows significant improvements to be introduced in the processes on which outsourcing is based, and moreover, it generally leads to a reduction in operating costs.

Economies of scope—Although it can be appealing to outsource a service or the production of goods to a specialized firm that operates simultaneously in several markets and benefits from economies of scale, it can also be efficient to conclude a contract with an operator that supplies multiple different goods or services and therefore benefit from economies of scope.

The literature on the economies of scope in the water sector is relatively scarce and inconclusive. Most often, these studies conclude that there are no or very few scope economies in the water sector, even if there might be an advantage in terms of negotiation power to outsource to one operator several services (e.g., water distribution and water sewage; see Marques and De Witte 2011; Desrieux et al. 2013).

The search for incentives—In addition to the technical reasons that were previously mentioned to provide an understanding of the advantages of outsourcing, other reasons that relate more to management issues are also alluded to in the economic literature. These scholars justify the use of outsourcing because it is a situation (1) where competition can fully play its part, (2) where management incentive practices are easier to establish, and (3) where risks are more likely to be shared between the public authority and the operator.

Outsourcing and competition—When a public authority decides to outsource a public service, it must choose a partner among a certain number of potential suppliers. For example, if a public authority decides to open a project to competition by organizing a call for tenders, then outsourcing is considered to guarantee a certain level of cost control that is more difficult to reach with in-house provisions, because internal services are generally not put in competition with potential external contractors. When performed properly, the opening to competition that is realized as part of the outsourcing process thus forces potential partners to disclose information regarding their costs by offering a price.

This is an important point, especially because it is raised systematically in the debate on PPP efficiency. With in-house provisions, it is not possible to enjoy the benefits of competition since such direct management does not compel the public body to organize competitive calls for tenders or to conduct a preliminary assessment to justify this organizational choice. In contrast, when outsourcing is the chosen solution, several operators are likely to compete in a call for tenders for a given contract. Therefore, adopting a direct management method deprives public authorities of the competitive pressures that are at play in the markets.

Incentives and management—Internal organization facilitates the control of production activities, whereas outsourcing—although it makes such control more difficult—presents the advantage of increasing the level of incentives for the operator that is in charge of supplying the goods or services. Because relationships within an organization are based on an employment contract, which establishes a relation of subordination that leaves little room for incentives (even if incentive wages can be established), these relationships do not encourage partners to be efficient to the same extent as market relations. Indeed, in the case of outsourcing, the relationship is based on a contract that can include strong incentives by describing the expected service (i.e., a higher degree of precision as to the service to be provided) and by introducing a range of incentive clauses (bonuses and penalties) that allow the operator to keep all the additional revenue that it can generate by being efficient.

Risk sharing—Finally, outsourcing can also present advantages regarding risk sharing. Outsourcing the production or the management of a service makes it possible to transfer some risks from the public party to the private party. Depending on the public procurement tool, this transfer can be focused on production risks, demand-related risks, and risks that are associated with operating costs (see Table 1.1). From this perspective, the advantage of outsourcing lies in the greater ability of operators to diversify their risks (because of their level and the potential diversity of their activities) and their expertise in managing different types of risks. Consequently, operators bear lower costs than a public entity when coping with risks.

Advantages of outsourcing the provision of services to a private partner—Aside from economic advantages, which are common to every transaction, the economic literature also identifies a range of additional reasons that justify the use of outsourcing and that are only relevant when studying the management of public services. These reasons mostly concern the difficulty of controlling and managing public organizations, as well as the risk of political interference that characterizes this difficulty. Outsourcing is then viewed as a way to reduce or eliminate the disadvantages that are specific to the in-house provision of public services.

Difficulties in controlling and managing public organizations—Organization theory has long since identified a limit that is specific to large companies and collectives: the delegation of decision-making power. Although delegating becomes indispensable when an organization increases in size and diversifies its activities, it raises the issues of incentives or the control of the decision-maker. This is a key point of agency theory (Jensen and Meckling 1976; Laffont and Tirole 1993), and this point has always received special attention in economics. When discussing company managers to whom owning shareholders delegate their decision-making power, Adam Smith already observed that "the directors of such companies, however, being the managers rather of other people's money than of their own, it cannot well be expected that they should watch over it with the same anxious vigilance with which the partners in a private copartnery frequently watch over their own" (Smith 1776).

Although this problem is not exclusive to public organizations, it is accentuated in their case by the differences between a public and a private organization; a public organization is more difficult to control when decisions are delegated (especially from citizens to managers). According to Laffont (2000), these particularities concern the fact that public entities generally are subjected to several controls that involve various "controllers" with objectives that are potentially in conflict or even erratic with little credibility. Consequently, incentives are limited, and it is impossible for public organizations—if it is even desirable—to replicate the incentive rules of the private sector, which is characterized by more efficient corporate governance mechanisms and a single and inviolable objective: profit maximization. Outsourcing is then a way to rationalize the production of a public good or the provision of a public service by restricting the intended objectives to only economic performance while implementing more efficient control mechanisms.

Outsourcing as a way to reduce political interference—In addition to their low incentives, public organizations are more subject to political interference, which potentially diverts them even further from the pursuit of the economic objectives of performance. This point is often advanced in the economic literature (see, e.g., Boycko et al. 1996). Outsourcing is then viewed as a way to reduce political interference by rendering them more difficult for governments.

The economic analysis therefore emphasizes that outsourcing provides many advantages regarding productive efficiency and production cost reductions for the provision of a public service. However, this analysis is only a partial insight. Indeed, the advantages that are mentioned above generally can be obtained only by signing a contract, often a long-term contract, which allows private operators to obtain a return on investment or achieve performance targets. This is an important point because the signature on a contract leads to various implementation and performance difficulties for PPPs, such as transaction costs, renegotiations, and adverse effects on quality. The theoretical advantages that are associated with public-private contractual arrangements must therefore be weighed against the costs that they entail, which will now be outlined.

The Costs of PPPs

PPPs generally lead to long-term contracting. The contract forms the basis for the "partnership" between the public and private parties. The contracting process would be quite straightforward if it were possible to draw up complete contracts by considering all future contingencies and specifying an efficient means to address them in a way that is understandable to both the contracting party and third parties (e.g., law courts that are in charge of enforcing them). Unfortunately, in the case of long-term contracts, when the context is uncertain or complex and the economic operators are characterized by different levels of information or expertise, contracting becomes much more challenging. Several theoretical approaches focus on the contracting problems that are encountered in such agreements and the ways to minimize them. These approaches emphasize the importance of information asymmetries and risk sharing, the issue of contractual incompleteness, and the influence of the political dimension of PPP contracts. In the end, the relative efficiency of PPPs compared with the direct in-house provision of public services depends on the parties' ability to limit the risks that are associated with PPP contracting.

Incomplete contracting and opportunistic behaviors—There is a need for long-term contracting to secure investors when private parties must invest in specific investments (i.e., investments whose value would be totally or partially lost in case of a contract breach) to provide public services. This lock-in situation is the cause of several difficulties, and the most obvious of these difficulties is that the parties may behave in an opportunistic manner (Klein et al. 1978; Williamson 1985), in particular by making cost effort reductions that might degrade quality (which is often not perfectly contractible), by trying to renegotiate the initial contractual terms, or by failing to completely fulfill their obligations (see Table 1.2). Such behaviors are helped by the fact that long-term contracts are incomplete; every transaction is characterized by a certain degree of uncertainty, that is, by future hazards that cannot be anticipated contractually by the parties. This uncertainty may arise from the economic context (i.e., demand shocks, the introduction of product or process innovations, etc.), the regulatory environment (i.e., a modification of the existing rules or the introduction of new rules), and the complexity of the transaction itself. Although a long-term contract is an option that can be considered without too many difficulties in a stable environment, the circumstances are different in an environment that is characterized by high uncertainty. Consequently, contract design has critical importance in the case of outsourcing. Table 1.2 shows that renegotiations in the water sector are particularly frequent and occur rapidly: 92% of contracts are renegotiated and the first renegotiation

Table 1.2 Renegotiations in PPPs

Geographical area	Industry	% of renegotiated contracts	Average time of the first renegotiation	References
Latin America and the Caribbean	All industries	68	1.8 years	Guasch (2004)
	Electricity	41	2.1 years	
	Transportation	78	2.9 years	
	Water	92	1.3 years	
United States	Transportation	40	–	Engel et al. (2011)
France	Urban parking	74	3.8 years	Beuve et al. (2013)
France	Motorways	50	–	Athias and Saussier (2007)
United Kingdom	All industries	55	–	NAO (2001)

Source: Estache and Saussier (2014)

occurs after 1.3 years (Guasch 2004). In a recent update, Guasch et al. (2014) found that this is 87% of water contracts that are renegotiated, on average 0.8 year after their signature.

Many complex public goods or services are regularly contracted out by public entities under complex contractual arrangements that favor the continuity of the relationship and the management of potential conflicts (through safeguard clauses, periodic review clauses, etc.). The main challenge here is to generate trust by reaching two contradictory objectives: securing the relationship by getting a steadfast commitment from the parties while retaining a certain level of flexibility that is necessary to allow contractual adaptations to the economic, financial, and statutory environment during the implementation phase.

Contestability and contract rigidity—The contracting costs that are associated with outsourcing can also be larger in the case of public contracts because of the political interference that we previously mentioned. The specificities of PPPs make them intrinsically different from private-private contracts (i.e., contracts that are conducted between two private partners). Some recent theories (Moszoro and Spiller 2012; Moszoro et al. 2016; Spiller 2008) indicate the importance of third-party opportunism in public contracts. Indeed, opportunism can come not only from the contracting parties (i.e., the private operator and public authority) but also from third parties such as interest groups (i.e., consumer associations, lobbies, the political opponents of the decision-maker, the rival companies of the operator, or citizens). A PPP contract relates to the provision of a public service, implies the investment of public funds, and therefore concerns society as a whole. This interest of third parties in PPP contracts may prove to be useful, especially when it plays a supervisory role by preventing the public party from straying from the announced political agenda or by precluding private operators from renegotiating the terms of the initial agreement in an opportunistic way (McCubbins and Schwartz 1984). However, third parties may also hinder the running of PPPs when they seek to pursue their own interests instead of the general interest. For example, it can be in the interest of third parties to question the integrity of the public party in relation to a public-private contract when they are competing with it in a "political market" (Spiller 2008). This type of opportunistic behavior may have significant consequences on contracting costs. Indeed, to protect themselves from political contestability, both the public body and the private party may be tempted to design more rigid public contracts. The idea is that public contracts include more clauses than the contracts that are concluded between two private parties, because private parties do not have to protect against third-party opportunism. Thus, the public parties will generally draw more rigid contracts to avoid contract renegotiations that could be costly in terms of image if there is a high level of political competition within a municipality (Beuve et al. 2018). However, establishing rigid contracts presents two drawbacks that should be explained. First, such contracts are particularly costly. Indeed, drafting rigid contracts generates additional transaction costs, both for the public authority (i.e., *ex ante* information retrieval, the drawing up of detailed specifications, etc.) and for the private party (i.e., longer and more complex procedures to participate in calls for tenders), which may result in cost overruns that would negate the potential advantages of a rigid contract (i.e., the limitation of third-party opportunism). Second, such rigidity can hinder the ability of contracts to adapt to changes in the environment (i.e., statutory, technological, and financial changes, etc.), while these evolutions are unavoidable in the case of long-term contracts.

There are thus many potential costs to contract with private firms in the delivery of complex public services like water. This complexity might lead to differences in performance.

2 The Impact of PPPs on Performance: Evidence from Water Industries

If economic theory can identify the advantages and drawbacks of private sector participation, it is then an empirical question to determine their efficiency in one specific sector. Water services clearly range from easy to complex to manage in areas where institutions are sometimes weak with significant or few needs for infrastructure investments. This diversity is clear considering different countries; the magnitude of the advantages and drawbacks of contracting with private firms or government capabilities are not the same when comparing developing and industrialized countries. This diversity is also clear within a given country. For example, the quality and scarcity of water and density of consumers might differ greatly in the north compared with the south of Italy, which explains the different choices that are made by the local public authorities concerning the management of water services. Empirical studies that examine the efficiency of water services might be helpful to determine if PPPs are appropriate or not.

Water Prices

Even if a large part of the public discontent that is translated in newspaper articles focuses on prices, there are few empirical studies that have examined the effect of governance structures on water prices.

The different chapters in the book show that the literature has not found clear conclusions for the French, German, or Spanish cases. For example, in Spain (see Chap. 2), studies have detected higher price levels under private provision (Martínez-Espiñeira et al. 2009; García-Valiñas et al. 2013). Similarly, Bel et al. (2015) found that private firms with a larger market share set higher water prices. In Germany (see Chap. 6),

Ruester and Zschille (2010) investigated the impact of governance structure on water retail prices with a database of 765 German water suppliers. Controlling for scale economies, technical and structural characteristics, and endogeneity issues, they found that private sector participation leads to higher prices. It seems that there is at least a correlation between private sector participation and prices.

In France, Chong et al. (2015) use longitudinal data on municipalities for 1998, 2001, 2004, and 2008.[1] The authors analyze the influence of private management, controlling for considerable information, such as the contract duration, the date of signature, water treatment types, leaks, population density, and sewage management, and considering the endogeneity of the management choice (public vs. private). They do not find any price difference between public and private management in municipalities of more than 10,000 inhabitants. A price difference to the advantage of in-house provision remains for municipalities under 10,000 inhabitants, in the range of 8%. It is important to emphasize that such price gaps cannot be explained by differences in water quality, which is controlled for. This study is important because it is the only one, to our knowledge, that is based on a representative dataset (more than 70% of French consumers are included in the study) and using panel data. This basis allows the authors to compare prices between municipalities that are comparable but differ only in the way that the water services are managed. However, it also allows the authors to compare one municipality with itself by using fixed effects as soon as one municipality changed the way that its water services are provided over the period. In this way, the unobserved heterogeneity that is constant over time and that might explain the cross-sectional results is controlled for.

Other French studies that have examined governance choices and water prices have demonstrated that elements other than governance choices might be central to this relationship. For example, Porcher (2017)

[1] The data come from the water survey of the French Environment Institute (IFEN)/Observation and Statistics Department (SOeS). Five thousand French municipalities were surveyed in 1998, 2001, 2004, and 2008, which provided a representative panel of French water services. The survey was abandoned after 2008 and replaced with a French Observatory (Sispea) that was supposed to collect information from municipalities on every water service in France on a voluntary basis. The observatory exists, but despite multiple efforts to improve the situation, the data are far from representative and reliable.

quantified the impact of the choice of contracting out the management of water public services on price. He used a unique dataset of utilities with unusually detailed financial indicators, such as the debt of the water public service. He found evidence that private management is associated with higher prices on average, *ceteris paribus*, but that this difference disappears when accounting for the "hidden costs" of water, that is, the price considering the debt refunding of the public service that could increase the price in the following years. Indeed, private management is characterized by higher tariffs but a lower debt level so that the price ensures the full-cost recovery, while under public management, prices are set at a lower level than under private management but with a higher debt from the public service.

Accordingly, the results from the studies that examine the impact of governance choices on prices are mixed. This discrepancy suggests that the influence of governance structures (i.e., public vs. private management) is not central to the final prices that are paid by consumers. Other elements such as corporatization, productive efficiency, competition level, regulations, and contractual choices are probably at least equally important.

Technical Efficiency

Efficiency can be measured as cost efficiency (decreasing costs for a given level of outputs) or total factor productivity (producing more for a given level of inputs). In industrialized countries, there is no clear relation between ownership and efficiency. Bhattacharyya et al. (1995) suggest that in the United States, publicly owned water utilities are more efficient. They apply a translog variable cost function to the data of 221 US water utilities in 1992. Shih et al. (2004) find that public utilities have lower costs than private utilities. They apply Data Envelopment Analysis (DEA) to two datasets, each with more than 1000 observations of water suppliers that were obtained through the Community Water System Survey that was conducted by the US Environmental Protection Agency. Garcıa-Sanchez (2006), by using DEA with the data of 24 water utilities, cannot find any efficiency differences between publicly and privately owned companies in Spain. Stiel (2017), with a newly constructed and unique dataset from the German Federal Statistical Office, investigates the link between organizational innovation and productivity by focusing on three elements, namely, corporatization, outsourcing, and partial privatization. The data comprised 2325 German state-owned firms for energy and water supply between 2003 and 2014 (German energy and water firms are multiproduct firms). Performance is measured as total factor productivity that is derived from a translog production function. These scholars find that corporatization and outsourcing are positively correlated with productivity and that fully state-owned firms outperform the firms with private minority shareholders. This paper does not directly compare the efficiency of PPPs versus direct public management. However, it sheds light on the fact that many organizational arrangements are possible and that corporatization is already one step toward the benefits of PPPs (i.e., the reduction of political interference).

In developing countries, some studies find a slight positive impact of private ownership on company efficiency. Kirkpatrick et al. (2006) use DEA and Stochastic Frontier Analysis (SFA) to determine the impact of ownership structure on the efficiency performance of 110 water utilities in African countries. Higher relative efficiency is shown for privately owned utilities when using the DEA method, whereas no statistically significant result for the impact of ownership is found with SFA. Estache and Kouassi (2002) estimate a Cobb-Douglas production function for 21 African water utilities for the period of 1995 to 1997. In a second stage, they use a Tobit model to relate the resulting inefficiency scores to governance and ownership variables. Their results indicate that private ownership significantly decreases inefficiency. However, their dataset contains only three privatized firms, and corruption and governance seem far more important than the ownership variable in explaining the efficiency differences between firms. No significant differences between efficiency under public and private ownership are observed by Estache and Rossi (2002), who estimate a stochastic cost frontier model on the data from 50 water utilities in developing and transition countries in the Asian and Pacific region.

Instead of comparing public and private water utilities that operate at the same point in time, another body of work focuses on the impact of privatization on the efficiency and productivity of the sector, mostly in the United Kingdom. Saal and Parker (2000) study the privatization of water utilities in England and Wales in 1989. They expect privatization to improve efficiency on the premise that it removes soft-budget constraints, eliminates any political or special interest group interference that is associated with public ownership, exposes utilities to the market for corporate control, and incentivizes management and employees with performance-pay structures and the market for managerial talent. Using cost function and total factor productivity (TFP) analyses on a panel of ten private UK companies, the authors conclude that there is no statistically significant reduction in the trend growth rate of total costs following privatization with cost function analysis and no changes in productivity after privatization with TFP.

Service Quality

Few papers use service quality as a performance indicator. Aside from the rates of compliance to water quality, there are no real measures of service quality. A good measure of service quality, such as the rate of complaints to capture the feeling of the user, is usually not available or largely unfulfilled. A global measure of quality is computed by the World Bank, but it aggregates different indicator and is a computed score instead of a real indicator.[2] The issue of water quality is particularly important in developing countries where access to safe water can largely decrease the disability adjusted life years (see Chaps. 8 and 9). Another significant dimension of water services quality is network losses (see especially Chaps. 3, 5, and 12 on Italy, France, and the United States).

While water quality is usually a regulatory measure implemented by the government, a large literature looks at the potential impact of private sector participation on water quality. Studies in France show rather small differences or no differences between public and private management

[2] The water utility performance index (WUPI) of the World Bank is a simple global index measuring how closely a utility company operates to accepted good practices. The index is based on ten dimensions generally accepted as key performance indicators in the industry and available from the IBNET dataset for most utility companies in the region. Those dimensions fall into three categories: coverage (water, sewer, and wastewater treatment); quality (service continuity and sewer blockage); and management (metering, NRW, staff productivity, collection rate, and operating cost recovery).

(see Chap. 5). In a recent study, Lyon et al. (2018) present the first large-scale empirical study of the ownership-service quality relationship that examines both privatization and municipalization in the US drinking water supply. Using a panel of 65,025 water systems from 2006 to 2014 and combining matching and difference-in-differences methods to construct comparable counterfactual control groups, they find privatization consistently produces quality improvements, but municipalization does not. Leveraging the different financing structures of public versus private water systems, the authors also find that privatization is more beneficial in communities with more government financing challenges as represented by higher government borrowing costs. Despite being highly regulated, water quality can differ between two governance modes.

Coverage and Access for the Poor

Another important issue in the debate about private sector participation relates to its impact on coverage and access for the poor. Studies on coverage usually occur in developing countries because coverage is largely assured in industrialized countries. A telltale story is that privatization decreases coverage because firms want to maximize their profits and do not want to serve unprofitable households.

Privatization is often linked to price increases (which, in turn, are often linked to productivity increases or to the evolution of regulatory norms), so one usually can expect a negative impact of privatization on water affordability for the poor. Similarly, private firms might not have any incentives to invest in the network to connect poor consumers who might not be able to afford paying for the service. Such a result is found in Lee (2011) for Malaysia. However, it is difficult to disentangle the results of privatization from the results of pro-access or tariff regulation government policies. Clarke et al. (2009) find no impact of private sector participation on coverage rates, most likely because the governments that were supporting privatization were also encouraging increasing coverage rates.

An important outcome of coverage is the impact on health of users. A famous study by Galiani et al. (2005) on privatization in Argentina shows that privatization, via an increased access to drinkable water, decreases child mortality. Chapters 8, 9, and 11 (looking at Latin American countries, African countries, and China) discuss the health issues related to water provision which are crucial, specifically in developing countries.

Increasing block tariffs (IBTs) are popular because they supposedly protect the poor and are environmental conscious (see Chaps. 8 and 9 for discussion on Latin American and the Caribbean (LAC) countries and sub-Saharan Africa). IBTs reveal multiple challenges for regulators who have to decide on the number of blocks, the unit price in each block, and the volume of water associated with each block. Chapter 8 reveals that in most LAC capitals, the average price per cubic meter increases as a function of the consumption level. Nauges and Whittington (2017) and Whittington et al. (2015) show that IBTs perform poorly in targeting subsidies to the poor. These tariff schemes also introduce price distortions that create economic efficiency losses, though welfare losses are relatively low. Increasing block tariffs might have lower redistributive effects, because water consumption, although it is positively correlated with income, is largely dependent on the size of the household (Smets 2004; Porcher 2014). Vital consumption of water is the same for all individuals; therefore, price elasticities between the rich and the poor do not differ much. In this case, IBTs should be used to promote water conservation. In practice, however, social tariffs and increasing block tariffs are implemented together. Social tariffs or measures such as energy paychecks are efficient redistributive measures to decrease the price of water for the poor because they directly target households with low incomes (see Chap. 5 for a discussion on social tariffs in France).

3 Other Governance Factors Influencing the Performance of Water Services

Empirical studies do not give a definitive answer concerning the efficiency of PPPs compared to direct public management in the provision of public services. This should not come as a surprise. After all, theory suggests that there is not one specific type of management that is the most efficient in all locations. One of the main lessons is that many dimensions are driving water management efficiency of which private participation is only one (not independent) dimension and is not necessarily the most important dimension. Characteristics of the service, corporatization, the level of competition in the market, the quality of institutions, risk transfers, contractual choices, and the degree of decentralization are many other dimensions of the problem that are driving the performance of these services and are mentioned as crucial in chapters of this book.

Regulatory Agencies, Benchmarking, and Information

The regulatory and institutional framework matter heavily since the quality of contract enforceability and governance are critical factors that affect water governance as a whole. There are mainly three types of regulation found worldwide. In the Anglo-Saxon model, there is no direct competition between private providers. Yardstick competition benchmarks the different providers. In the French model, competition for the limited operation of local water supply monopolies takes place in the form of a bidding process. In contrast, in Germany, there are natural monopolies at the regional level. Public authorities influence the operation of municipal water networks through rights of ownership as the municipalities are involved in the supply companies. Information is critical to empower local governments and to run a proper benchmarking: in the Anglo-Saxon model, information is needed to benchmark utilities; in the French model, information asymmetry can lead to poor performance and monitoring; in the German model, information asymmetry can be smaller as municipalities are directly involved in the supply of water.

Designing an independent regulatory institution requires maintaining an appropriate balance between independence and accountability. Independence implies financial autonomy and protection from political interference. Accountability demands effective administrative appeals process and the creation of vehicles for consumer participation. Moreover, regulation agencies should have proper instruments to regulate the industry: such instruments (e.g., tariff or output regulation) were discussed above; others (e.g., regulatory information) are discussed in the chapters of the book (e.g., see Chaps. 2, 5, or 6 on Spain, France, and Germany).

Many voices advocate for regulatory agencies in order to strengthen governments' and companies' commitments. As noted in Chap. 9, the decision to create separate regulatory agencies for water and sanitation utilities in sub-Saharan African countries is one of the main innovations of the 1990s. The idea is that regulatory agencies would not only protect investors and private operators in risky political and economic environments but they would also protect consumers and taxpayers including subsidies as needed. Unfortunately, the authors also mentioned that when put in place; they were rarely given training and instruments to perform their mandate with competence and even lacked political support from the government. Moreover, in some cases, the government had political control over them, which raises concerns regarding autonomy and accountability (Estache 2006). In some other cases, regulators can be captured by private firms.

There are few international studies showing the impact of regulatory agencies on performance. Bertoméu et al. (2018), looking at more than 123 developing countries, suggest that, for developing countries, the creation of economic regulatory agencies, separated from the ministry in charge of water and sanitation, may not be a necessary or sufficient condition to stimulate large-scale private involvement in the sector.

Public Participation and Stakeholder Engagement

Stakeholder engagement is important in water because this is a highly decentralized and fragmented sector. Multiple, independent players co-exist at different levels. While traditional stakeholders—governments, utilities, business, river basin organizations, legislators, and so on—are well included in the management of water, some groups such as poor people or property developers are now getting a more important voice in the governance of public services (OECD 2015b). Governments are more aware of the need to engage different stakeholders in the management of public services to improve their quality.

Stakeholder engagement is discussed in some of the chapters of the book (e.g., Chap. 5 on France, Chap. 9 on Africa, Chap. 11 on China, and Chap. 12 on the United States). In Chap. 11 on China, Jensen shows that the level of transparency and the scope for public participation in China's water sector appear to be very limited. While public hearings are mandated as part of the tariff review process, these are not always held, the representatives selected to participate are restricted, and the level of information available to members of the public on performance is limited, making it impossible for the public to engage in the deliberations in a meaningful way. Stakeholders' engagement, in industrialized and developing countries, is a key issue.

4 Conclusion

This book provides a cutting-edge, comprehensive overview of the challenges facing water governance and regulatory choices. Drawing together empirical studies from different countries all around the world, each chapter reviews the individual challenges facing each country/geographical area, from the supervision of autonomous regulatory bodies to the question of centralization and the influence of local utility companies.

Throughout the book, it appears that water challenges are mainly governance ones. This has already been pointed out by OECD's work on water since a long time (Akhmouch et al. 2018). The OECD Principles on Water Governance (OECD 2015a) identified three mutually reinforcing and complementary dimensions of water governance—effectiveness, efficiency, and trust and engagement—and developed 12 principles on water governance around which indicators could be developed in order to follow the efficiency of water governance. The principles are related, among other things, to the division of responsibilities, stakeholder engagement, financing, data and information, and evaluation (see Fig. 1.1). One question still open is the extent to which those governance indicators are connected to each other's (complement or substitutes) and which ones are the most important ones.

This book illustrates the usefulness of those principles and, depending on the country or the geographical area that is studied, stresses which dimensions are a priority. If each country is unique, some common conclusions emerge concerning the importance of governance issues as well as the need for innovative governance choices in order to respond to the fast and accelerating urbanization process, changes in consumption, and demographic factors.

Fig. 1.1 The Water Governance Principles (Source: OECD 2015a)

References

Akhmouch, A., Clavreul, D., & Glas, P. (2018). Introducing the OECD Principles on Water Governance. *International Water, 43*, 5–12.

Athias, L., & Saussier, S. (2007). Contractual Flexibility or Rigidity for Public Private Partnerships? Theory and Evidence from Infrastructure Concession Contracts. *Revue Economique, 58*(3), 565–576.

Bel, G., González-Gómez, F., & Picazo-Tadeo, A. J. (2015). Does Market Concentration Affect Prices in the Urban Water Industry? *Environment and Planning C, 33*(6), 1546–1565.

Bertoméu, S., Camos, D., & Estache, A. (2018). Do Economic Regulatory Agencies Matter to Private-Sector Involvement in Water Utilities in Developing Countries. *Utilities Policy, 50*, 153–163.

Beuve, J., de Brux, J., & Saussier, S. (2013). Renégocier pour durer : une analyse empirique des contrats de concessions. *Revue d'économie industrielle, 141*(1), 117–148.

Beuve, J., Moszoro, M., & Saussier, S. (2018). Political Contestability and Public Contract Rigidity. *Journal of Economics and Management Strategy*, Forthcoming.

Bhattacharyya, A., Harris, T. R., Narayanan, R., et al. (1995). Specification and Estimation of the Effect of Ownership on the Economic Efficiency of the Water Utilities. *Regional Science and Urban Economics, 25*(6), 759–784.

Boycko, M., Shleifer, A., & Vishny, R. W. (1996). A Theory of Privatization. *The Economic Journal, 106*, 309–319.

Chong, E., Saussier, S., & Silverman, B. S. (2015). Water Under the Bridge: Determinants of Franchise Renewal in Water Provision. *Journal of Law, Economics and Organization, 31*(1), 3–39.

Clarke, G. R. G., Kosec, K., & Wallsten, S. (2009). Has Private Participation in Water and Sewerage Improved Coverage? Empirical Evidence from Latin America. *Journal of International Development, 21*(3), 327–361.

Desrieux, C., Eshien, C., & Saussier, S. (2013). Putting All One's Eggs in One Basket: Relational Contracts and the Provision of Local Public Services. *Journal of Economic Behavior & Organization, 89*, 167–186.

Engel, E., Fischer, R., & Galetovic, A. (2011). *Public-Private Partnerships to Revamp U.S. Infrastructure* (pp. 1–26). The Hamilton Project—Brookings.

Estache, A., & Kouassi, E. (2002). *Sector Organization, Governance, and the Inefficiency of African Water Utilities*. Policy Research Working Paper Series 2890. Washington, DC: The World Bank.

Estache, A., & Rossi, M. (2002). How Different Is the Efficiency of Public and Private Water Companies in Asia? *World Bank Economic Review, 16*, 139–148.

Estache, A., & Saussier, S. (2014). Public-Private Partnerships and Efficiency: A Short Assessment. *CESifo DICE Report, 12*(3), 8–13.

Estache, A. (2006). PPI Partnerships vs. PPI Divorces in LDCs. *Review of Industrial Organization, 29*(1), 3–26.

Galiani, S., Gertler, P., & Schargrodsky, E. (2005). Water for Life: The Impact of the Privatization of Water Services on Child Mortality. *Journal of Political Economy, 113*(1), 83–120.

Garcia-Sanchez, I. M. (2006). Efficiency Measurement in Spanish Local Government: The Case of Municipal Water Services. *Review of Policy Research, 23*(2), 355–371.

García-Valiñas, M. A., González Gómez, F., & Picazo-Tadeo, A. J. (2013). Is the Price of Water for Residential Use Related to Provider Ownership? Empirical Evidence from Spain. *Utilities Policy, 24*, 59–69.

Guasch, J. L., Benitez, D., Portabales, I., & Flor, L. (2014). *The Renegotiation of PPP Contracts: An Overview of Its Recent Evolution in Latin America*. International Transport Forum Discussion Papers 2014/18, OECD Publishing.

Guasch, J.-L. (2004). *Granting and Renegotiating Infrastructure Concession: Doing It Right*. Mimeo: The World Bank.

Jensen, M. C., & Meckling, W. H. (1976). Theory of the Firm: Managerial Behavior, Agency Costs and Ownership Structure. *Journal of Financial Economics, 3*(4), 305–360.

Kirkpatrick, C., Parker, D., & Zhang, Y.-F. (2006). State Versus Private Sector Provision of Water Services in Africa: An Empirical Analysis. *The World Bank Economic Review, 20*(1), 143–163.

Klein, B., Crawford, R. G., & Alchian, A. (1978). Vertical Integration, Appropriable Rents, and the Competitive Contracting Process. *The Journal of Law and Economics, 21*(2), 297–326.

Laffont, J.-J., & Tirole, J. (1993). *A Theory of Incentives in Procurement and Regulation*. Cambridge: MIT Press.

Laffont, J.-J. (2000). Etapes vers un Etat moderne : une analyse économique.

Lee, C. (2011). Privatization, Water Access and Affordability: Evidence from Malaysian Household Expenditure Data. *Economic Modelling, 28*(5), 2121–2128.

Lyon, T. P., Montgomery, A. W., & Zhao, D. (2018). *A Change Would Do You Good: Privatization, Municipalization and Drinking Water Quality*. Working Paper.

Marques, R., & De Witte, K. (2011). Is Big Better? On Scale and Scope Economies in the Portuguese Water Sector. *Economic Modelling, 28*(3), 1009–1016.

Martínez-Espiñeira, R., García-Valiñas, M. A., & González-Gómez, F. (2009). Does Private Management of Water Supply Services Really Increase Prices? An Empirical Analysis in Spain. *Urban Studies, 46*(4), 923–945.

McCubbins, M. D., & Schwartz, T. (1984). Congressional Oversight Overlooked: Police Patrols Versus Fire Alarms. *American Journal of Political Science, 28*(1), 165–179.

Mizutani, F., & Urakami, T. (2001). Identifying Network Density and Scale Economies for Japanese Water Supply Organizations. *Papers in Regional Science, 80*(2), 211–230.

Moszoro, M. W., & Spiller, P. T. (2012). Third-Party Opportunism and the Nature of Public Contracts. NBER w18636.

Moszoro, M. W., Spiller, P. T., & Stolorz, S. (2016). Rigidity of Public Contracts. *Journal of Empirical Legal Studies, 13*(3), 396–427.

National Audit Office. (2001). *Managing the Relationship to Secure a Successful Partnership in PFI Contracts.* Report by the Controller and Auditor General.

Nauges, C., & Whittington, D. (2017). Evaluating the Performance of Alternative Municipal Water Tariff Designs: Quantifying the Tradeoffs Between Equity, Economic Efficiency, and Cost Recovery. *World Development, 91*, 125–143.

OECD. (2012). Strategic Transport Infrastructure Needs to 2030. OECD Publishing. Retrieved from https://doi.org/10.1787/9789264114425-en.

OECD. (2015a). Principles on Water Governance. Retrieved from http://www.oecd.org/governance/regional-policy/OECD-Principles-on-Water-Governance.pdf.

OECD. (2015b). Stakeholder's Engagement for Inclusive Water Governance. Retrieved from https://read.oecd-ilibrary.org/governance/stakeholder-engagement-for-inclusive-water-governance_9789264231122-en.

Porcher, S. (2014). Efficiency and Equity in Two-Part Tariffs: The Case of Residential Water Rates. *Applied Economics, 46*(5), 539–555.

Porcher, S. (2017). The 'Hidden Costs' of Water Provision: New Evidence from the Relationship Between Contracting-Out and Price in French Water Public Services. *Utilities Policy, 48*, 166–175.

Ruester, S., & Zschille, M. (2010). The Impact of Governance Structure on Firm Performance: An Application to the German Water Distribution Sector. *Utilities Policy, 18*(3), 154–162.

Saal, D. S., & Parker, D. (2000). The Impact of Privatization and Regulation on the Water and Sewerage Industry in England and Wales: A Translog Cost Function Model. *Managerial and Decision Economics, 21*(6), 253–268.

Saussier, S., & de Brux, J. (2018). *The Economics of Public-Private Partnerships.* Springer International Publishing.

Shih, J.-S., Harrington, W., Pizer, W. A., & Gillingham, K. (2004). *Economies of Scale and Technical Efficiency in Community Water Systems.* Resources for the Future, Discussion Paper 04-15.

Smets, H. (2004). *La solidarité de l'eau potable: aspects économiques.* Paris: Report, Académie de l'eau.

Smith, A. (1776). *An Inquiry into the Nature and Causes of the Wealth of Nations.* London: Strahan and Cadell.

Spiller, P. T. (2008). *An Institutional Theory of Public Contracts: Regulatory Implications*. NBER Working Paper (14152).
Stiel, C. (2017, December). *Modern Public Enterprises: Organisational Innovation and Productivity*. Working Paper 1713, DIW Berlin.
Whittington, D., Nauges, C., Fuente, D., & Wu, X. (2015). A Diagnostic Tool for Estimating the Incidence of Subsidies Delivered by Water Utilities in Low-and Medium-Income Countries, with Illustrative Simulations. *Utilities Policy, 34*, 70–81.
Williamson, O. E. (1985). *The Economic Institutions of Capitalism*. The Free Press.

2

Spain

Water Governance in Spain: The Role of Federalism and Public-Private Partnerships

María A. García-Valiñas

1 Introduction

Water has been traditionally considered as a merit good that has significant implications in economic, social, and environmental terms (OECD 2003, 2013). This renewable natural resource is characterized by a strong heterogeneity in terms of spatial distribution. Water scarcity, stress, and quality problems are key issues that local, national, and supranational institutions have to deal with. Water management and governance schemes become strategic when designing public policies.

Spain is a European Union (EU) member country facing strong weather differences among regions. Both quality and availability of water experience broad variations depending on the area, and the country is subject to severe droughts (Lopez-Gunn et al. 2012) and quality problems (Willaarts et al. 2014). Moreover, it is a federal country, where sub-central governments actively design and develop public policies related to

M. A. García-Valiñas (✉)
Oviedo Efficiency Group, Universidad de Oviedo, Oviedo, Spain
e-mail: mariangv@uniovi.es

S. Porcher, S. Saussier (eds.), *Facing the Challenges of Water Governance*,
Palgrave Studies in Water Governance: Policy and Practice,
https://doi.org/10.1007/978-3-319-98515-2_2

water resources. However, decentralization is not always combined with coordination among different government levels, as explained in latter sections. As a result, economic and regulatory tools show a strong heterogeneity. Additionally, Public-Private Partnerships (PPPs) have emerged as a usual framework to manage water resources. The expanding trend of PPPs in the last decades has generated a public and academic discussion about the superiority of this kind of organizational schemes, which will be also addressed in this chapter.

The chapter is organized as follows. First of all, a comprehensive picture of water sector in Spain is presented. Next, some key issues in water governance in Spain are described, such as the river basin organizational framework, the decentralization of economic instruments, or citizens' participation channels. The fourth section includes a broad discussion on the role of PPP schemes in the water sector. Finally, the chapter finishes with a brief summary of the main conclusions and future challenges.

2 Water Sector in Spain: Outstanding Problems

Since ancient times, water services provision in Spain has been a difficult task to drive. Several pressures have been identified. On the one hand, Spain is the most semi-arid country in the EU (Lopez-Gunn et al. 2012), and droughts and other water scarcity issues frequently affect some regions of the country. Moreover, climate change emerges as a major threat for water management in Spain. An increase of the average temperatures up to +1.9 °C has been predicted for 2040 (CEDEX 2011), jointly with a reduction in rainfall (up to 6%). In some specific areas, such as Canary Islands and the southern basins, significant changes in rainfall patterns are expected (Garrido et al. 2013). Southern basins could experience reductions of up to 13% of surface water runoff and around 15% of aquifers' recharge, while the reductions at Eastern Mediterranean basins would be below 10%. Northern basins will also experience reductions, but their relative profuse resource availability will likely mitigate the impact (Garrido et al. 2013).

On the other hand, significant population increases have been registered in the last decades. Thus, population has extended from 38.8 million in 1990 to 46.6 million in 2016 (INE 2017a). Despite a negative population growth rate predicted for the next decades (INE 2017a), there are still a high number of people keeping the pressure up on the water supply and sewerage systems. Moreover, the slight GDP growth rate increase forecasted for the next years (IMF 2017) is an additional factor with impact on water resources management. Water is considered as an additional input in the production process (Arbués et al. 2010; Renzetti 2015). As a consequence, recovering economic activity could lead to intensify both industrial and agricultural water use and waste.

Pollution and consumption are significant key drivers of water quality. A good monitoring system to control the quality of water bodies emerges as an essential instrument for water governance. Under the enactment of the EU Water Framework Directive (WFD, 2000/60/EC), the surveillance programme for water bodies developed in Spain has allowed the creation of a broad network of monitoring stations. Willaarts et al. (2014) conclude that around 50% of surface water bodies in Spain are in poor ecological status. Moreover, saline intrusion and non-source pollution have been the two main sources lessening the quality of underground water bodies (Custodio 2014).

In spite of water availability and quality problems, water prices in Spain do not register a high level (Calatrava et al. 2015). Moreover, inefficient price schemes are still applied.[1] The basis for water pricing in Spain was set at the 1985 Water Law (Garrido and Calatrava 2009). Since then, only partial changes have been observed, the most important of which resulted from the transposition of the WFD into

[1] For example, the majority of irrigation water users' associations face prices based on land surface. Water tariff usually consists of a fixed amount per hectare. These pricing structures are intrinsically linked to overconsumption and efficiency losses. Additionally, some urban water tariffs include free allowances. A free allowance is a minimum water consumption that is charged at zero marginal price. Usually, when the user pays the fixed charge, he/she gets the right to consume a basic amount of water. The use of free allowances has been criticized because they lead to significant efficiency losses. In general, it has been proven that this kind of elements provides strong incentives to water overuse, being against water conservation (Castro et al. 2002), and not very equitable (OECD 2003).

Spanish national legislation. However, current water prices for different users in Spain have not achieved the goals established at the WFD, especially when looking at cost recovery and environmental issues (Calatrava et al. 2015).

Although water infrastructures have improved in the last decades, there is still an urgent need to invest in both urban wastewater treatment and water supply systems. On the one hand, average water leakage rate in Spain is around 25% (INE 2014). What is most worrying is that there are many cities where the water loss rate is much higher than the average rate, particularly in the south of Spain (González-Gómez et al. 2012).

On the other hand, Spain has not got a full compliance with the Council Directive 91/271/EEC of 21 May 1991 concerning urban wastewater treatment (Environment Agency Austria 2015). Aldaya and Llamas (2012) pointed out that many wastewater treatment plants do not operate correctly, and most that do only perform secondary treatment (not tertiary). The cost of restoring all these treatment plants and upgrading those with secondary treatment has been estimated at around €19 billion. That figure would amount to 4€ per m^3, which annualized in ten years would be about 0.45–0.5€ per m^3. This surcharge would represent an average increase of 20–40% of the urban tariff. This is not an extremely large increase but it could generate high political costs (Calatrava et al. 2015). On top of that, water reuse, closely linked to wastewater management, is a top-listed priority area in the Strategic Implementation Plan of the European Innovation Partnership on Water (EC 2012). However recycling water processes are far from being generalized. Actually, Spain shows by far the highest reuse potential, the calculations suggesting a value of over 1,200Mm^3/yr (EC 2014).

Definitively, water management in Spain is far to be a straightforward process. Moreover, it has not generated the expected outcomes. There are unfortunately still many issues that need to be resolved both in the short and long run. A mixture of scarce non-controllable inputs, overpressures on water resources, and a non-adequate design of institutional and regulatory framework are some of the reasons explaining the current situation of water resources in Spain.

3 Water Governance in Spain

Water governance is the set of rules, practices, and processes (formal and informal) through which decisions for the management of water resources and services are taken and implemented, stakeholders articulate their interest, and decision-makers are held accountable (OECD 2015). Regulatory issues emerge as a key factor of governance in the water sector. However, formal norms and regulation are not the only aspects that matter in this field. Other non-necessarily formal aspects that articulate coordination, participation, and information are also significant when attempting to improve water sector outcomes management. This section is especially devoted to describe the regulatory framework in Spain, with special attention to federalism and economic instruments.

An Overview

Spain is a federal country, whose bases are set at the Spanish Constitution (1978). The current vertical structure of government includes both central and sub-central levels. The Constitution recognizes explicitly the existence and right to self-governance of sub-central governments. Spanish institutional framework is composed of the central government (CG), 17 autonomous communities (regional governments, CCAA) and 2 autonomous cities at the intermediate level, and 50 provinces and 8,124 municipalities (MUN) at the local level (INE 2017b).

This framework makes water management more complex in terms of governance, since different levels of government develop several tasks related to water policies. As a consequence, coordination and negotiation processes emerge as substantial issues in this context. In the next subsection we will analyse the most significant topics related to water planning, economic and social aspects through the integrated water cycle.

Water Planning: River Basins

River basins are the basic managerial units set at the WFD (Arts. 3 and 13). These organisms are in charge of water planning and policy design strategies (OECD 2011). Basin-based governance unit has been defended as the best scale option that can be adopted to manage water resources. In this respect, some advantages such as the adaptation to physical or hydro-ecological environment have been remarked (Cohen 2012). However, there are some critical voices that claim higher flexibility and dynamism in demarcating the optimal hydrological unit (Moss 2012).

Spain has a long tradition of using the river basins as basic organizational and managerial units. The germ of the current Confederaciones Hidrográficas (CCHH) was found at the end of the nineteenth century. In 1865, some preliminary efforts were made in order to set a basin-based management organizational framework. During the Primo de Rivera Dictatorship (1923–1930), these units were endowed with legal and administrative contents, while in 1934 they were called using their current denomination (Del Moral Ituarte 2016). During the Franco Dictatorship (1939–1975), water policy was strongly centralized through these organisms. However, coinciding with the return of democracy, the Constitution of 1978, and the State of Autonomies, sub-central governments emerged as significant agents in the development of water policies in the country.

Actually, responsibilities and decisions on water policies are taken depending on the water body considered. On the one hand, there are inter-communitarian river basins, in charge of planning and managing water bodies that cross several CCAA. In this case, the responsibility of water management lies on the central state (Spanish Ministry of Environment). On the other hand, there are a few intra-communitarian water bodies, where the distribution of functions between the central state and each region is bilaterally negotiated and set in the so-called autonomy statutes. Those could be reformed upon the request of the regional government. In any case, regional governments are assuming higher responsibilities than in the case of inter-communitarian water bodies.

Since past many years, there has been a long discussion about the decentralization level that is more convenient in this kind of situations. Thus, fiscal federalism theory recommends central public services provision when it comes to control for externalities and/or spillover effects in contexts where cooperative management is not possible (Oates 2001; Banzhaf and Chupp 2012). However, several problems such as pollution, ecosystem degradation, and water pressures are still significant in the inter-communitarian water bodies (Thiel 2015). Moreover, central provision imposes uniform public policies, neglecting sub-central heterogeneity (Banzhaf and Chupp 2012).

In this context, more flexibility could be demanded, since the regional differences in Spain are remarkable. However, the Spanish Constitution and the European regulation strongly constrain the possibility of decentralization in this context. Actually, some attempts to extend the decentralization process on the CCHH have been aborted. That is the case of the Guadalquivir CH, in charge of an inter-communitarian river basin, where water bodies are shared by Andalusia, Extremadura, and Castilla-La Mancha territories.[2]

Economic Tools and Decentralization

The WFD sets requirements for member states to take account the principle of cost recovery of water services. In this respect, cost recovery should include both financial recovery of *operational and maintenance costs* (and ideally costs of capital renewal and extension as well) for water and sanitation services and also the recovery of the *environmental and resource costs* including costs related to water abstraction and costs induced by the pollution of water bodies. In this section, we will study the charges/taxes that have been designed to get the aim of full cost recovery, and which are commonly included in final users' water bills. We refer here to the water abstraction charge and the water pollution charge.

The Spanish system combines both central and sub-central economic tools. Thus, the 1999 Spanish Water Act requires that water users pay different charges to river basin authorities, since they are recipients of different

[2] In 2007 Andalusia gained exclusive responsibilities on the Guadalquivir water bodies that flowed through its territories and did not affect another autonomous community (Junta de Andalucía 2007). A combination of different factors, such as the high weight of Andalusia in the basin (this region covers the most part of the Guadalquivir River Basin area), the fact that other communities were downstream, and some political alignments in that period, led to make that decision. However, the remaining regions submitted claims to the Court, and finally in 2011, responsibilities fell back to the central government (Thiel 2015).

water services. First, users of the public hydraulic domain are charged a levy to protect and improve the domain's conditions. Second, urban and industrial users pay an "Effluent Control Levy" (*Canon de Vertido*), a levy on point source pollution. Third, users of surface water should pay a "Regulation Levy" (*Canon de Regulación*) to compensate the basin authority for the costs linked to building, operating, and maintaining public water regulation infrastructures. Finally, the "Water Use Tariff" (*Tarifa de Uso del Agua*) is set to fund investment, operation, and maintenance costs of specific infrastructures that are not regulation infrastructures (Calatrava et al. 2015).

Moreover, regional governments are in charge of developing green taxation. The majority of them have set additional charges/taxes on water resources use and pollution. For instance, the *Canon de Saneamiento* is a usual charge set by regional governments. The main aim of this tax is to get revenues to finance sewage services. However, there are a broad variety of green taxes/charges related to the hydrological cycle. Table 2.1 shows the taxes/charges that autonomous communities have set on water abstraction, consumption, and pollution. Except for Castilla-La Mancha, all the regional governments have established taxes at different stages of hydrological cycle. Most of them have set taxes on water use and pollution, where the tax base is the real or potential (if estimated) water consumption. However, a broad variety of tax rate/bill schemes are observed.

On the one hand, it is surprising that there are still free allowances in some Spanish regions, where a minimum consumption level is obligatorily billed. For instance, Catalonia sets a free allowance of 6 m^3 per month for domestic users, a mandatory inclusion in the final water bill. Other alternative schemes are also observed, where an exemption threshold is set, but taxpayers should pay a fixed charge (which could be seen as equivalent to a free allowance). This is the framework applied in some regions such as Andalusia, Extremadura, or Galicia.[3] On the other hand, only seven regions decide to set increasing tax rates, while the remaining regions set a linear variable tax rate. On top of that, the majority of regional governments apply pollution-based adjustments in the calculation of final tax bill, and some of them consider agglomeration factors.[4]

---

[3] In Basque Country, a minimum water consumption of 130 litres per person and day is established, but no fixed charge is set.

[4] For instance, Catalonia considers the population size in the calculation of final tax bill, assuming that higher water pressure and pollution levels are generated in bigger municipalities.

Table 2.1 Regional water taxation in Spain, 2016

Region	Tax (Spanish name)	Tax event	Tax base	Free allowance/ exemption threshold	Increasing tax rate	Pollution-based adjustments
Andalusia	Impuesto sobre vertidos a las aguas litorales	Emissions to coastal water bodies	Pollution units (1000 m^3)	No	No	Yes
	Cánon de mejora de infrestructuras hidráulicas de depuración	Water use and availability	Water consumption (real or potential, m^3)	Yes	Yes	No
Aragon	Impuesto sobre la contaminación de las aguas	Water use and pollution	Water consumption (real or potential, m^3)	No	No	Yes
	Impuesto Medioambiental sobre determinados usos y aprovechamientos de agua embalsada	Environmental damage when using water from reservoirs	Reservoir storage (hm^3) + reservoir height (m)	No	No	No
Asturias	Impuesto sobre las afecciones ambientales del uso del agua	Water use and pollution	Water consumption (real or potential, m^3)	No	Yes	No
Balearic Islands	Cánon de saneamiento de aguas	Water use and pollution	Water consumption (real or potential, m^3)	No	Yes	No
Canary Islands	Cánon de vertidos	Water use and pollution	Pollution units	No	No	Yes

(*continued*)

Table 2.1 (continued)

Region	Tax (Spanish name)	Tax event	Tax base	Free allowance/ exemption threshold	Increasing tax rate	Pollution-based adjustments
Cantabria	Cánon de agua residual	Water use and pollution	Water consumption (real or potential, m^3)	No	No	Yes
Castilla-León	Impuesto sobre la afección medioambiental causada por determinados aprovechamientos del agua embalsada[a]	Environmental damage when using water from reservoirs[a]	Reservoir storage (hm^3) + reservoir height (m)	No	No	No
Castilla-La Mancha						
Catalonia	Canon del agua	Water use and pollution	Water consumption (real or potential, m^3)	Yes	Yes	Yes
Valencia	Cánon de saneamiento	Water use and pollution	Water consumption (real or potential, m^3)	No	No	Yes
Extremadura	Cánon de saneamiento	Water availability	Water consumption (m^3)	Yes	Yes	No
Galicia	Cánon de saneamiento	Water use and pollution	Water consumption (real or potential, m^3)	Yes	Yes	Yes

(*continued*)

Table 2.1 (continued)

	Impuesto sobre el daño medioambiental causado por determinados usos y aprovechamientos del agua embalsada	Environmental damage when using water from reservoirs	Reservoir storage (hm^3)	No	Yes	No
Madrid	Tarifa de depuración de aguas residuales	Water use and pollution	Water consumption (m^3)	No	Yes	Yes
Murcia	Cánon de saneamiento	Water use and pollution	Water consumption (real or potential, m^3)	No	No	Yes
	Impuesto sobre vertidos a las aguas litorales	Emissions to coastal water bodies	Pollution units	No	No	Yes
Navarra	Cánon de saneamiento de las aguas residuales de Navarra	Water use and pollution	Water consumption (real or potential, m^3)	No	No	No
Basque Country	Cánon del agua	Water use and pollution	Water consumption (m^3)	Yes	No	No
La Rioja	Cánon de saneamiento	Water use and pollution	Water consumption (real or potential, m^3)	No	No	Yes

Source: Own elaboration from http://www.minhafp.gob.es

[a]This tax is also levied on other environmental damages (i.e. the visual impact of wind farms)

Moreover, it is also noteworthy that only a few regions have levied charges on environmental damages linked to water abstraction. Thus, Aragón, Castilla-León, and Galicia have designed a tax to internalize the environmental damage when abstracting water from different natural sources, storing it in reservoirs or dams. Water abstraction has impact on the ecological water bodies' flow, generating significant environmental problems. Reservoirs' storage and height are usual dimensions considered in the calculation of tax base.

Finally, we should not forget the key role of local governments in the management of water economic tools. Local governments are in charge of the water services provision to non-agricultural users. As mentioned before, Spain is an extremely atomized country, since its territory is split in more than 8,000 municipalities (INE, 2017b). Almost 5,000 of them have less than 1,000 inhabitants (INE 2017a). Paradoxically, the number of municipalities has registered a slight increasing trend during the last decades (INE 2017b). Retail urban water prices are set at the local level. Moreover, there is no central government institution or regulatory body coordinating retail water prices in Spain. As a result, a wide range of water structures and price levels emerges (Calatrava et al. 2015). In any case, Spanish water prices have been traditionally among the lowest in the European Union (OECD 2010, 2013). Moreover, the country is poorly complying with the polluter-pay principle set at the European WFD (EEA 2013).

Civil Society Participation

Public participation is considered to have a positive and significant impact on the quality of water governance (De Stefano et al. 2013). The European WFD and other EU water directives oblige Spain to involve civil society in the decision-taking process (Osbeck et al. 2013). Public participation should involve three different issues: information, consultation, and active participation. Despite the information and consultation stages have been developed, however active participation processes are relatively new in Spain and "collide heavily with Spanish political and institutional tradition" (Ruiz-

Villaverde and García-Rubio 2017, p. 2490). Civil society participation is mainly conducted through the Water Basin Councils (institution under the river basin authorities). Several stakeholders are involved in the design of water policies. However, different groups have different weights in the decision-making process, being user groups (e.g. agriculture, hydropower, urban users) have much stronger weight. Moreover, campaigns and demonstrations (informal participation tools) could also have impact on the final policy structure. Actually, informal participation has emerged especially in the last decade, since "there is greater social awareness of water as a common good" (Ruiz-Villaverde and García-Rubio 2017, p. 2491).

4 Urban Water Management Services: The Role of PPPs

Spanish regulatory framework[5] establishes that local governments are responsible for guaranteeing urban water services. However, municipalities are flexible to choose the legal regime to provide local public services. The local government may choose to either manage the service in-house or externalize it. When the last option is chosen, management may be transferred either to a public[6] or private company. Moreover, there are some municipalities opting for joint management model, in order to take advantage of economies of scale.[7] In this section, the attention is focused on the role of PPPs in the water sector, showing some figures and outcomes of this kind of management in Spain.

[5] Law 7/1985 on the Regulation of Local Government Terms and Conditions and Law 57/2003 on Local Government Modernization Measures and the Royal Decree 2/2000 establish the legal regimen related to the provision of municipal services.

[6] Public companies are a usual scheme to provide public services in Spain. On the one hand, they provide more flexibility than in-house provision. On the other hand, many sub-central governments have decided to set public companies to provide public services as a strategy to elude the legal constraints on public deficit and debt (Fernandez-Llera and García-Valiñas 2013).

[7] According to González-Gómez et al. (2014) and based on FEMP (2012), 325 associations of municipalities provide wholesale of retail water services in Spain. Private companies are also looking for the optimal provision size, since they design growth strategies to provide the service to a group of municipalities close to each other.

PPPs in Spanish Water Sector: Schemes and Figures

Public-private partnerships are a usual way of collaboration in water services management (World Bank 2006). In terms of PPP schemes, the management of the service may be either fully privatized (contractual PPP) or partially privatized to a mixed company (institutionalized PPP), including both public and private capital (García-Valiñas et al. 2013). It is worth highlighting that Spanish legislation only contemplates privatizing the management of the service, as the infrastructure remains in public hands.

Statistics show that there has been a progressive process of water services privatization in Spain during the last decades. In 2008 private companies supplied 46% of the Spanish population (AEAS[8] 2010). During the last economic crisis, an increasing privatization trend is also observed, since local governments needed extra resources to reduce their level of public debt and achieve the budgetary stability goals (García-Valiñas et al. 2015). Thus, in 2014 private corporations provided urban water services to 56% of the Spanish population (AEAS/AGA 2016). Moreover, Spanish water industry is highly concentrated. *Aqualia* and *Aguas de Barcelona* are the two main private operators[9] (García-Valiñas et al. 2013; González-Gómez et al. 2014; Bel et al. 2015).

Contractual public-private partnerships are one of the most widespread forms of public services privatization in Spain. In 2014, fully private companies provided water services to the 34% of Spanish population (AEAS/AGA 2016). Concessions are the usual way through which the local government entrusts an individual or legal entity with the management. They are awarded following a public tender and for a limited period of time. In the case of water supply companies, contracts that involve building infrastructures and operating the service must be no longer than 50 years, while

[8] The Spanish Association of Water Supply and Sanitation (AEAS) is a private organization carrying out periodic surveys on water services and pricing in Spain. Nowadays, it is the only source from which it is possible to get statistical information on PPPs in Spain, since no official institution is collecting such information. However, these surveys are not census, since they represent approximately 40% of the population and 88% of the municipalities (González-Gómez et al. 2014).

[9] Acciona, Valoriza, Gestagua, Aguas de Valencia, FACSA, Agua & Gestión, PRODAISA, Espina & Delfín, and Hidrogestión are examples of minor companies (at the national level) operating in the Spanish water sector (González-Gómez et al. 2014).

those that only imply running the service have a 25-year constraint.[10] At the end of the contract, local governments must again decide how they wish the service to be managed for the next years.

Institutionalized PPPs are another alternative scheme to allow private sector sharing in urban water services management (González-Gómez et al. 2011; García-Valiñas et al. 2013), whereby the board is shared between the private and public sector. In those companies, local government participation is usually sufficiently significant to guarantee that public objectives will be accomplished successfully. In 2014, those entities served to the 22% of Spanish population (AEAS/AGA 2016). This form of management has significantly gained weight in the last years (AEAS/AGA 2016), since it makes possible to merge public interests (universal access, quality standards) with the industry know-how of private firms. In this sense, the private partner is mainly responsible for the daily water services management, while the public partner makes political decisions.

Whatever the PPP scheme is, when local governments allow private initiative coming into the management of water services, there is a bidding announcement with some requirements to operators in order to attend to the competition. Those specifications include, among others, a minimum entrance and/or annual fees, and/or a certain level of investment during the concession period. Private operators might submit a bid trying to improve those minimum requirements. Once all the bids are registered, local governments make a decision based on several criteria (García-Valiñas et al. 2015).

PPPs in Spanish Water Sector: Outcomes

There is not a wide body of scientific literature comparing public and private management in the Spanish water sector. One of the main reasons is the difficulty to collect data on the industry, since no official institution

[10] Contracts' length is nowadays under discussion. Spanish parliament is currently debating a draft law to implement the transposition of some European Directives related to public contracts (2014/23/UE; 2014/24/UE; 2014/25/UE) into national law. According to the European regulation, contracts could not be longer than five years, and only extensions would be accepted if these are justified by new investments. Further information on the draft bill could be checked at http://transparencia.gob.es.

provides micro-data and/or disaggregated information. That fact makes complex the economic analysis of water sector in the country. The majority of published papers consider partial- and small-sized samples, since capturing information is a difficult task. In any case, there are three key dimensions where empirical studies have focused on: efficiency, prices, and quality. Tables 2.2, 2.3, and 2.4 summarize the main contributions in the three aforementioned items.

Efficiency has been a substantially controversial issue. From a theoretical point of view, it has been traditionally argued that private companies could be more efficient, since they are specialized in the provision of water services, while local public entities have to deal with different kind of services. However, the empirical evidence in Spain does not show clear conclusions on this particular issue. Among others, García-Sanchez (2006), González-Gómez and García-Rubio (2008), and González-Gómez et al. (2013) show that there is no evidence of the superiority of private management with respect to public one. Spain is a highly heterogeneous country, where climatic and regional differences emerge. Some papers have shown that, once controlled for non-controlled inputs and environmental factors, efficiency differences between private and public management tend to disappear (García-Rubio et al. 2010; González-Gómez et al. 2013).

Another interesting conclusion is that private corporations seem to be more efficient in managing specific inputs. Using non-parametric techniques, Picazo-Tadeo et al. (2009a, b) and Suárez-Varela et al. (2017) found that private firms are more efficient when dealing with labour. These findings are basically explained by the higher flexibility of the regulatory and legal context for private firms. As Suárez-Varela et al. (2017, p. 2366) mentioned, "the superiority of the technology used by private units in the management of labour might be due to certain regulatory and institutional restrictions faced by public management units that could reduce their flexibility in adjusting this production factor. In general, public managers are constrained by more stringent labour regulation which makes it more difficult to fire employees, and they also face higher levels of absenteeism (…). In addition, local governments, particularly those ruled by left-wing parties, tend to develop policies to promote employment stability (…). Finally, creating overemployment when public services are delivered in-house might also form part of local politicians' rent-seeking strategy (…)".

Table 2.2 PPPs and water services efficiency in Spain

Reference	Sample	Area and period	Main variables	Methodology	Key results
García-Sanchez (2006)	24 municipalities (over 50,000 inhabitants: 21.24% national population)	National scope, 1999	Inputs: staff; treatment plants; delivery network; total costs Outputs: water supplied; number of connections; quality analyses	*DEA*; non-controllable inputs; second stage (Tobit model)	No significant differences between public and private management were found
Picazo-Tadeo et al. (2009a)	34 municipalities (50% regional population)	Andalusia, 2001	Inputs: delivery network; sewer network; labour; operational costs Outputs: water delivered; collected sewage; treated sewage	*SFA*; second stage to capture the effect of environmental variables	Private utilities outperform public companies in the management of labour
Picazo-Tadeo et al. (2009b)	34 municipalities (50% regional population)	Andalusia, 2001	Inputs: delivery network; sewer network; labour; ground, surface, and purchased water Outputs: water delivered; collected sewage; treated sewage	*DEA*; second stage to capture the effect of environmental variables	Private utilities are superior to public companies in the management of labour

(*continued*)

Table 2.2 (continued)

Reference	Sample	Area and period	Main variables	Methodology	Key results
García-Rubio et al. (2010)	20 water utilities	Andalusia, 1993–2006	Inputs: labour costs; operational costs; hydraulic yield Outputs: water delivery; connections; delivery network; quality	*DEA*	Differences between public and private governance disappear when hydraulic yield (proxy of the degree of renovation network) is included in the analysis
González-Gómez et al. (2013)	80 rural water utilities	Andalusia, 2009	Inputs: delivery network; wastewater treatment capacity; staff costs; operational costs Outputs: population; sewage treated	*DEA*, controlling for environmental factors	No significant differences between public and private management were found when controlling for environmental factors
Suárez-Varela et al. (2017)	70 municipalities (under 50,000 inhabitants)	National scope, 2013	Inputs: labour; operational costs; distribution network Outputs: water delivered; population served	*DEA*; metafrontiers and directional distance functions	Private utilities are superior to public companies in the management of labour

Source: Own elaboration
DEA: Data envelopment analysis, *SFA:* stochastic frontier analysis

Table 2.3 PPPs and residential water prices in Spain

Reference	Sample	Area and period	Main variables	Methodology	Key results
Martínez-Espiñeira et al. 2009	53 municipalities (over 100,000 inhabitants; 33.5% of national population)	National scope, 2006	Dependent variable: monthly water bill for several consumption levels (3, 5, 10, 15, 20, 25, and 50 m^3) Independent variables: weather conditions; water tariff characteristics; water quality and treatment; geographical, housing, and sociodemographic factors; political orientation of local government; private/public management dummies	*OLS, RE, FILM*	Private management of water services leads to set higher prices
Garcia-Valiñas et al. (2010)	301 municipalities (79% of regional population)	Andalusia, 2005	Dependent: household expense on basic water threshold (% of municipal average income) Independent variables: income; geographical, housing, and sociodemographic factors; water tariff characteristics; provincial and river basin dummies; private/public management dummies	*OLS*	More affordable residential water tariffs under in-house provision regime

(*continued*)

Table 2.3 (continued)

Reference	Sample	Area and period	Main variables	Methodology	Key results
García-Valiñas et al. (2013)	396 municipalities (60% of regional population)	Andalusia, 2009	Dependent variable: average representative bill (15 m^3/month) Independent variables: population; housing stock; water treatment; different management regimes; market concentration index; water corporation dummies	*OLS*, *FILM*	Public companies supplying water services set higher prices than those establish under any PPP scheme
Bel et al. (2015)	715 municipalities (93% of municipalities in the region)	Andalusia, 2009	Dependent variable: average representative bill (15 m^3/month) Independent variables: population; population density; water treatment; different management regimes; water corporation dummies; market concentration indices	*FILM*	Private firms with a larger market share make their dominant position effective by setting higher water prices

Source: Own elaboration

OLS: Ordinary least squares, *RE:* random effects model, *FILM:* Heckman selection models based on full information maximum likelihood

Table 2.4 PPPs and water services quality in Spain

Reference	Sample	Area and period	Main variables	Methodology	Key results
González-Gómez et al. (2012)	133 municipalities (50% of regional population)	Andalusia, 2005	Dependent variable: percentage of water losses Independent variables: population; population growth; water treatment; abstraction and network characteristics; local financial burden; local government ideology; water price levels; different management regimes	*WLS*	Water services outsourcing is linked to higher water losses. Private suppliers seem to lack incentives to devote resources to the reduction of water losses
García-Rubio et al. (2016)	1,023 individuals, covering information is on 64 Spanish cities (population over 100,000 inhabitants)	National scope, 2011	Dependent variable: respondent's satisfaction with tap water quality on a scale 1 to 4 Independent variables: gender, age, education, income; ideology; water price; public services satisfaction; water chemical characteristics; geographical and climatic characteristics; private/public management dummies	*OLM*	User's tap water quality perceptions are worse when a private company is managing water services

Source: Own elaboration
WLS: Weighted least squares, *OLM:* ordered logit model

Regarding water prices, previous literature has not found clear conclusions for the Spanish case. All the studies have focused on residential water tariffs, since it is a controversial sector where different policy aims merge (OECD 2003). On the one hand, some previous studies have detected higher price levels under private provision (Martínez-Espiñeira et al. 2009). Similarly, Bel et al. (2015) found that private firms with a larger market share set higher water prices. On the other hand, some studies conclude that public management does not necessarily lead to lower prices. Based on a sample of southern Spanish municipalities, García-Valiñas et al. (2013) found that public companies supplying water services set higher prices than those established under any PPP scheme. However, when looking at the price paid for low water consumption levels, conclusions could change. In this respect, Garcia-Valiñas et al. (2010) showed that in-house provision leads to set more affordable water prices in Andalusia. They argued that the proportion of income spent in a minimum water threshold is lower when the service is not outsourced.

Finally, only a few studies have focused on the quality dimension of water services. A priori, it is expected that private firms looking for profits could reduce water services quality standards, in order to keep/increase their markup. Network losses emerge as a significant dimension of water services quality. Some authors have mentioned that private companies have no incentive to repair water losses, since raw water abstraction costs are lower (Garcia and Thomas 2003). González-Gómez et al. (2012) found more significant water leakage levels under any PPP scheme. Moreover, tap water quality is another significant dimension. García-Rubio et al. (2016) analysed users' perceptions on tap water quality, finding that users' perceptions are poorer when there is a private company supplying water.

5 Conclusions

Throughout these pages, a broad picture of water sector governance in Spain has been provided. An extremely heterogeneous country, Spain is dealing with serious water stress and quality problems. In this context, the design of a good institutional and regulatory framework could contribute to get an equitable and efficient use of water bodies. Then, governance emerges as a key issue to improve water resources allocation.

We have also explained how this country is characterized by a federal structure of regional and local governments that participate in the design and development of water policies. As a consequence, water sector management is felt in general to be complex, in terms of the agents and organizations involved. The strong variety of economic and regulatory tools make difficult the coordination and cooperation among different institutions. Moreover, the exceptionally atomized map of sub-central governments planning water policies is not always aligned with the European Union requirements. On top of that, the weight of PPP schemes in the management of water resources in Spain has increased in the last decades. However, checking the empirical literature on efficiency, prices, and quality, no intensively positive outcomes have been detected in the last years.

Some authors have claimed the urgent need to coordinate water policies in Spain, through the establishment of a regulatory agency with jurisdiction throughout the national territory (González-Gómez et al. 2014). However, this option is a very sensitive topic in a highly decentralized nation, where sub-central governments are trying to conquer new policy areas to develop their autonomy, instead of looking for joint or central actions.

Moreover, as De Stefano et al. (2013, p. 224) pointed out, "the most compelling challenge is possibly ensuring the reliability and the consistency of the information made available by public administrations". Despite the efforts made in the last years, more public institutional support and transparency is needed when it comes to collect and publish information related to the Spanish water sector.

References

AEAS. (2010). *XI Encuesta de Suministro de Agua Potable y Saneamiento en España*. Madrid: Asociación Epañola de Abastecimientos de Agua y Saneamiento.

AEAS/AGA. (2016). *XIV Estudio Nacional de Suministro de Agua Potable y Saneamiento en España*. Madrid: Asociación Española de Abastecimiento y Saneamiento/Asociación Española de Empresas Gestoras de los Servicios de Agua Urbana.

Aldaya, M. M., & Llamas, M. R. (Eds.). (2012). *El Agua en España: Bases para un Pacto de Futuro*. Madrid: Fundación Botín.

Arbués, F., García-Valiñas, M. A., & Villanua, I. (2010). Urban Water Demand for Service and Industrial Use: The Case of Zaragoza. *Water Resources Management, 24*(14), 4033–4048.

Banzhaf, H. S., & Chupp, B. A. (2012). Fiscal Federalism and Interjurisdictional Externalities: New Results and an Application to US Air Pollution. *Journal of Public Economics, 96*(5–6), 449–464.

Bel, G., González-Gómez, F., & Picazo-Tadeo, A. J. (2015). Does Market Concentration Affect Prices in the Urban Water Industry? *Environment and Planning C, 33*(6), 1546–1565.

Calatrava, J., García-Valiñas, M. A., Garrido, A., & González-Gómez, F. (2015). Water Pricing in Spain: Following the Footsteps of Somber Climate Change Projections. In *Water Pricing Experiences and Innovations* (pp. 313–340). Cham, Switzerland: Springer.

Castro, F., Da-Rocha, J. M., & Delicado, P. (2002). Desperately Seeking θ's: Estimating the Distribution of Consumers Under Increasing Block Rates. *Journal of Regulatory Economics, 22*(1), 29–58.

CEDEX. (2011). *Evaluación del Impacto del Cambio Climático en los Recursos Hídricos en Régimen Natural*. Madrid: CEDEX.

Cohen, A. (2012). Watersheds as Boundary Objects: Scale at the Intersection of Competing Ideologies. *Environment and Planning A, 44*, 2207–2224.

Custodio, E. (2014). *Trends in Groundwater Pollution: Loss of Groundwater Quality & Related Services*. Groundwater Governance Thematic Paper.

De Stefano, L., Hernández-Mora, N., López-Gunn, E., Willaarts, B., & Zorrilla-Miras, P. (2013). Public Participation and Transparency in Water Management. In L. De Stefano & R. Llamas (Eds.), *Water, Agriculture and the Environment in Spain: Can We Square the Circle?* (pp. 218–225). Taylor and Francis Group.

Del Moral Ituarte, L. (2016). Nuevos debates sobre escalas en política de aguas. Estado, Cuencas Hidrográficas y Comunidades Autónomas en España. *Ciudad y Territorio: Estudios Territoriales, 190*, 563–583.

EC. (2012). European Innovation Partnership Water. Strategic Implementation Plan. Retrieved from www.eipwater.eu.

EC. (2014), Water Reuse in Europe. *Relevant Guidelines, Needs for and Barriers to Innovation*. Joint Research Centre, Institute for Environment and Sustainability, European Commission.

EEA. (2013). *Assessment of Cost Recovery Through Water Pricing*. European Environment Agency Technical Report No 16/2013.

Environment Agency Austria. (2015). *Technical Assessment of the Implementation of Council Directive Concerning Urban Waste Water Treatment* (91/271/EEC), Umweltbundesamt GmbH.

FEMP (Federación Española de Municipios y Provincias). (2012). Radiografía de las Mancomunidades en España, Madrid.

Fernandez-Llera, R., & García-Valiñas, M. A. (2013). The Role of Regional Public Enterprises in Spain: Room for a Shadow Government? *Hacienda Pública Española, 205*, 9–32.

Garcia, S., & Thomas, A. (2003). Regulation of Public Utilities Under Asymmetric Information. *Environmental and Resource Economics, 26*, 145–162.

García-Rubio, M. A., González-Gómez, F. J., & Guardiola, J. (2010). Performance and Ownership in the Governance of Urban Water. *Municipal Engineer, 163*(1), 51–58.

García-Rubio, M. A., Tortajada, C., & González-Gómez, F. J. (2016). Privatising Water Utilities and User Perception of Tap Water Quality: Evidence from Spanish Urban Water Services. *Water Resources Management, 30*(1), 315–329.

García-Sanchez, I. M. (2006). Efficiency Measurement in Spanish Local Government: The Case of Municipal Water Services. *Review of Policy Research, 23*(2), 355–371.

Garcia-Valiñas, M. A., Martínez-Espiñeira, R., & González-Gómez, F. J. (2010). Affordability of Residential Water Tariffs: Alternative Measurement and Explanatory Factors in Southern Spain. *Journal of Environmental Management, 91*(12), 2696–2706.

García-Valiñas, M. A., González-Gómez, F. J., & Picazo Tadeo, A. (2013). Is the Price of Water for Residential Use Related to Provider Ownership? Empirical Evidence from Spain. *Utilities Policy, 24*, 59–69.

García-Valiñas, M. A., González-Gómez, F. J., Suarez-Pandiello, J., & Zaporozhets, V. (2015). *Private Sector Involvement in Water Services: Theoretical Foundations and Empirical Evidence*, TSE Working Paper, n. 15-590.

Garrido, A., & Calatrava, J. (2009). Trends in Water Pricing and Markets. In A. Garrido & M. R. Llamas (Eds.), *Water Policy in Spain* (pp. 131–144). Leiden: Taylor & Francis.

Garrido, A., Willaarts, B., López-Gunn, E., & Rey, D. (2013). Considerations on Climate Variability and Change in Spain. In L. De Stefano & M. R. Llamas (Eds.), *Water, Agriculture and the Environment in Spain: Can We Square the Circle?* (pp. 191–202). Leiden, The Netherlands: CRC Press/Balkema, Taylor and Francis Group.

González-Gómez, F., & García-Rubio, M. A. (2008). Efficiency in the Management of Urban Water Services. What Have We Learned After Four Decades of Research? *Hacienda Pública Española, 185*(2), 39–67.

González-Gómez, F., Picazo-Tadeo, A. J., & Guardiola Wanden-Berghe, J. (2011). Why Do Local Governments Privatize the Provision of Water Services? Empirical Evidence from Spain. *Public Administration, 89*(2), 471–492.

González-Gómez, F. J., Martínez-Espiñeira, R., García-Valiñas, M. A., & García-Rubio, M. A. (2012). Explanatory Factors of Urban Water Leakage Rates in Southern Spain. *Utilities Policy, 22*, 22–30.

González-Gómez, F., García-Rubio, M. A., Alcalá-Olid, F., & Ortega-Díaz, M. I. (2013). Outsourcing and Efficiency in the Management of Rural Water Services. *Water Resources Management, 27*(3), 731–747.

González-Gómez, F. J., García-Rubio, M. A., & González-Martínez, J. (2014). Beyond the Public-Private Controversy in Urban Water Management in Spain. *Utilities Policy, 31*, 1–9.

IMF. (2017). World Economic Outlook, October. Retrieved from www.imf.org.

INE. (2014). *Encuesta sobre el Suministro y Saneamiento del Agua*. Instituto Nacional de Estadística. Retrieved from www.ine.es.

INE. (2017a). Padrón. Población por Municipios, Instituto Nacional de Estadística. Retrieved from www.ine.es.

INE. (2017b). *Relación de Municipios y sus Códigos por Provincias*. Retrieved from www.ine.es.

Junta de Andalucia. (2007). *Estatuto de Autonomía para Andalucía*. Sevilla: Texto aprobado por el congreso de los diputados, Junta de Andalucia.

Lopez-Gunn, E., Zorrilla, P., Prieto, F., & Llamas, M. (2012). Lost in Translation? Water Efficiency in Spanish Agriculture. *Agricultural Water Management, 108*, 83–95.

Martínez-Espiñeira, R., García-Valiñas, M. A., & González-Gómez, F. (2009). Does Private Management of Water Supply Services Really Increase Prices? An Empirical Analysis in Spain. *Urban Studies, 46*(4), 923–945.

Moss, T. (2012). Spatial Fit, from Panacea to Practice: Implementing the EU Water Framework Directive. *Ecology and Society, 17*(3), 2.

Oates, W. E. (2001). *A Reconsideration of Environmental Federalism*. Discussion Paper 01–54, Resources for the Future, Washington.

OECD. (2003). *Social Issues in the Provision and Pricing of Water Services*. Paris: OECD.

OECD. (2010). *Pricing Water Resources and Water Sanitation Services*. Paris: OECD.

OECD. (2011). *Water Governance in OECD Countries: A Multi-level Approach*. Paris: OECD.

OECD. (2013). *Environment at a Glance 2013. OECD Indicators*. Paris: OECD.

OECD. (2015). *OECD Principles on Water Governance*. Paris: OECD.

Osbeck, M., Berninger, K., Andersson, K., Kuldna, P., Weitz, N., Granit, J., & Larsson, L. (2013). *Water Governance in Europe: Insights from Spain, the UK, Finland and Estonia*. Stockholm: Stockholm Environment Institute.

Picazo-Tadeo, A. J., González-Gómez, F., & Sáez-Fernández, J. (2009a). Accounting for Operating Environments in Measuring Water Utilities' Managerial Efficiency. *Service Industries Journal, 29*(6), 761–773.

Picazo-Tadeo, A. J., González-Gómez, F., & Sáez-Fernández, J. (2009b). The Role of Environmental Factors in Water Utilities' Technical Efficiency. Empirical Evidence from Spanish Companies. *Applied Economics, 41*(5), 615–628.

Renzetti, S. (2015). Non-household Water Demand: Industrial and Commercial Sectors. In R. Q. Grafton, K. Daniell, C. Nauges, J. Rinaudo, M. Ward, & W. Chan (Eds.), *Understanding and Managing Urban Water in Transition* (pp. 297–310). Springer Publishing.

Ruiz-Villaverde, A., & García-Rubio, M. A. (2017). Public Participation in European Water Management: From Theory to Practice. *Water Resources Management, 31*(8), 2479–2495.

Suárez-Varela, M., García-Valiñas, M. A., González-Gómez, F., & Picazo-Tadeo, A. J. (2017). Ownership and Performance in Water Services Revisited: Does Private Management Really Outperform Public? *Water Resources Management, 31*(8), 2355–2373.

Thiel, A. (2015). Constitutional State Structure and Scalar Re-organization of Natural Resource Governance: The Transformation of Polycentric Water Governance in Spain, Portugal and Germany. *Land Use Policy, 45*, 176–188.

Willaarts, B. A., Ballesteros, M., & Hernández-Mora, N. (2014). Ten Years of the Water Framework Directive in Spain: An Overview of the Ecological and Chemical Status of Surface Water Bodies. In P. Martínez-Santos, M. M. Aldaya, & M. R. Llamas (Eds.), *Integrated Water Resources Management in the 21st Century: Revisiting the Paradigm* (pp. 99–120). Leiden: CRC Press/Balkema.

World Bank. (2006). *Approaches to Private Participation in Water Services: A Toolkit*. Washington, DC: The World Bank and Public-Private Infrastructure Advisory Facility.

3

Italy

Asymmetric Regulation and Incentives for Quality Provision: The New Phase of Italian Water Services

Lorenzo Bardelli and Paola Valbonesi

1 Introduction

In this chapter, we present the new design for the water sector in Italy, resulting from reforms adopted in the last six years. We also discuss the current and relevant issues concerning the water services governance and

The views expressed in this chapter are those of the author and do not necessarily represent those of ARERA.

L. Bardelli
ARERA, The Italian Regulatory Authority for Energy, Networks and Environment, Milan, Italy

P. Valbonesi (✉)
Department of Economics and Management "Marco Fanno", University of Padova, Padova, Italy

National Research University, Higher School of Economics (NRU-HSE), Moscow, Russian Federation
e-mail: paola.valbonesi@unipd.it

S. Porcher, S. Saussier (eds.), *Facing the Challenges of Water Governance*, Palgrave Studies in Water Governance: Policy and Practice,
https://doi.org/10.1007/978-3-319-98515-2_3

its provision. In so doing, we contribute to the literature on the regulation of water services, with a focus on the recent Italian case.[1]

It is noteworthy to stress here that, different from other national settings, the empirical literature on the Italian water sector is scant; this is because until the water sector Regulator started its activities by the end of 2011, there was no institutional and regular data collection on the sector.[2] Data on tariffs, investments, governance, and management of the water service were collected and presented since 2005 in the so-called Blue Books by Utilitatis—a research centre of firms active in the Italian energy and water sectors—and, in the period 2006–2011, in the annual reports by Conviri—the national committee supervising the use of water resources. However, these databases were largely on samples, often focussed on specific issues, and lacked regular, long-lasting collection.[3] Similarly, the existing empirical studies on the Italian water sector refer to selected samples. Antonioli and Filippini (2001), Fabbri and Fraquelli (2000), and Guerrini et al. (2018), on different samples of water Italian utilities collected in different periods,[4] investigated economies of scale and economies of density. These studies adopted diverse kinds of empirical strategies/methods and—not surprisingly—found conflicting results. Focussing on the effect of ownership on the firm's performance, on the one side, by exploiting a dataset on 65 Italian water companies, Di Cosmo (2012) found no relationship between them. On the other side, Guerrini et al. (2011) on a the annual financial statements of 80 Italian water utility companies between 2004 and 2008 highlighted that

[1] For discussions on policies carried out in the past in the Italian water sector, see among others: Muraro and Valbonesi (2003), Bardelli and Robotti (2009), Massarutto and Ermano (2013) and Muraro (2008).

[2] The sectoral Italian Regulator has changed its name since the institution in 1995: first, it was set to regulate the electricity and gas sectors and named "AEEG"; then, since December 2011, it included the regulation of the water sector and changed its name to "AEEGSI"; and recently, in January 2018, having also collected competences to regulate the waste sector, it became "ARERA".

[3] In the Blue Book 2005, 2007, 2008, 2011 are presented data on the needed investments as recorded in a sample of "Piani d'Ambito" (the official plan which should be set at local level, according to the Galli Law); in the Blue Book 2014 and 2017 are presented data on realised investments according to the ARERA Decision 347/2012 and on the planned investments according to ARERA Decision 643/2013, referring to different regulatory periods.

[4] Antonioli and Filippini (2001) run their analysis on a sample of 32 Italian water distribution firms over the period 1991–1995; Fabbri and Fraquelli (2000) refer their analysis to 173 Italian water utilities in 1991; Guerrini et al. (2018) study 43 Italian water utilities in the period 2009–2014.

ownership—along with size, diversification, and geographical location—had an impact on the performance of water utility companies, although with different degrees of significance.

The low availability of extended datasets concerning the Italian water sector points out the relevance of these studies. As described in the next sections, ARERA has drawn the new regulatory setting along with large data collections, which—in the near future—should increasingly support the regulation itself and, hopefully, would allow developing new and robust research.

The present chapter is organised as follows. Section 2 concisely reports the evolution of Italy's legal and regulatory framework for the water service, along with its main issues. Section 3 introduces some element of the Italian water governance, clarifying its multilevel nature; Sect. 4 illustrates briefly the new tariff regulation adopted by ARERA and reports the first evidence that emerged from the implementation phase; and Sect. 5 focusses on quality regulation, both contractual and technical. The concluding remarks are presented in Sect. 6.

2 Historical Background: the Period Before 2012

The governance and management of water services in Italy finds its origin at the beginning of the last century with the Giolitti Law n. 103, 29 March 1903, and the Consolidated Law n. 2578, 15 October 1925. These acts set the normative framework for the supply of municipal services and for their positive externalities in the vastly enlarged main towns.[5] A unified framework for tariffs was later set with the Law n. 347, 19 October 1944. Such act also established one national and several local committees to monitor and manage pricing activities (the so-called CIP and CPP, "Comitato Interministeriale Prezzi", Interministerial Committee for Prices; "Comitati Provinciali Prezzi", Provincial

[5] Between 1891 and 1901, the largest Italian towns faced an incredible fast growth in their population: Roma increased by 54%, Milano by 43%, Torino by 32%, Palermo by 27%, and Napoli by 14% (Bardelli and Muraro 2003, p. 380).

Committee for Prices, respectively). During a period of high inflation, the CIP adopted new rules allowing for increases in water tariffs. The ratio for these new rules was to take into account (1) the inflation dynamic, (2) new investments and management costs necessary to provide the water service, and (3) differences in consumption levels (i.e., the adoption of higher tariffs for larger consumption, as compared to standard ones).

At the beginning of the 1970s, a new and extended debate addressed the issues regarding the optimal size of water services as referred to both the horizontal and vertical boundaries. The definitions of the optimal size of the network (i.e., number of consumers supplied) and the level of integration among water services (i.e., water supply, water treatment, and sewage) were widely discussed as a result of the Law n. 183, passed on 18 May 1989. This Law allowed for the aggregation for the supply of water services in Italy (i.e., from the provision by the single municipality to the provision by larger management units). Shortly after this, the Law n. 142, passed on 8 June 1990, established that the water service, considered as a local public utility, could be managed by joint-stock companies.

During the same period, further incentives for a reform at national level of the water services were provided by the European Directive n. 91/271/CE on wastewater treatment. Such Directive, throughout the adoption and implementation of strict regulations, controls, and procedures, aimed to protect the environment from the adverse effects of wastewater discharges from large urban agglomerations and from some specific industrial sectors.[6]

On these grounds, a new design of the Italian water sector was defined by the Law n. 36, 5 January 1994, the so-called Galli Law. This new Law established a clear-cut separation between the activities related to the service provision and those related to the administration of the service itself. In particular, this Law provided the definition of:

---

[6] In particular, this Directive defines rules for (1) the collection and treatment of wastewater in all agglomeration larger 2000 population equivalent; (2) the secondary treatments of all discharges from agglomeration of larger 2000 population equivalent and more advanced treatment; and (3) the monitoring and the performance of the water's treatment plants, the treated water re-use, and the sewage sludge disposal.

1. an integrated water system, which included all the phases of the water cycle (i.e., water purification, sewage, and supply);
2. a geographical aggregation for the service provision called Optimal Territorial Areas (i.e., Ambiti Territoriali Ottimali, ATOs) managed by autonomous authorities with a legal status;
3. a tariff covering all the costs of the service and, in so doing, overcoming the practice of cross-subsidies among utilities at municipal level;
4. a national committee for the monitoring of the sector (i.e., "Comitato per la Vigilanza sull'Uso delle Risorse Idriche", Conviri) and of an observatory for the data collection on the sector aiming at supporting its regulation (i.e., "Osservatorio sui Servizi Idrici").

The implementation of the Galli Law was very difficult, being the reforming process often blocked in one of the mandatory sequential steps required by the Law itself. According to the Law, indeed, once the Authority of the Optimal Territorial Area has been settled by the regional government, a survey on the existing sewage, water, and sanitation infrastructures must be conducted. Then, by exploiting the information collected in the survey, a business plan on the locally provided services must be issued. On these grounds, the operator supplying the local integrated water service in the Optimal Territorial Area has to be appointed and the tariffs must be defined according to the *Regulation 01 August 1996* (i.e., the so-called "Metodo Normalizzato").[7]

The difficulties in adopting and enforcing the Galli Law are well documented. In 2004, ten years after the Law's approval, only 38 out of 91 planned Authorities of Optimal Territorial Areas were effectively settled.[8] Such slow and often ineffective implementation of the Galli Law could be explained—on the one hand—by the vast heterogeneity of the local water sector provisions. In terms of efficiency by the local public administrations, a sort of *case-by-case approach* was adopted to enable the reform. This was very much dependent on local interactions, and it did not acti-

[7] Specifically, the "Normalised Method" defines the cost components and determines the reference tariffs to be adopted.

[8] In these 38 ATOs, operators were appointed to manage and operate the local water integrated service; in terms of governance, these operators were mainly mixed joint-stock companies (in 25 ATOs) and fully public joint-stock companies (in 12 ATOs); only in one case a concession contract was chosen (Conviri Report 2005).

vate worthy effects in implementing the reforming process. On the other hand, the reduced action by the Conviri (a national "committee"), which—by construction—lacked the power of an independent regulatory sectoral authority in setting and enforcing the Law, determined such regulatory failure.[9] In the light of these facts and according to Petretto (2017), it could be concluded that, in the period 2001–2012, the main changes in the Italian water sector belong to the EU legislation on the services of general economic interest (SGEI).

In this phase, referring to the regulation drivers of the model centred on local water authorities ("Autorità d'Ambito Territoriale Ottimale", AATO, Authorities of the Optimal Territorial Areas), we observe that contracts among parties define water tariff and agents' risk sharing. In particular, tariffs were determined on projected variables such as capital expenditure (Capex), operating expenditure (Opex), and water consumption. There was also an obligation to run an investigation every three years to verify whether any differences in projected values for tariffs were due to end users or operators (e.g., reduced consumption and cost inefficiency, respectively). In general, the implementation of this obligation was highly unsatisfactory, and it resulted in numerous disputes between consumers (unwilling to pay for investments planned, but not started) and service operators (asking for full recovery of actual costs). During this blurred and complex regulatory period, in June 2011, a national referendum was run. Its result established that (1) a fair rate of return should not be included in the water tariff calculations and (2) the obligation either to partially sell 100% publicly owned companies or to award concession contracts for the water service through auction procedures was to be abolished.[10] All these elements contributed to increase the uncertainty

---

[9] The coordination between the Authorities of the Optimal Territorial Areas (AATO), on the one hand, and the Conviri (i.e., the sectoral committee), on the other hand, was often very difficult, because such Authorities had the ambiguous role of an independent local Regulator, often affected by local political interests. Moreover, the regulatory failure of this model could also be referred to the large relevance in the Conviri's action of ex ante regulation and the lack of (1) attention to ex post monitoring and (2) tools to address ex post renegotiation.

[10] Specifically, the referendum posed to vote Yes or No on the following two questions: (1) the abolition of the obligation to use the call for tenders as ordinary procedure to contract out the local public services included in the so-called Ronchi Decree and (2) the abolition of the 7% profit rate in the tariff of the water service. The referendum's result was in favour to abolish both.

over the future of the Italian water sector, leading both the private and public decision-makers involved in such economic activity to adopt an inertial strategy.

3 The New Regulatory Design

A few months after the referendum, in November 2011, with the Law n. 214/2011, the Italian government assigned the new design and enforcement of the water sector to the already existing independent Regulator for electricity and gas, that is, the "Autorità per l'Energia Elettrica e il Gas" (AEEG).[11] Accordingly, the Regulator's name was then amended as "Autorità per l'Energia Elettrica, il Gas e il Settore Idrico" (AEEGSI).[12] For the water sector, the Regulator was specifically appointed to: (1) define a new mechanism for designing water tariffs based on full cost recovery and on "the polluter pays" principle and to investigate a single tariff proposal; in order to have final approval, (2) set general and specific thresholds (standards) for the provision of service qualitative; (3) verify the plans for investment defined at local level; and (4) define a scheme of standard contract for the awarding of the water service.

The reformed regulation for the Italian water sector presents two main levels of decision-making (see Fig. 3.1). The first level refers to the Italian independent sectoral Regulator, the AEEGSI (now ARERA)—which being equipped with the technical skills and resources,—acts in the light

[11] The AEEG was established by the Law 481/1995 as an Italian independent sectoral authority, ruled by a board of five members appointed by the national government and confirmed by the Parliament. The legal mandate of this Regulator for the electricity and gas sectors was: (1) to balance the protection of the consumers' interest with the guarantee of industry viability; (2) to operate in an accountable and open manner, also in order to inform the public debate and ensure public participation during the regulation process. In its operational activity, such sectoral authority had wide autonomy. With the Italian Law Decree 201/2011 and the change of AEEG in AEEGSI with the Italian Law Decree 145/2013, all these regulatory activities were also referred to the regulation of the water service. Recently, the Law 205/2017 allocated the Authority regulatory and control functions over the waste cycle, including sorted, urban, and related waste, and the Authority's name was changed to Italian Regulatory Authority for Energy, Networks, and the Environment (Autorità di Regolazione per Energia Reti e Ambiente, ARERA).

[12] Note that this relevant institutional change from a national committee for the water sector (Conviri) to an independent sectoral regulatory authority (in the first period AEEG, then AEEGSI, and finally ARERA) was also in line with the 2012 OECD's recommendations on the reform of the Italian water sector (see OECD 2012a, b).

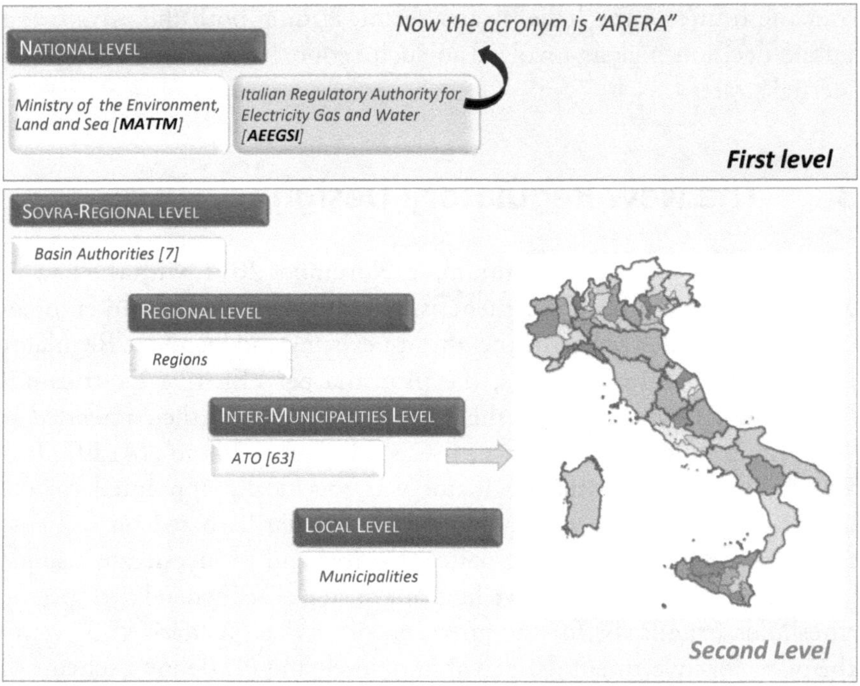

Fig. 3.1 The new regulatory design: the two levels of decision-makers

of EU Directives affecting the sector and sets—at the rules for the national water sector concerning:

- revenue and tariff calculation
- contractual quality discipline
- technical and infrastructural quality regulation
- unbundling and information feedback
- consumer protection
- enforcement, monitoring the conditions under which the services are provided, with powers to demand documentation and data, applying sanctions, and determining those cases in which operators should be required to provide refunds to consumers

The second level refers to the local territorial government, the so-called EGA (i.e., "Ente di Governo di Ambito"), which acts as a local Regulator in charge of interacting with the Municipalities' Committee and population.[13] Such interactions are aimed at ensuring a wide public participation in the regulatory process at a local level. The EGA should appoint the water operator(s) in charge of providing the service in the optimal territorial area.[14] In doing so, the EGA should also choose the governance for the service provision among those allowed by the EU legislation (see Box 3.1).[15]

Finally, the EGA should present the ARERA with a proposal for the water tariff, coherent with the new regulation (see Section 4 and Fig. 3.3).[16] Note that the local operator(s) can directly communicate decision about tariff proposal to the Regulator if the EGA does not act, i.e. if it does not proceed with the tariff proposal and the related investment/financial plan. In addition, if even the local operator(s) does(do) not act, the Regulator can move on the decision-making process and also apply a 10% penalty on the tariff.

In the new local governance of the water sector, EGAs and water operators play a relevant role. Recently, both have recorded relevant changes as follows:

- EGAs have been substantially rationalised and improved, bringing about a reduction in their number (they have decreased from 91 in 2011 to 63 in 2017). In so doing, in many cases (i.e., in 12 out of 20 regions),

---

[13] The Italian Law 133/2014 (so-called "Sblocca Italia") includes mandatory rules for the setting of EGAs and for their role in the awarding of the water service in the optimal territorial area.

[14] Note that, as recalled by the Italian Law 133/2014, the EGAs should only have one operator providing the water service in the optimal territorial area. The aim was to exploit economies of scale and scope which are relevant to foster new investments in the sector. However, the same Law also defines a transitional period during which more than one operator could be active in the optimal territorial area. Accordingly, in some optimal territorial area, we still temporarily find a few operators providing the water service.

[15] Specifically, the EGA should choose the governance of the water service among three options: a joint-stock company to which the service is awarded by a competitive tender; a mixed joint-stock company in which the private firm is chosen by a competitive tender; and a fully public company, that is, the so-called in-house option.

[16] Note that each EGA is monitored and authorised by the Regulator in different steps of the application of the reformed regulatory design; and each EGA should transmit to the Regulator the requested and updated information on the optimal territorial area (including those collected by the water operator).

Box 3.1 The Local Management of the Water Services

In terms of management and considering the forms allowed by the Law, in below Fig. 3.2, which refers to December 2016, we can observe that the Italian water services are locally provided in 48% of cases by in-house operators; in 29% of cases by joint-stock company, of which 12% are in the stock exchange; in 2% of cases by concessionaires; and the remaining cases, which are largely located in the South and Centre Italy, are managed through simplified forms inherited from the past regulatory settings.

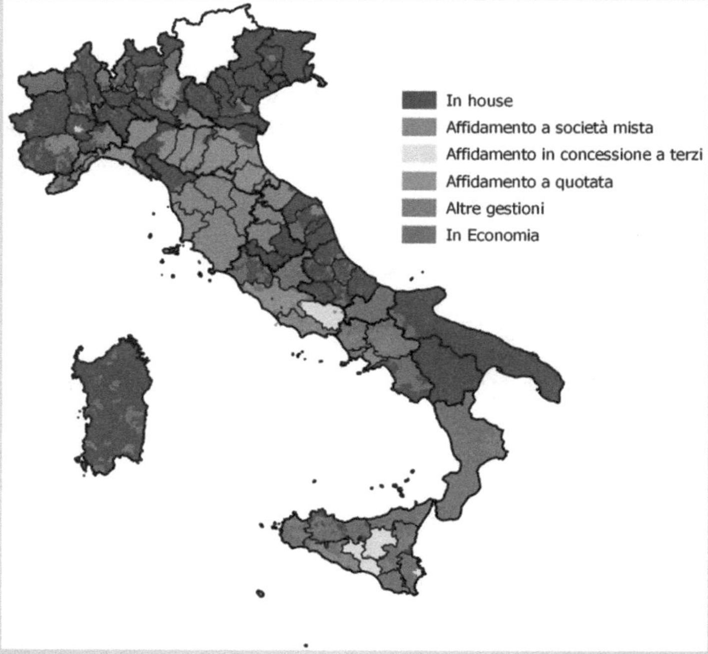

Fig. 3.2 Local Management of the water services. Source: Blue Book (2017), p. 69

the model of one EGA per region has been adopted with the aim to reinforce its instutional functions and competences and to upgrade the quality of each EGA's technical staff.

- Water operators have been considerably restructured, with their numbers decreased from more than 2600 in 2011 to nearly 2100 in 2017

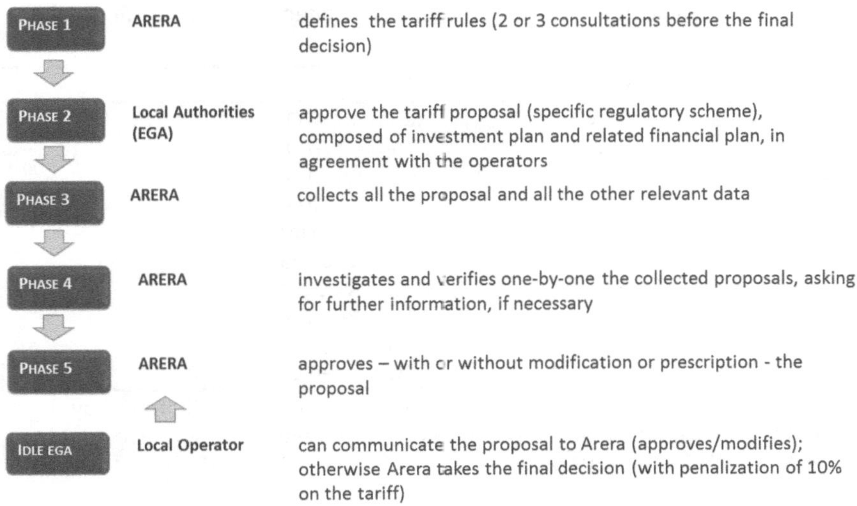

Fig. 3.3 Tariffs' decision-making process

(most of them, about 1300, are still present in the South of the country), within a process of continuous rationalisation and acquisition of infrastructures and their direct management. Such restructuring process leads to the vertically integrated supply (i.e., only one operator supplying the water services); as for the remaining not-restructured provisions, according to the Law, mergers among local suppliers of the water services should be planned in the short run.

As we can see in the AEEGSI's Report (2017a, b), n. 898/2017/I/IDR, addressed to the Parliament on the local governance of the water sector, the largest part of the EGAs in the North-East, the North-West (with the exception of the Valle d'Aosta Region), and in the Centre of Italy had already chosen the operator(s) for the water service. And just a few EGAs were very close to its selection. However, in the South of Italy and the Islands, a limited number of EGAs have chosen the water operator(s) (see Table 3.1).

According to this empirical evidence, on the one hand, the adoption of the new regulatory design in the North and Central part of Italy can be considered a success, at least if compared to the previous slow and weak adoption of the Galli Law. On the other hand, the different implementation

Table 3.1 Cases in which water operator has not been chosen according to legislative provisions

Region	ATO[a]
Valle d'Aosta	Regional ATO
Campania	Regional ATO
Molise	Regional ATO
Calabria	Regional ATO
Sicilia	ATO 1 Palermo
	ATO 2 Catania
	ATO 3 Messina
	ATO 4 Ragusa
	ATO 7 Trapani
	ATO 8 Siracusa

[a]ATO ("Ambito Territoriale Ottimale") is a specific territory, delimitated by regional legislation following criteria of homogeneity.
Source: AEEGSI Report (2017a)

of the same new regulatory design in the two areas (i.e., in the North-Central Italy and in the South Islands) highlights the well-known "Italian divide" phenomena: such different implementation is not only related to technical and infrastructural features of the service provision, but it strongly belongs to institutional elements. In the attempt to better manage such relevant territorial differences and, hopefully, move towards improvements and convergence on overall Italy in the medium-run period, the ARERA designed the new water tariff (see its description in Sect. 4) on an asymmetric base.

4 Tariff Regulation

The new regulation for the Italian water sector includes a revenue-cap approach method that limits the operator's revenue arising from tariffs; in turn, tariffs should cover the efficient costs of the service. These costs are directly defined by the Regulator; they are calculated by the EGAs on information collected from the water operators with two-year lag (see Fig. 3.3). The cap on the operator's revenue in the period $t + 1$ is applied to guarantee social sustainability, and it is calculated as a limit of the revenue at time t, related to the total allowed cost at time t, which in turn depends on the realised cost at time $t - 2$.

The Italian water Regulator's focus on the service's revenues has been largely driven by the vast amount of investment required in the sector: in 2012, such investment was estimated to be five billion Euros per year.[17] However, investment needs result not uniformly distributed among the different Italian areas. Accordingly, the new tariffs' regulatory design includes the definition of an "asymmetric revenue cap" in that it specifically indicates different price limits for the operator(s) providing the water's service in each Italian optimal territorial area.

The tariff design also establishes other novel elements[18]: (1) a cost-recovery component for the invested capital; (2) a component referring to environmental aims; (3) a component, in the form of incentives, addressing the quality element of the service provision; (4) and a component allowing to specifically address vulnerable consumers.

The full implementation of the new tariff design requires sequential regulatory periods and, at the time of writing, we are in the second one.

In 2012, the Regulator published several consultation papers regarding the ongoing tariff regulation reform. By the end of that year, *Decision 585* was adopted, introducing the first relevant change with respect to the past[19]: the new method of tariff calculation results based not anymore on *planned costs*, but on *effective costs* that were collected from the operators providing the water service. Moreover, in order to promote efficiency,[20] an overall cap on revenues was adopted; the out-turn operative cost (Opex) based on a base year was rolled forward, and some standardised parameters for the "reimbursement" of fiscal and financing costs to operators were established on a reference basis.[21] These new rules issued by the new national water Regulator largely contribute to increasing the regulatory transparency and accountability in the Italian water sector, also

[17] See AEEGSI (2013) consultation document 339/2013. The evaluations of the Regulator were subsequently shared also by operators, even comparing data collected for other countries, namely, France and Germany (see http://www.festivalacqua.org/cartella-stampa-edizione-2015/).

[18] See Box 3.2.

[19] AEEGSI (2012), Decision 585/2012/R/IDR.

[20] See: Laffont and Tirole (1993); Leibenstein (1996).

[21] In the same period, the Regulator concluded a procedure to reimburse consumers with the difference between the allowed return on capital included in the tariff (charged by water service operators under pre-referendum regulation) and the reference fiscal and financing costs.

transferring the risk of delayed investment from the customers to the water operator.

In 2013, the Regulator began a consulting phase to collect relevant information for the creation of the long-term regulation and tariff design. At the end of the year, *Decision 643* (the "MTI") was adopted. It included the "Regulatory Matrix",[22] a framework allowing the new regulation to vary and directly depend on the operative characteristics of the water service setting. The Regulatory Matrix was designed to introduce a set of innovative and asymmetric rules that, considering the heterogeneous service provisions in the optimal territorial areas, would provide different incentives to foster different levels of investments. Specifically, such Regulatory Matrix MTI defined four regulatory schemes according to which, each operator—on the basis of its own reported characteristics—is provided with the precise rules to be applied in the definition of its local water tariff. The two key entries of Regulatory Matrix MTI were:

1. the ratio between the planned investment expenditure and the regulatory asset base (RAB). Accordingly, if this ratio results above a certain threshold, it would be possible to apply rules to achieve higher cash flows.
2. the size of activities to be managed by the water service operator. Therefore, if there are mergers among operators in the same optimal area and/or new infrastructures have to be managed, it would be possible to recover the costs through allowed revenues.

As discussed in Bardelli (2018), the Regulatory Matrix MTI—by relating to investments' need and size of the area/activities supplied—defines four different types of operators as follows:

1. operators with low planned investments and no changes in the area/activities supplied (Scheme I, limit price at 6.5%)
2. operators with low planned investments and changes in the area/activities supplied (Scheme II, limit price at 6.5%)

[22] AEEGSI (2013), Decision 643/2013/R/IDR.

3. operators with high planned investments and no change in the area/activities supplied (Scheme III, limit price at 9%)
4. operators with high planned investments and changes in the area/activities supplied (Scheme IV, limit price at 9%)

In December 2015, the Regulator approved the regulatory framework for the period 2016–2019, the so-called MTI-2,[23] which includes a set of new rules we present in details in Box 3.2. Specifically, the constraint to the operator's revenue (i.e., the so-called *VRG*, Vincolo ai Ricavi del Gestore) is defined following the principle of full cost recovery and corresponds to the sum of Capex (capital costs), Opex (operative costs), FoNI (a component related to new investments), *ERC* (a component specifically addressing environmental issues), and RC_{TOT} (a component related to coverage of previous years costs).

The tariff's dynamic is regulated by the price limit as follows:

$$\frac{\vartheta^a}{\vartheta^{a-1}} \le \Gamma^j \equiv \left[1 + rpi + \left(1 + \gamma_{K,j}\right)K - \left(1 + \gamma_{X,j}\right)X\right] \quad j = 1,2\ldots6$$

where *rpi* is the inflation rate; *K* is a factor taking into account the investment needs and it is equal to 5%; *X* is a sharing component equal to 0.5%. Thus, tariffs will grow accordingly to the multiplier ϑ and its dynamic. Values of parameters γ_K and γ_X depend on the selected scheme of the matrix, as indicated in Table 3.2.

In the "Regulatory Matrix MTI-2", the price limit is asymmetrically determined for six regulatory schemes. Therefore, Fig. 3.4 presents the six maximum values (on the top in each box) for the period 2016–2019 and the formula determining such values and referring to level of Opex (relative to the national average value, OPM) and to the level of planned investment (IP) over the regulatory value of infrastructures (RAB).

Note that, considering the ratio between the planned investments (IP) and the regulatory value of the existing infrastructure (RAB), for the period 2016–2019, on the one hand, Schemes IV, V, and VI are intended

[23] AEEGSI (2015). Decision 664/2015/R/IDR.

Box 3.2 Definition of the Cost Components [MTI-2]

The Constraint to the Operator's revenue (Vincolo ai Ricavi del Gestore, VRG) is defined as:

$$VRG^a = Capex^a + FoNI^a + Opex^a + ERC^a + RC^a_{TOT}$$

In what follows we present each component of the VRG.

[1] Infrastructures costs:

$$Capex^a = AMM^a + OF^a + OFisc^a + \Delta CUIT^a_{Capex}$$

where:

AMM^a is the cost component representing depreciation, which is allowed to increase under Schemes IV, V, and VI

OF^a is the financial cost

$OFisc^a$ is the fiscal cost

$\Delta CUIT^a{}_{Capex}$ is the cost component defined in excess of the allowed cost for the use of existing public infrastructure

[2] Component in support of specific investment objectives:

$$FoNI^a = FNI^a_{FoNI} + AMM^a_{FoNI} + \Delta CUIT^a_{FoNI}$$

where:

$FNI^a{}_{FoNI}$ is the cost component for financing new investments (allowed under Schemes IV, V, and VI), determined by the competent local authority within the limit of the component $FNI^{new,a}$

$AMM^a{}_{FoNI}$ is the cost component representing depreciation of non-repayable public transfers

$\Delta CUIT^a{}_{FoNI}$ is the cost component defined in excess of the allowed cost for the use of existing public infrastructure

[3] Operating costs:

$$Opex^a = Opex^a_{end} + Opex^a_{al}$$

where:

$Opex^{a}_{end}$ are the endogenous costs, among which, in case of Schemes III and VI, the Op^{new} can be recovered (i.e., quality improvements or social sustainability issues)

$Opex^{a}_{al}$ are the updatable costs

[4] Environmental and resource costs:

$$ERC^{a} = EnvC^{a} + \mathrm{Re}sC^{a}$$

where:

$EnvC^{a}$ are the environmental costs; they represent the costs of damage that water uses impose on the environment and ecosystems

$ResC^{a}$ are the resource costs; they represent the costs of foregone opportunities that other uses suffer, due to the depletion of the resource beyond its natural rate of recharge or recovery

[5] Component to recover costs related to previous years' balance:

$$Rc^{a}_{TOT} = \left(Rc^{a}_{VOL} + Rc^{a}_{EE} + Rc^{a}_{ws} + Rc^{a}_{ERC} + Rc^{a}_{ALTRO}\right) * \prod_{t=a-1}^{a}\left(1 + I^{t}\right)$$

Table 3.2 Values of γ_K and γ_X parameters

		γ_K	γ_X
Scheme	I	0	0
	II	0	1
	III	0	−1
	IV	0.5	0
	V	0.5	1
	VI	0.5	−1

Source: AEEGSI (2015)—Decision 664/2015/R/IDR

		No Changes in the Water Operator's Activities		*Changes in the Water Operator's Activities:* • *water system integration* • *quality improvements*
		$\frac{OPEX}{pop} \leq OPM$	$\frac{OPEX}{pop} > OPM$	
Investments	$\frac{\Sigma_{2016}^{2019} IP^{exp}}{RAB_{MTI}} \leq \omega$	**SCHEME I – price limit:** 6.0% $\frac{\vartheta^a}{\vartheta^{a-1}} \leq [1 + rpi + K - X]$	**SCHEME II -- price limit:** 5.5% $\frac{\vartheta^a}{\vartheta^{a-1}} \leq [1 + rpi + K - 2X]$	**SCHEME III-- price limit:** 6.5% $\frac{\vartheta^a}{\vartheta^{a-1}} \leq [1 + rpi + K]$
	$\frac{\Sigma_{2016}^{2019} IP^{exp}}{RAB_{MTI}} > \omega$	**SCHEME IV – price limit:** 8.5% $\frac{\vartheta^a}{\vartheta^{a-1}} \leq [1 + rpi + (1.5)K - X]$	**SCHEME V -- price limit:** 8.0% $\frac{\vartheta^a}{\vartheta^{a-1}} \leq [1 + rpi + (1.5)K - 2X]$	**SCHEME VI -- price limit:** 9.0% $\frac{\vartheta^a}{\vartheta^{a-1}} \leq [1 + rpi + (1.5)K]$

Fig. 3.4 Price limits for water tariffs, 2016–2019. Source: Our translation from AEEGSI (2015), Decision 664/2015/R/IDR

for operators addressing relevant investment needs; on the other hand, Schemes I, II, and III are designed for operators addressing lower investment needs.

Moreover, as for the Opex component, the "Regulatory Matrix MTI-2" distinguishes Schemes I and IV for operators reporting per-capita Opex below the national average value, that is, the OPM, and Schemes II and V for operators with per-capita Opex greater than OPM.[24] Note that Schemes III and VI are applicable in the event of structural changes in the activities, both in terms of served area and scope of supplied services. The different rules for Schemes I, II, IV, and V, on the one hand, and Schemes III and VI, on the other, are motivated by the aim of fully exploiting economies of scale and scope in water service management (i.e., to have only one water operator in each EGA). Thus, the Regulatory Matrix MTI-2 addresses such an objective by providing the most favourable rules for merging operators in those EGAs where the management of the service is still fragmented.

As recently highlighted by the AEEGSI Report (2017b) and presented in Fig. 3.5, 47 operators have chosen regulatory Schemes I, II, and III, which correspond to limited levels of investments. The population served in the corresponding EGAs, about 11,500,000 residents (see Fig. 3.5), will have average tariff increases of 5.0% in 2016, 3.5% in 2017, 0.7% in

[24] Note that it is very difficult to have efficiency comparisons, since the operating situation may vary in relevant terms. The underlying idea was to set thresholds only for sharing purposes, distinguishing situations with higher costs from the others.

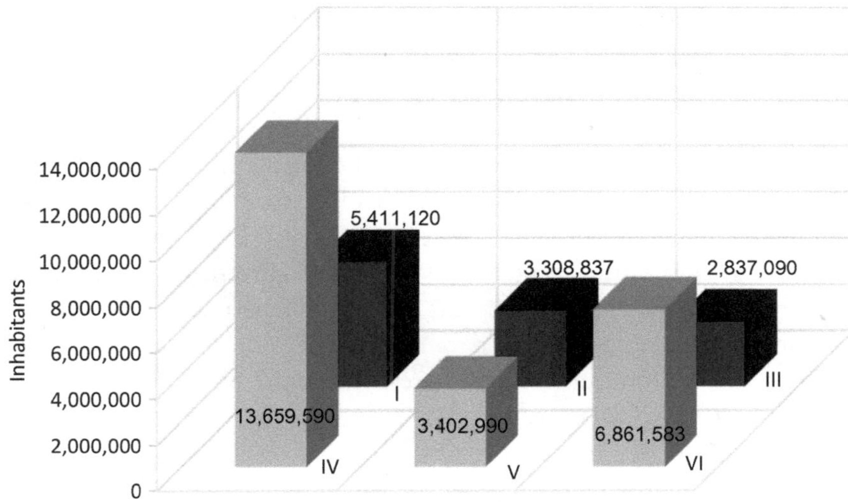

Fig. 3.5 Distribution of the Italian inhabitants according to the regulatory groups, 2016–2019. Source: Our translation from AEEGSI Report (2017b)

2018, and a decrease of −1.0% in 2019. Moreover, 62 operators—about 24,000,000 residents (see Fig. 3.5)—have chosen regulatory Schemes IV, V, and VI. These schemes are going to generate a large amount of investments and a tariff increase of 4.4% in 2016, 3.7% in 2017, 3.3% in 2018, and 2.3% in 2019.

In 2016, the water tariffs in Italy increased on average by 4.57%, with heterogeneous value in the different areas (i.e., 6.09% in the South, 5.39% in the North-East, 4.51% in the North-West, and 2.38% in the Centre of Italy).

At the end of the first regulatory period, investments (net of public contributions) for the years 2014 and 2015 reported a considerable growth with an increase of 55% recorded in 2015 compared with 2012.

Referring to the information collected from 130 operators who provide water service to 46,663,760 users, for the second regulatory period (2016–2019), tariffs are going to finance 7.8 billion Euros of investments, corresponding to an average of 167 Euros for each Italian user (i.e., 207 Euros for each user in the Central Italy, 176 Euros for each user in the North-East, 164 Euros for each user in the North-West, and 121 Euros for each user in the South of Italy). In the same period (2016–2019), once we also consider the public funds required to finance the water infrastructures, the average increases to 214 Euros for each Italian user, and the total amount of investments will be 9.9 billion Euros in the four years (i.e., 2.2 billion Euros in 2016, 2.5 billion Euros in 2017, 2.6 billion Euros both in 2018 and 2019). Considering the aims for which these investments are planned, the Regulator Report (AEEGSI 2017b, Figure 5.16, p. 256) highlights that infrastructures for wastewater and sewage represent the largest recipients.

The Regulator estimates that once the planned investments are considered for the Italian population and not only for that belonging to the sample of 130 operators, the need for investment in the period 2016–2019 is about 12.7 billion Euros which correspond to 3.2 billion Euros in each year from 2016 to 2019 (see Fig. 3.6).[25]

The AEEGSI Report (2017b) also shows, investigating in detail tariffs for domestic use of 102 operators who supply 37,931,369 inhabitants, that—not surprisingly—in 2016 the higher water tariffs are to be referred to local services for which the larger investments are planned. Given a national average expenditure for the water service provision of 291 Euro/year (considering a three-component family with a consumption of 150 m^3 of water), in Central Italy the average becomes 344 Euro/year. However, as highlighted by the AEEGSI Report (2017b, p. 254), there is large heterogeneity among geographical areas and within each area.

As for the tariff's weights of domestic water, wastewater, and sewage expenditure, we can see that, at national level, a three-component family pays on average 115 Euro/year (39% of the total) for the consumption of 150 m^3 of water, 38 Euro/year for the sewage service, and 85 Euro/year for the wastewater service. For the same consumption, a similar family living in the South of Italy will record a lower expenditure for sewage and wastewater service (29 Euro/year and 72 Euro/year, respectively) because of the larger coverage by public resources of the ongoing investments.[26]

---

[25] Accordingly, the value of the Italian water sector infrastructures is going to increase: the AEEGSI Report (2017a) estimates an increase of 17% from 2016 to 2019 (i.e., from 15.7 billion Euros in 2016 to 18.4 billion Euros in 2019).

[26] For a study about Italian water bills, regional heterogeneity, and distributional effect, see Miniaci et al. (2008).

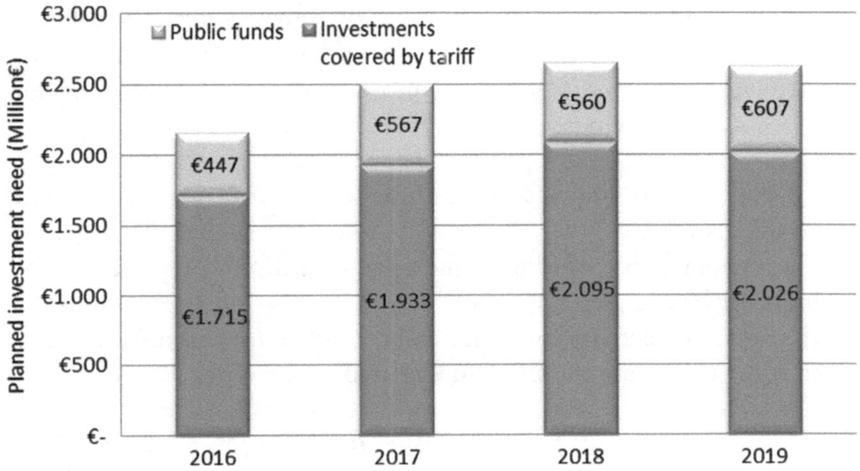

Fig. 3.6 Planned investments for 2016–2019 period (estimate at national level). Source: AEEGSI Report (2017b)

5 Quality Regulation

The output-based approach implemented by the Italian Regulator's activities becomes evident in the field of quality regulation. Indeed, in the reformed setting, quality regulation consists of the definition of both general and specific standards for quality: the former refers to thresholds for the performance in service provision to end users as a whole, while the latter refers to thresholds for the performance in service provision addressing single users. In this new setting, local authorities are free to (1) choose thresholds that are above the ones defined by the Regulator and (2) provide within established cost-reimbursement rules, for resources coherent with the selected objectives.

According to the asymmetric regulation for the period 2016–2019, in the aim to reduce local differences in the water services supply, the Italian Regulator has also added a relevant set of rules relating to "contractual" (CQ) and "technical" (TQ) quality of water services. Specifically, the Regulator has designed detailed frameworks for each one. Incentives to improve the "contractual quality" are set in the Decision 655/2015/R/IDR[27]

[27] AEEGSI (2015), "Regolazione della qualità contrattuale del servizio idrico integrato ovvero di ciascuno dei singoli servizi che lo compongono". Decision 655/2015/R/IDR.

where the Regulator refers to a number of activities performed by the water operator in its interactions with end users. The Regulator has defined *minimum quality standards,* in particular for the following specific items, among which:

1. the rules and timing according to which the contract for the water services is valid
2. the type and time of billing, the operator's time to reply to written inquiries by the user
3. the operator's activity through a call centre and/or through offices
4. the operator's time to solve issues related to the water service.

Accordingly, the additional costs arising from improvements in quality beyond such minimum standards are to be included in the calculation of the revenue cap. Moreover, an incentive mechanism defines compensation, penalties, and rewards as referred to the difference between the supplied quality and the defined minimum quality standards. These rules have been applied since July 2016. The ratio below this regulation is to encourage the water operators to map their contractual performance in the aim of highlighting the critical elements and their solution, also considering the incentives provided by the adopted regulation. In so doing, the water operators have to refer their activities to the required minimum level defined by the Regulator when monitoring their contractual quality. Moreover, the water operators should update their information system to collect all the data needed to measure the required performance and to verify their performance relative to the minimum standards set by the Regulator.[28] Note that, in this area, the role of the EGAs is crucial: indeed, they possess larger amounts of information on the local service provision than any other national institution. Such information allows the EGAs to identify the needs of local communities and to set the path to improve the quality of the water service. EGAs can achieve these improvements through the incentive mechanism, that is, rewards for operators paid for by users benefiting from higher standards.

[28] In the AEEGSI's report (2017a, b), a first evaluation about the "contractual quality" on data collected by the national Regulator is included.

As for the "technical quality", the Italian Regulator has proceeded in defining a set of macro-indicators referring to different measurable outputs and somehow related to infrastructural elements of the water service provision. Specifically, these macro-indicators are:

1. leakages, both in their linear measure and in their percentage on the total water provided;
2. service interruptions, taking into account the different sources used to provide the service;
3. quality of the water supplied for human consumptions, considering both tests for drinking water and other controls;
4. sewage overflow due to storms, also referring to their frequency and to the contingent sewage network conditions;
5. treatments for sludge, considering its environmental and sustainability impact;
6. treatments for wastewater, considering its environmental and sustainability impact.

For each macro-indicator, the technical quality regulation defines a relationship between ex ante measured value and the target to be achieved during the period that follows. In case of ex ante good performance, the target is its maintenance and the cluster associated is called "A". In all the other cases, the targets should be to improve the ex ante values and their magnitude should depend on the distance of the ex ante values from the cluster A.

Each water operator, on the ground of its macro-indicators' measures, will then receive a relative evaluation based on the Technique for Order of Preference by Similarity for Ideal Solutions (TOPSIS); see Fig. 3.7. Moreover, given the announced tasks in terms of technical quality and each water operator's relative evaluation, from 2020 onwards, ARERA will implement a system of penalties and awards. This approach aims to provide incentives for better water infrastructures' management, for a progressive reduction of impacts of wastewater and sludge treatment, for a progressive increase of drinking water quality, etc.

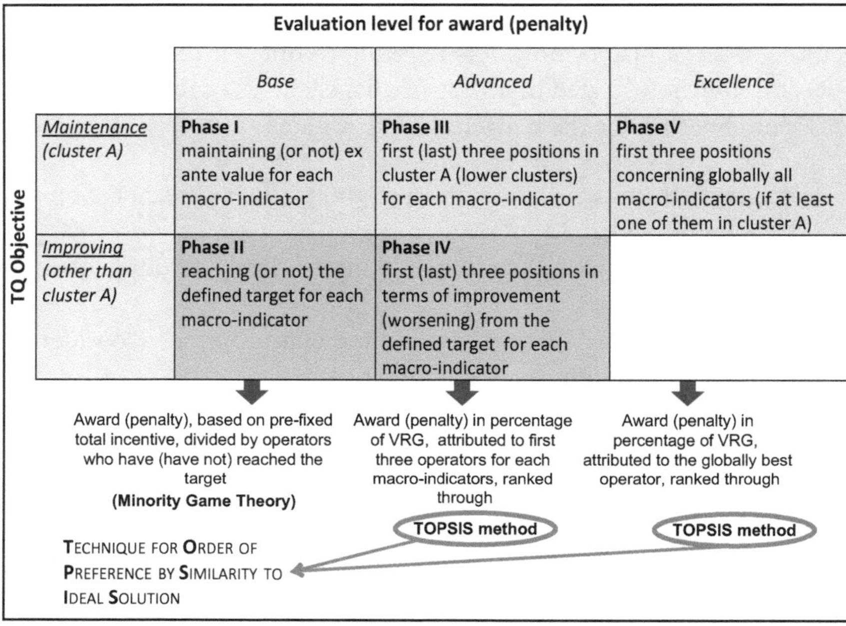

Fig. 3.7 Award/penalty incentive mechanism

6 Concluding Remarks

The new water regulation in Italy is coherent with a multilevel governance design, and it directly refers to all the relevant functions and objectives of the interested entities, being industrial operators or public institutions. The underlying idea of the new regulation design was to promote steady improvements in the local decision-making processes, clarifying the relationships among objectives and instruments, with a national Regulator in charge of managing, controlling, and enforcing the overall process.

The new water regulation in Italy is composed of both input-based and output-based mechanisms, with a tendence to reduce the incidence of the first and to develop the relevance of the second.[29] It consists in a

[29] Such a setting will esplicitate the link among contractual and technical regulation, on one side, and accounting unbundling, on the other. In particular, a properly developed performance-based mechanism (based on the monitoring through macro-indicators will be connected with accounting unbundling, thus enabling decision-makers to evaluate the direct costs related to the improvement of an indicator.

revenue-cap approach where regulatory instruments—referring to investments' needs and operators' activities to provide the water services in the local area—are asymmetrically defined by the Regulator and referring to real data. A crucial asset in this new regulation is the extended data collection from the local operators and the EGAs the Regulator has implemented since starting its activities. This data collection has permitted—on the one hand—to better address the asymmetric regulatory design (e.g., MTT, MTI, and MTI-2) and will permit, on the other hand, to effectively monitor the effects of the new regulation. It is worth noting that from 2019 the first data stream coming from the implementation of the entire new regulatory setting will be available.

By using the available data and following the Regulator official reports and decisions, in the previous sections, we have presented the current picture of the new water regulation in Italy, with a focus on the governance of the sector, the tariff's design, the incentives for investments, and the rules for contractual and technical quality.

The evidence from the recently reformed regulation of the Italian water sector highlights an extended compliance by local operators and institutions in the adoption of the new regulation. This is a relevant result, in particular if referred to the weak implementation of previous sector regulations (i.e., the Galli Law). In this achievement, the reputation and the skills of ARERA, the Italian Regulator (i.e., related also to the Regulator's previous activity as AEEG and as AEEGSI) have played a key role. Moreover, the significant increases in investments recently recorded in the Italian water sector seem to highlight that such regulatory reform is moving the sector on the right path. Relevant steps by the Italian Regulator towards the design and implementation of rules on the contractual and technical quality of the service provided represent further relevant signals of a challenging regulatory process.

References

AEEGSI, Autorità per l'energia elettrica il gas e i servizi idrici. (2012). Regolazione dei servizi idrici: approvazione del metodo tariffario transitorio (MTT) per la determinazione delle tariffe negli anni 2012 e 2013. Decision 585/2012/R/IDR.

AEEGSI, Autorità per l'energia elettrica il gas e i servizi idrici. (2013). Metodo tariffario idrico. Retrieved from https://www.autorita.energia.it/allegati/docs/13/643-13all.pdf.

AEEGSI, Autorità per l'energia elettrica il gas e i servizi idrici. (2015a). Regolazione della qualità contrattuale del servizio idrico integrato ovvero di ciascuno dei singoli servizi che lo compongono—RQSII. Decision 655/2015/R/IDR.

AEEGSI, Autorità per l'energia elettrica il gas e i servizi idrici. (2015b). Metodo tariffario idrico (2016–2019). Retrieved from https://www.autorita.energia.it/allegati/docs/15/664-15all.pdf.

AEEGSI, Autorità per l'energia elettrica il gas e i servizi idrici. (2017a). Relazione Annuale sullo stato dei servizi e dell'attività svolta. Volume 1: Stato dei servizi.

AEEGSI, Autorità per l'energia elettrica il gas e i servizi idrici. (2017b). Relazione Annuale sullo stato dei servizi e dell'attività svolta. Volume 2: Attività svolta.

Antonioli, B., & Filippini, M. (2001). The Use of a Variable Cost Function in the Regulation of the Italian Water Industry. *Utility Policy, 3–4*, 181–187.

Bardelli, L. (2018). Water Regulation: An Innovative Approach. In A. Gilardoni (Ed.), *The Italian Water Industry*. Springer International Publishing.

Bardelli, L., & Muraro, G. (2003). L'offerta e la regolamentazione dei servizi idrici italiani: l'esperienza italiana. In Muraro and Valbonesi (Ed.), *I servizi idrici tra mercato e regole* (pp. 347–383). Milano: Carocci.

Bardelli, L., & Robotti, L. (2009). *The Water Sector in Italy*. CIRIEC Working Paper N.1/2009.

Blue Book. (2017). Rome: Utilitatis.

Conviri Report. (2005). Relazione al Parlamento sullo stato dei servizi idrici 2004.

Di Cosmo, V. (2012). Ownership, Scale Economies and Efficiency in the Italian Water Sector. *Journal of Industry Competition and Trade, 13*(3), 399–415.

Fabbri, P., & Fraquelli, G. (2000). Costs and Structure of Technology in the Italian Water Industry. *Empirica, 27*(1), 65–82.

Guerrini, A., Romano, G., & Campedelli, B. (2011). Factors Affecting the Performance of Water Utility Companies. *International Journal of Public Sector Management, 24–26*, 543–566.

Guerrini, A., Romano, G., & Leardini, C. (2018). Economies of Scale and Density in the Italian Water Industry: A Stochastic Frontier Approach. *Utilities Policy, 52*, 103–111.

Laffont, J. J., & Tirole, J. (1993). *A Theory of Incentives in Procurement and Regulation*. The MIT Press.

Leibenstein, H. (1996). Allocative Efficiency vs. "X-Efficiency". *The American Economic Review, 56*(3), 392–415.

Massarutto, A., & Ermano, P. (2013). Drowned in an Inch of Water. How Poor Regulation Weakened the Italian Water Reform. *Utilities Policy, 24*, 20–31.

Miniaci, R., Scarpa, C., & Valbonesi, P. (2008). Distributional Effects of Price Reforms in the Italian Utility Markets. *Fiscal Studies, 9*(1), 135–163.

Muraro, G. (2008). Water Services and Water Policy in Italy. In C. Clini, I. Musu, & M. L. Gullino (Eds.), *Sustainable Development and Environmental Management: Experiences and Case Studies*. Berlin: Springer.

Muraro, G., & Valbonesi, P. (2003). *I servizi idrici tra mercato e regole*. Milano: Carocci.

OECD. (2012a). Economic Survey: Italy, 2011. Retrieved from http://www.oecd-ilibrary.org/economics/oecd-economic-surveys-italy-2011_eco_surveys-ita-2011-en.

OECD. (2012b). Water Governance in OECD Countries: A Multilevel Approach. Retrieved from http://www.oecd.org/cfe/regional-policy/water-governanceinoecdcountriesamulti-levelapproach.htm.

Petretto, A. (2017). La regolamentazione del servizio idrico integrato: struttura tariffaria e investimenti. In S. Antoniazzi & F. Merusi (Eds.), *Vent'anni di regolazione accentrata di servizi pubblici locali*. Collana Cesifin, Firenze: Giappichielli.

4

Portugal

Water Governance Diversity: The Progress of Water Services in Portugal, a Silent Revolution

Jaime Melo Baptista

1 The New Public Policy for Water Services

The water services involve many areas, which are distinct but must be interconnected. It is necessary to bring together in a suitable way institutional, governance, management, planning, technical, economic, legal and environmental, public health, social, and ethical tools to ensure the suitable provision of these services. In 1993, when it was decided in Portugal to define a new public policy, which would ensure suitable water services to the population, an integrated approach was adopted involving various components:

- Approval of the strategic plan for the sector
- Definition of the legislative framework
- Definition of the institutional framework
- Definition of the governance of the services

J. M. Baptista (✉)
LNEC/LIS-Water, Lisboa, Portugal
e-mail: jmbaptista@lnec.pt

S. Porcher, S. Saussier (eds.), *Facing the Challenges of Water Governance*, Palgrave Studies in Water Governance: Policy and Practice,
https://doi.org/10.1007/978-3-319-98515-2_4

- Definition of the access targets and the quality of service goals
- Definition of the tariff and tax policy
- Provision and management of financial resources
- Construction and asset management of the infrastructures
- Improving structural efficiency
- Improving operational efficiency
- Human resources capacity building
- Promotion of research and development
- Development of entrepreneurship
- Increasing of competition
- Protection, awareness, and involvement of users
- Making information available

The successful implementation of this new public policy resulted from the ability to manage the implementation of all these components at relatively the same time, ensuring an effective global and integrated approach.

This policy was naturally based on a careful analysis of the evolution of the sector in the past, learning from successes but mainly from the errors. Between others, a large study (Baptista et al. 1995) was carried out in 1993, providing support for public policies that were defined for the water services. Another study was published in 2011 (Pato 2011) that carried out a reconstitution and historic analysis of water supply and sanitation in Portugal from the end of the nineteenth century until the first decade of this century. A critical analysis of the facts was also made to support improvements to be introduced in the design, definition, implementation, and assessment of policies for the sector. A complementary study was published in 2014 (Pato 2014) with chronology and testimonies.

Adoption of Strategic Plans for the Sector

The Portuguese situation concerning water services before 1993 was inacceptable regarding the expectations of the population and the country's development ambitions, without a clear strategy to resolve these issues. Only 81% of the dwellings were covered by public water supply service, and only 50% were provided with safe water, 61% by public wastewater collection services, and 31% with collection and treatment. As a result, only 57% of coastal beaches and 17% of river beaches had good water quality and some waterborne diseases occurred.

The political, social, and economic importance attributed led to a political commitment to reorganise the sector, involving appropriate strategies, expressed within a strategic plan for drinking water supply and wastewater management, nationwide in scope and for the medium term.

The strategic plan for drinking water supply and wastewater management involved stages characterising and diagnosing the situation, defining intended goals, evaluating the corresponding investment requested, identifying the measures needed, defining measures to implement strategies, and specifying the monitoring instruments.

One of the success factors was the stability over time of this public policy, in spite of the natural rotation of parties in government, with their different policy options. Although subject to natural minor adaptations in terms of its path and evolution, the public policy remained quite stable during those 25 years, without any forward or backward movement.

Definition of the Legislative Framework

Insofar as public policy needs to be reflected in legislation, Portugal set up a new and modern legislative framework to cover the legal regime for water services and their regulation, as well as tariff legislation, quality of service, quality of water, and also other technical aspects. There was essentially a first generation of legislation in 1993 and a second generation in 2009. Additionally, the country improved its modern legislation on water resource management, environmental management, waste management, consumer protection and competition.

The regulatory authority, described in more detail below, has drawn up proposals for new legislation and the altering of existing legislation and the issuing of regulations and recommendations to contribute towards a clarification of the rules for the provision of these services.

Definition of the Institutional Framework

A new public policy should be supported by a well-structured administrative organisation. Portugal has established an appropriate institutional framework, with a clear assignment of responsibilities for the public entities involved, especially the regulatory authority for water services and authorities for the environment, water resources, waste management, public health, consumer protection and competition. This definition was absolutely fundamental and essential for a good performance by the sector, since it has enabled the responsibilities of stakeholders to be specified, along with clear rules of operation and the articulation between the close and complementary sectors mentioned above, without overlaps or important gaps.

Definition of the Governance Models for the Services

Portugal decided on the governance models that could be used in the sector, in accordance with political options. They are direct management, delegated management, and concessions for water services, with provision by State-owned, municipal, and private entities. Currently there are around 80% of cases involving direct management, 10% of cases involving delegated management, and another 10% involving concessions, with about 300 utilities. In 15% of the utilities, private management is involved. The introduction of these different models of governance, all of them having clear cases of success, has meant that the option adopted varies from municipality to municipality and from region to region and there is a comparative analysis of their performance and thus the stimulus to improve services.

The regulatory authority ensures the legal and contractual monitoring of the utilities throughout their life cycle, whatever their model of governance, particularly through analysing the tendering and contracting processes, contract modifications, contract terminations, and reconfigurations and mergers of systems, accompanying the carrying out of contracts and intervening where necessary in reconciliation activities between parties.

Definition of the Access Targets and the Quality of Service Goals

The access targets and the quality of service were specified and have been continually monitored with regard to the goals specified, based on indicators relating to the user interface, the sustainability of utilities, and environmental sustainability.

The regulatory authority has ensured the regulation of the quality of service provided to users by the utilities, assessing their performance and comparing the utilities among themselves, through the application of a suitable selection of performance indicators, to promote effectiveness and efficiency. It has also ensured drinking water quality regulation, assessing the quality of water supplied to the users, comparing the utilities among themselves (metric benchmarking), and monitoring any non-compliances in real time.

Definition of the Tariff and Tax Policy

Portugal has set out a tariff policy for public water services with the goal of promoting a gradual trend towards cost recovery, consistent with the economic capacity of the population. Although somewhat distant from a full application, the recovery of the costs of these services has gradually been carried out through tariffs paid by the users, supplemented where possible through transfers of European funding and, if necessary, through rates from the levying of taxes. The country has thus sought to evolve from a situation of low tariffs to the gradual full recovery of costs, and although the process is not yet fully completed, significant successes have been achieved.

The regulatory authority has ensured gradually the economic regulation of the utilities, thus promoting the regulation of prices to ensure efficient tariffs, which are socially acceptable to users without prejudice to the necessary economic and financial sustainability of the utilities, within an environment of efficiency and effectiveness in the provision of their service.

At the same time an attempt has been made to introduce tax instruments that encourage desirable behaviour, for example, the rational use of water as a primary material or as final destination, through a water resources usage tax.

Provision and Management of Financial Resources

To implement a new public policy and carry out coverage of service goals in Portugal, it was necessary to ensure the availability of important financial resources, both national resources and those originating from European funds. Indeed, in the last 25 years, there has been and continues to be a significantly higher level of investment in infrastructure for water (about ten thousand million euros). It was also necessary to create capacity for the efficient management of these so important financial resources, knowing how to apply the resources where they could provide greater profits and benefits for society.

The regulatory authority has contributed by supporting the government in defining suitable criteria for the awarding of this financing and the ongoing monitoring of the results obtained.

Construction and Asset Management of the Infrastructures

Portugal has throughout this period promoted the construction of vast infrastructure for the provision of public water services, using appropriate technologies, naturally implying substantial costs, both in terms of initial investments and at the operational level. This infrastructure has enabled an enormous increase in the level of compliance with European legislation in this matter. Nowadays the focus is more directed towards on the rehabilitation of existing assets, for instance, trying to increase the network rate of replacement from the existing 0.6% to 1–4%, namely, to reduce the leakage from 126 per connection.day to less than 100 per connection.day.

The regulatory authority has contributed by monitoring and even approving new investments by utilities within the scope of economic regulation.

Improving the Structural Efficiency

Portugal has been promoting the improvement of the structural efficiency of the public water services, to reduce costs to users and to society in general.

An optimised territorial organisation was specified for the management of these services, making use of economies of scale at the regional level. The utilities are therefore encouraged to promote, as much as possible, the physical integration and aggregation of the systems at a technically and economically appropriate scale, promoting the implementation of joint solutions with similar entities. In fact, nowadays there are clear benefits in the bulk service resulting from the physical aggregation of systems, from 300 to 5 utilities, not only in economic terms but also in terms of the quality of the services provided. The physical aggregation of retail systems is not yet concluded and is a national priority. Another important measure is the promotion of economies of scope, joining the services of water supply to those for wastewater. Process economies have also been assessed and discussed, with the aim of providing vertical integration of bulk systems (water production, treatment of wastewater) with retail systems (water distribution and collection of wastewater).

The regulatory authority has contributed towards the improved organisation of the sector.

Improving the Operational Efficiency

Portugal has been promoting the improvement of the efficiency of the operation by the utilities, to reduce costs to users and to society in general.

The utilities should seek to adopt, given existing legislation, the most advisable organisation-type, particularly at the staffing level, functional content, information circuits, administrative procedures, planning, budget, control, and measures which will tend to ensure quality assurance.

The regulatory authority has contributed to improve the efficiency of the operation of the utilities.

Human Resource Capacity Building

A major effort regarding human resources capacity building has been carried out in Portugal in terms of quantity and competencies, as this is an essential factor to guarantee the general quality of the sector. There has been a strengthening of technological and traditional technical courses to overcome any shortages of personnel with relevant academic qualifications to carry out the existing roles in the sector. There are training and updating courses for specific human resources in the sector, involving training activities for managers and technicians at various levels. There has been a considerable increase in technical publications.

The regulatory authority has provided support to the utilities through the production of technical publications in partnership with knowledge centres, the direct and indirect promotion of seminars and conferences, and support for training events by third parties. In this way, it contributed to an improvement in the technical capacity building of the utilities. The country currently has around 15,000 well-prepared professionals, with solid training in planning, design, financing, construction, and operation of water services, with recourse to advanced technology.

Promotion of Research and Development

Portugal has promoted research and development in areas associated with public water services, creating endogenous knowledge. A research and development covers pre-normative applied research projects, for the development and strengthening of the technological infrastructure on which they can base these projects. It has been possible to gradually promote innovation and technical support to the utilities, with research centres moving closer to industry, and ensure greater national economy in terms of knowledge and technologies. The Portuguese research centres have increased their participation in international research and development projects, particularly European ones, and in some areas even heading such initiatives.

The regulatory authority has supported innovation and the promotion of research and development in the sector, in partnership with the research centres. The annual information provided by the regulator about the sector and the utilities, easily available and free of charge, was and still is very useful for many academic works, at master and doctoral levels.

A recent initiative of the National Laboratory of Civil Engineering (LNEC) was the creation of the Lisbon International Centre on Water (LIS-Water), an international centre of excellence for leading edge R&I on water services and water resources with a high impact on public policies, management, and regulation in Portugal, Europe, and worldwide. It is funded by the Teaming Programme H2020 of the European Commission and aggregates the knowledge of many stakeholders, namely, the Cranfield Water Science Institute at Cranfield University and the Sorbonne Business School at Université Paris 1 Panthéon-Sorbonne.

Development of the Entrepreneurship

Portugal has sought to improve the development of the economic sector, taking advantage of the implementation of strategies to develop water services, which has created exceptional conditions to promote the development of national knowledge and hence strengthen the capacity of the business sector in the domestic and international market, generating new activities with the creation of employment and wealth.

The regulatory authority has provided technical support to the sector and particularly the utilities through the production of technical publications in partnership with knowledge centres, the direct and indirect promotion of seminars and conferences, and support for training events by third parties. In this way, it can also encourage the consolidation of the national business sector.

Increasing of Competition

In the case of natural or legal monopolies, and where there is therefore no competition in the market, Portugal has promoted virtual competition, introducing benchmarking between utilities, and, in the case of private involvement, competition in the market, for example, through tender procedures for the allocation of delegations, concessions, and the provision of services. Indeed, the introduction of different models of governance has enabled competition to increase, as well as a comparative analysis of their performance and thus an ongoing stimulus to improve services. Currently, five major groups compete for the concession contracts in Portugal, where about 30 contracts have been signed in the last 25 years, and for the participation in delegated management, where about 15 contracts have been signed in the same period.

The regulatory authority was the main agent to increase competition in the sector. This encourages innovation and technical progress and, therefore, increases the efficiency and quality of the provision of these services.

Protection, Awareness, and Involvement of the Users

In Portugal the tools for the protection of users have been promoted, especially for the most economically disadvantaged, as well as awareness and involvement with regard to public water services. The users now have rights reinforced through specific legislation and increasingly involvement about water services, particularly regarding physical and economic access to services, their quality, the quality of drinking water, information about services, complaints about services, and their participation in decisions. Complaints have become a powerful instrument for the defence of users.

The regulatory authority continually ensures compliance by the utilities with consumer protection legislation and, in particular, undertakes an analysis of any complaints and promotes their resolution between users and the service provision utilities. It has also fostered the participation of the users of the services, creating advisory mechanisms and disseminating information. The users are naturally represented on the regulatory authority's advisory board.

Provision of Information

A very complete and transparent information system has been established in Portugal for the water services. It makes use of reliable information, both to support the definition of public policies and business strategies and to evaluate the service that is actually provided to society, to be able to convey an overview of the sector in a reliable and regularly updated manner.

The regulatory authority has the responsibility to regularly collect, validate, file, and process information about the situation of the sector, following a standardised format, which is sufficiently complete. This corresponds to about one million data per year and is available online, together with the performance indicators calculated based on that data (http://www.ersar.pt/pt/publicacoes/relatorio-anual-do-setor).

2 The Evolution of Water Services Regulation in Portugal

Portugal may be compared with its European counterparts in the area of regulation of network services, having indeed been one of the first countries to create independent regulatory bodies in the 1990s, following an international trend for the liberalisation of traditional monopolist markets. This was the case with the telecommunications, electricity, and gas sectors, which, after liberalisation, needed regulatory bodies to supervise these recently created competitive markets. These bodies now have around two decades of existence and the experience has been, overall, positive.

This has also been the case with the water services from the beginning of the present century, in which Portugal was one of the pioneering European countries in setting up a regulatory body, with its intervention recognised in the country and abroad due to the rationality and robustness of its regulation model (RITA-ERSAR). This context leads to the consolidation of independent, credible, strong, and universal regulation in Portugal, defending the general interest and the uses of regulated services, without prejudice to safeguarding the economic viability of the utilities and their legitimate interests.

Evolution of the Regulatory Authority

In Portugal, the setting up of a regulatory authority had from the outset been envisaged in the reform of the drinking water supply and wastewater management, undertaken from 1993 onwards, with a new public policy for these services.

The first attempt to regulate these services was taken in 1995 with the setting up of the Concessions Monitoring Committee. It was a positive initiative but without enough capacity to face all the challenges.

Shortly afterwards, the National Observatory for Municipal and Multi-municipal Systems was set up, but it was closed three years later. It was demonstrated not being easy to effectively monitor the sector from outside, without deep involvement and effective power in the day-to-day activities of the utilities and all the correspondent complexity.

As a third attempt, the Institute for the Regulation of Water and Solid Waste (IRAR) was established in 1997, with the duties of the water and waste services regulator in Portugal. It was a public institute endowed with administrative and financial autonomy but subject to the responsibility of the minister for the environment, distinct, therefore, from the model of the independent regulatory body existing at that date in other network services. Its powers were limited and it acted more using its powers of influence through constructive interaction with the other sector stakeholders. According to its bylaw, as far as municipal concessions were concerned, IRAR would issue an advice on the tendering process but would not intervene in the process of establishing the tariff, which constituted a determining element in the choice of the concession holder, with this element being regulated by the contract, as well as its updating. On the other hand, IRAR would also issue an advice on the tariffs for multi-municipal concession systems, annually proposed to the State by the concession holder. However, as far as the management models not operated under a concession, no regulatory activity was foreseen.

From 2003, IRAR was given the responsibility of the national authority for drinking water quality. With this new duty, there were a vast additional number of functions to be covered, with reinforced powers. Of note were the approval of the water quality control programmes presented annually by the utilities; the appraisal, validation, and supervision of the methods used and the laboratories undertaking analyses which provide this service to the utilities; the monitoring, inspection, and possible sanctioning of the utilities; and the drawing up of annual reports on the situation concerning compliance with legislation.

A resolution of the Council of Ministers in 2004, concerning the restructuring of the water sector, would consider regulation an essential part for the development of the water sector, within a preferentially competitive market. This was the real indicator of the changeover in Portugal from the infrastructure stage to a stage characterised by the safety and quality of the service provided. It was thus recommended to carry out a redefinition of the regulatory model, particularly in terms of goals, administrative nature, organic and functional independence, neutrality, and universality of the regulated utilities. However, this did not take place that time yet. The organic law of the ministry for the environment decided that the IRAR remained within the indirect administrative sphere of the State with its statute as a public institute, changing its designation to Water Services Regulation Authority (ERSAR).

The legal system for the local business sector of 2006 extended ERSAR's regulation to municipal companies, supplemented by the new Law of Local Finances and which would attribute responsibility to the regulatory authority for the verification of provisions regarding the prices of the services provided by municipal or intermunicipal direct management bodies, including those provided in the form of municipalised or intermunicipalised services, and by municipal and intermunicipal companies. This extended, once again, its scope of intervention. It was published in 2009 the legal systems for the municipal services and multi-municipal services for the public drinking water supply and wastewater management, which confirmed and intensified the powers of the regulator.

The legal regulation framework was reinforced in 2009 with the new bylaw of ERSAR, leading to the extension of the scope of intervention to all the utilities of these services, independent of their management model, as well as a greater uniformity of the regulatory body's procedures with regard to all of them. Therefore, in accordance with this legal framework, the direct management of services involving State and municipal ownership (municipal, municipalised, and intermunicipalised services) became subject to quality of service regulation. Economic regulation then covered all the municipal services managed by contract (concessions, delegations, and partnerships) and municipal services managed without a contract (municipal and municipalised services), following a mechanism involving supervision of tariffs, in the first case, and tariff verification mechanisms through sampling, in the second. However, the effective power to decide the tariffs was with the holder of the service. This diploma maintained once again the administrative nature of the regulatory body as a standard public institution, being governed by the legal system for public institutions.

Transition to an Independent Regulation Authority

In 2011, the Portuguese Government took on the goal of transforming the bodies with regulatory functions dependent on government, where the regulated markets justified this due to their importance, into independent administrative authorities, reintegrating the others into the traditional administrative service.

It was considered that ensuring a balanced market is not possible if one of the parties, the State, takes on conflicting roles, being both an active agent and the entity that establishes the rules regarding the functioning of the sector. This duality normally has the consequence of imbalance in the market function, to the detriment of appropriate performance.

It is therefore important to emphasise the autonomy of the regulator with regard to the executive power, granting it this status and the means that enable it to defend what is essential, the general interest, and the interests of the users of the regulated services, without prejudice to safeguarding the economic viability of the utilities and their legitimate interests. Effective regulation was used to establish an incentive to increase the efficiency and effectiveness of utilities, thus avoiding the risk of the prevalence of these with regard to the users and the subsequent risk of these users receiving services of lower quality and at a higher price.

As such in 2013 a framework law was approved for independent administrative bodies with functions involving the regulation of the economic activity of the private, public, and cooperative sector, which standardised the legal framework of operation of these bodies. This also envisaged the alteration of the legal nature of ERSAR, placing it on the same level as other independent regulatory bodies.

Following this, and with a view to achieving all these goals, the Portuguese Parliament in 2014 approved the current bylaw of ERSAR. This diploma seeks to recognise and accentuate its autonomy with regard to the executive power, granting it this status and the means to defend the general interest and the interests of the users of the regulated services, as well as to safeguard the economic viability of the utilities and their legitimate interests. The aforementioned diploma reinforces the independence of the regulatory authority in carrying out its respective functions, particularly through the reduction of the powers to government. It also alters the statute of the members of its board of administration concerning their appointment process, period of mandates, guarantees of immovability, and the rules for the termination of mandates. The powers of authority and the sanctuary and legislative powers of the regulatory body were reinforced, to increase its capacity to act within the regulated sector, through the attribution and strengthening of essential instruments for the regulatory and supervisory activity of the entities providing water services. ERSAR finally became an independent regulatory authority.

A Strong Regulatory Model

The regulatory authority developed and applied for more than one decade a model for the public water supply and wastewater management services based on a regulatory integrated approach (known in the abbreviated form as RITA-ERSAR model) (Baptista 2014, 2017). This regulatory approach was implemented through two main levels of intervention.

The first level is aimed generically at the sector as a whole, designated as structural regulation, which involves a contribution towards the better organisation of the sector, through the clarification of its operational rules and legislation, the drawing up and regular dissemination of information, and capacity building and innovation. The regulatory authority is not focused on any utility in particular, but on the sector as a whole, helping to create organisation, rules, and tools for its good functioning. It therefore corresponds to macro-regulatory intervention.

The regulatory authority contributes to the formulation of better public policies, towards their rationalisation and the resolution of any malfunctions regarding the regulated services and towards the organisation of the sector, promoting, for example, an increase in the efficiency and effectiveness of the water services and the search for economies of scale, scope, and process. It monitors the national strategy adopted for the sector, by accompanying its implementation and regularly reporting on any evolution or constraints.

It should draw up proposals for new legislation or the alteration of existing legislation, for example, at the level of the legal framework governing the systems, technical legislation regarding the water services, and the legislation governing regulation. It contributes towards a clarification of the rules for the provision of these services through proposed legislation and the issuing of regulations and recommendations. It monitors the application of legislation in force along with regulations and recommendations, assessing their effectiveness and the need for any improvements or replacements.

The regulatory authority makes available and regularly discloses thorough and accessible information to all sector stakeholders—through the collection, validation, processing, and disclosure of information regarding the sector and the respective utilities—as well as the general public, leading to increased public interest in consulting this information. It contributes to consolidating an actual culture of concise and credible information that can be easily interpreted by all, extendable to all utilities, regardless of the forms of management adopted for the provision of such services.

It also provides technical support to the utilities through the production of technical publications in partnership with knowledge centres, the direct and indirect promotion of seminars and conferences, and support for events by third parties, conducting opinion studies and promoting research and development in the sector, thus motivating the academic sector in this field. It should also make itself available to answer the various questions put to it by the different sector stakeholders. In this way, it can contribute to an improvement in the technical capacity building of the utilities and to encouraging the consolidation of the national business sector.

The second level of intervention, designated as behavioural regulation of the utilities, consists of its legal and contractual monitoring throughout the life cycle, economic regulation, quality of service regulation, drinking water quality regulation, and user interface regulation. In contrast to structural regulation, the regulator is focused here on each of the about 300 utilities operating in this sector. It therefore, in a supplementary manner to the previous level, corresponds to micro-regulatory intervention, multiplied by the number of regulated utilities.

The regulatory authority ensures the legal and contractual monitoring of the utilities throughout their life cycle, specifically through analysing the tendering and contracting processes, contract modifications, contract terminations, and reconfigurations and mergers of systems, accompanying the carrying out of contracts and intervening where necessary in reconciliation activities between parties. Legal and behavioural regulation should thus contribute to ensuring public interest and legality.

The regulatory authority ensures the economic regulation of the utilities, thus promoting the regulation of prices to ensure efficient tariffs that are socially acceptable to users without prejudice to the necessary economic and financial sustainability of the utilities, within an environment of efficiency and effectiveness in the provision of their service. Economic regulation also includes the assessment of investments to be undertaken by the utilities. As monopoly prices tend to be higher than those resulting from competing markets, obtaining lower prices that will allow the economic and financial viability of the utilities and will correspond to a fairer system for users requires major intervention from the regulatory authority.

It ensures the regulation of service provided to users by the utilities, assessing annually their performance and comparing the utilities among themselves (metric benchmarking), through the application of a suitable selection of performance indicators, to promote effectiveness and efficiency, which represents an improvement in their levels of service. Quality of service regulation is a way of regulating performances, which is inseparable from economic regulation. It constrains the permitted performances of the utilities regarding the quality of service they provide to users, steering the utilities in terms of effectiveness and efficiency and embodying in this way a basic right of users. This contributes to promoting an improvement in the levels of service provided to users.

The regulatory authority ensures the drinking water quality, assessing the quality of water supplied to the users, comparing the utilities among themselves, and monitoring any non-compliances in real time. As the quality of drinking water is an aspect of quality of service, and as it has a major interaction with economic regulation, there is a clear rationale for it to be regulated by the same body, although it is not obligatory to follow this model and this is not the case in various countries. It thus contributes to promoting an improvement in water quality and public health.

It should ensure compliance by the utilities with consumer protection legislation and, in particular, undertake an analysis of any complaints and promote their resolution between users and the service provision utilities. It should also foster the participation of service users, creating advisory mechanisms and disseminating information.

This regulatory model encompasses all the water and wastewater utilities, about 300, and is implemented by the regulator with a staff of 70 people and an annual budget of 8M€, with an average impact of 0.7% on the tariffs.

3 Results of the Implementation of This Policy and Regulation

This section summarises the results of the implementation of this policy in Portugal during the last two and half decades, not only regarding the accessibility of the drinking water supply and wastewater management service and the quality of drinking water but also regarding the impact on environmental quality and on public health.

Evolution of the Public Water Supply Policies

In 1993, only 81% of the dwellings in mainland Portugal were covered by the public water supply service. Currently, 95% of dwellings are covered by the service, which means that the coverage target envisaged in the strategic plan for the drinking water supply and wastewater management has been attained. The remaining 5% of dwellings are served by the individual solutions, such as individual holes or wells. There are, however, significant differences at the regional level between urban areas and rural areas, and it can be seen that around 99% of dwellings in urban areas have access to public water supply services, while in the rural areas access is in the order of 92%.

As for the water quality, in 1993 only 50% of the dwellings in mainland Portugal were provided with safe water, in accordance with national and European legislation. These public services currently ensure a high quality level of distributed water, with more than 99% complying with legislation, with the remaining non-compliant situations being subject to immediate corrective intervention. It can be noted that the quality of water has evolved in an extraordinary manner. The target envisaged in the strategic plan for the drinking water supply and wastewater management has been detained. This is a remarkable example of a successful strategy, with a very positive impact on public health, on the reduction of diseases and deaths, and on the reduction of days absent from work.

Quality of service is now monitored to all the about 300 water supply utilities, by the regulatory authority, using a comprehensive assessment system with 14 performance indicators, based on the best international references (Alegre et al. 2017), and reported every year (Relatório Anual dos Serviços de Águas e Resíduos em Portugal 2017a, b, c).

For utilities providing bulk service (water production), the average values of quality of service (2017) are:

- Physical accessibility of the service: 94%—target [100]
- Economic accessibility of the service: 0.18%—target [0; 0.25]
- Occurrence of supply interruptions: 0.02/(delivery point × year)—target [0.00]

- Quality of supplied water: 99.75%—target [98.50; 100]
- Reply to written suggestions and complaints: 88%—target [100]
- Coverage of total expenses: 1.05—target [1.0; 1.1]
- Effective connection to the service: 93.8%—target [100]
- Non-revenue water: 5.2%—target [0.0; 5.0]
- Mains rehabilitation: 0.3%/year—target [1.0; 4.0]
- Mains failures: 10/(100 km × year)—target [0; 15]
- Adequacy of human resources: [2.4/(10^6 m^3 × year)—target [1.0; 2.5]
- Water losses: 6.6 m^3/(km × day)—target [0.0; 5.0]
- Energy efficiency of pumping installations: 0.42 kWh/(m^3 × 100 m)—target [0.27; 0.4]
- Disposal of sludge from the water treatment: 100%—target [100]

For water supply utilities providing retail service (water distribution), the current average values of quality of service are:

- Physical accessibility of the service: 95%—target [95–100]
- Economic accessibility of the service: 0.40%—target [0; 0.50]
- Occurrence of supply interruptions: 0.8/(1000 connections × year)—target [0.0–1.0]
- Quality of supplied water: 98.88%—target [98.50; 100]
- Reply to written suggestions and complaints: 89%—target [100]
- Coverage of total expenses: 1.08—target [1.00; 1.10]
- Effective connection to the service: 86.5%—target [95.0; 100]
- Non-revenue water: 29.8%—target [0.0; 20.0]
- Mains rehabilitation: 0.6%/year—target [1.0; 4.0]
- Mains failures: 40/(100 km × year)—target [0; 30]
- Adequacy of human resources: 2.1/1000 consumers—target [2.0; 3.5]
- Water losses: 126 l/(water connections × day)—target [0; 100]
- Energy efficiency of pumping installations: 0.49 kWh/(m^3 × 100 m)—target [0.27; 0.4]
- Disposal of sludge from the water treatment: 97%—target [100]

Evolution of Wastewater Services

In 1993, only 61% of the dwellings in mainland Portugal were covered by public wastewater collection services. Currently 81% of dwellings are covered by these services. However, what is intended is to ensure not only the collection of wastewater but also its suitable treatment before it is discharged into the environment, and this is where the situation was clearly worse earlier. In 1993, only 31% of the dwellings in mainland Portugal were covered by public wastewater services involving collection and treatment. Currently 80% of dwellings are covered by these services.

This means that there has been a major evolution. However, the goal of 90% envisaged in the strategic plan has not yet been attained. The remaining dwellings (20%) are serviced by private solutions, such as septic tanks.

The geographical distribution of the physical accessibility of the wastewater service shows that the differences are quite significant at regional level. There are greater differences between urban areas and rural areas, given that, despite municipalities with an urban characteristic having an access percentage of 97% for dwellings with a wastewater management service, above that established as the national target (90%), in the case of more rural areas this is still situated at around 70%. These disparities also result from the fact that it can be seen that in the rural areas these public systems often have excessively high material costs, with the option for individual solutions being preferable.

Quality of service is now monitored every year to all the about 300 bulk and retail wastewater management utilities, by the regulatory authority, using a comprehensive assessment system with 14 performance indicators (Baptista 2017; Alegre et al. 2017; Relatório Anual dos Serviços de Águas e Resíduos em Portugal 2017a). For utilities providing bulk service, the current average values of quality of service (2017) are:

- Physical accessibility of the service: 90%—target [100]
- Economic accessibility of the service: 0.18%—target [0; 0.25]
- Flooding occurrence: 7.8/(100 km of pipes × year)—target [0; 0.5]

- Reply to written suggestions and complaints: 84%—target [100]
- Coverage of total expenses: 1.11—target [1.00; 1.10]
- Effective connection to the service: 90.7%—target [100]
- Sewerage rehabilitation: 0.3%/year—target [1.0; 4.0]
- Sewer collapses: 1.3/(100 km × year)—target [0]
- Adequacy of human resources: 3.1/(10^6 m^3 × year)—target [3.0; 4.5]
- Energy efficiency of pumping installations: 0.6 kWh/(m^3 × 100 m)—target [0.27; 0.45]
- Appropriate disposal of collected wastewater: 100%—target [100]
- Emergency control discharges: 57%—target]90; 100]
- Compliance with discharge parameters: 92%—target [100]
- Disposal of sludge from the wastewater treatment: 100%—target [100]

For wastewater management utilities providing retail service, the current average values of quality of service are improving, covering social, utility, and environmental sustainability:

- Physical accessibility of the service: 81%—target [85; 100]
- Economic accessibility of the service: 0.29%—target [0; 0.50]
- Flooding occurrence: 5.28/(1000 connections × year)—target [0; 0.25/(1 000)]
- Reply to written suggestions and complaints: 88%—target [100]
- Coverage of total expenses: 0.91—target [1.00; 1.10]
- Effective connection to the service: 88.7%—target [100]
- Sewerage rehabilitation: 0.4%/year—target [1.0; 4.0]
- Sewer collapses: 1.6/(100 km × year)—target [0.0]
- Adequacy of human resources: 7.9/(100 km × year)—target [5.0; 11.0]
- Energy efficiency of pumping installations: 0.65 kWh/(m^3 × 100 m)—target [0.27; 0.45]
- Appropriate disposal of collected wastewater: 99%—target [100]
- Emergency control discharges: 30%—target]90; 100]
- Compliance with discharge parameters: 92%—target [100]
- Disposal of sludge from the wastewater treatment: 99%—target [100]

The assessment of the quality of water and wastewater services in Portugal is mostly based on sound information collected and validated by the Portuguese regulatory authority ERSAR, which gives priority to the accuracy of data, essential to the confidence in the interpretation (Cabrera et al. 2018). Some academic research has been developed based on this regulatory information made publicly available annually (Baptista et al. 2008; Amaral 2017).

Impact on the Environment

The water services have a strong impact on environmental quality, particularly as regards the discharge of wastewater into water resources. An indicator to assess this impact is the quality of surface waters, insofar as it is obviously affected by any pollution coming from wastewater, even diffuse pollution and other sources can contribute significantly. It can be realised that its evolution in the last two decades has been very positive, going from 19% to 78% quality for surface waters good enough to be used for drinking water supply after treatment, in accordance with European legislation.

Other indicator is the quality of coastal bathing waters and its transition, insofar as it is heavily impacted by any pollution coming from wastewater. Its evolution in the last two decades has been extraordinarily positive, going from 57% to 99% for beaches with good water quality, in accordance with European legislation.

Another indicator is the quality of river bathing waters, insofar as it is also more strongly affected by any pollution, particularly from the discharge of wastewater. Its evolution in the last two decades has been extraordinarily positive, going from 17% to 95% for river beaches with good water quality, in accordance with European legislation.

Other indicator is the number of beaches with blue flag or classified as gold beaches, distinctions annually awarded by non-governmental organisations. The number of beaches and marinas distinguished has been increasing in a very significant manner, going from about 100 to 300 with a blue flag and similar figures with gold quality.

Impact on Public Health

The water services naturally have a strong impact on public health, particularly through the quality of drinking water, which may transmit waterborne diseases, particularly cholera, typhoid and paratyphoid fever, other salmonelloses, shigellosis, leptospirosis, Legionnaires' disease, and hepatitis A. Using this last indicator, it can be verified that its evolution in the last two decades has been very positive, going from 630 cases to 8 cases annually. All the other waterborne diseases have no more relevance in Portugal.

4 Conclusion

From the analysis carried out, it can be seen that the new public policy instituted in 1993 in Portugal for the water services was implemented in a global and integrated manner, with greater stability over time. It brought together institutional, governance, management, planning, technical, economic, legal and environmental, public health, social, and ethical instruments to ensure the suitable provision of these services. In that new policy, regulation has played an important role. As a consequence, a major evolution in the provision of these essential public services of drinking water supply and wastewater management can be noted, and also in the impact on environmental quality and on public health.

It can be also realised that the setting up of a regulatory authority for the water services resulted from the new public policy, and that this was seen as an indispensable component for its promotion. It has also been possible to define, develop, and gradually implement an integrated regulatory model which includes structural regulation of the sector, through contributing to its organisation, legislation, information, and capacity building, and the behavioural regulation of the utilities, through legal and contractual, economic, quality of service, quality of drinking water, and user interface regulation.

References

Alegre, H., et al. (2017). *Performance Indicators for Water Supply Services* (3rd ed.). London: International Water Association Publishing.

Amaral, R. (2017). *Strategic Investment Planning in Water Services in Portugal*. PhD thesis, University of Lisbon, Lisbon.

Baptista, J. M. (2014). *The Regulation of Water and Waste Services: An Integrated Approach (Model RITA-ERSAR)*. London: IWA.

Baptista, J. M. (2017). Portuguese Regulatory Model for Water and Waste Services: An Integrated Approach. In H. Cabrera & E. Cabrera Jr. *Regulation of Urban Water Services. An Overview* (Chapter 2). London.

Baptista, J. M., et al. (1995). *Management of Basic Sanitation Systems*. Lisbon: LNEC.

Baptista, J. M. et al. (2008). A Programmatic Perspective on Asset Management, from the National to the Utility Level—The Case of Portugal. *Journal Urban Water*. Water Asset Management International, Abingdon.

Cabrera Jr., E., et al. (2018). Adequacy of DEA as a Regulatory Tool in the Water Sector. The Impact of Data Uncertainty. *Environmental Science and Policy, 85*, 155–162. Elsevier, Amsterdam.

Pato, J. (2011). *História das políticas públicas de abastecimento e saneamento de águas em Portugal* [History of Public Policies for Water Supply and Sanitation in Portugal]. Lisbon: ERSAR.

Pato, J. (2014). *História das políticas públicas de abastecimento e saneamento de águas em Portugal—cronologia e depoimentos* [History of Public Policies for Water Supply and Sanitation in Portugal—Chronology and Testimonies]. Lisbon: ERSAR.

Relatório Anual dos Serviços de Águas e Resíduos em Portugal. (2017a). Volume 1: Caraterização Geral do Setor de Águas e Resíduos. Report, ERSAR, Lisbon.

Relatório Anual dos Serviços de Águas e Resíduos em Portugal. (2017b). Volume 1, anexo 1: Benchmarking. Report, ERSAR, Lisbon.

Relatório Anual dos Serviços de Águas e Resíduos em Portugal. (2017c). Volume 1, anexo 2: Matriz de Avaliação. Report, ERSAR, Lisbon.

5

France
In Hot Water? Issues at Stake in the Regulation of French Water Public Services

Simon Porcher

In France, as in most industrialized countries, municipalities must provide local public services that have natural monopoly characteristics such as water or sanitation. Water public service refers to production and distribution of water, while sanitation refers to collection and treatment of used water. A particularity in water and wastewater in France is that service provision is much decentralized: there are more than 14,000 drinking water services producing and distributing water to 64 million inhabitants, while in England, for example, there are 12 water-only companies and 12 water and wastewater companies. This large number of services is due to the fact that France counts more than 36,000 municipalities. Another particularity of France is that local governments monitor prices, control entry and exit of firms into the market, organize competition, and ensure uninterrupted service.[1] This is different from

[1] Note that water quality is defined at the national level by the central government, following several European directives (European directive 2000/60 on water).

S. Porcher (✉)
IAE Paris - Sorbonne Business School, University of Paris 1 Panthéon-Sorbonne, Paris, France
e-mail: Simon.Porcher@univ-paris1.fr

S. Porcher, S. Saussier (eds.), *Facing the Challenges of Water Governance*, Palgrave Studies in Water Governance: Policy and Practice,
https://doi.org/10.1007/978-3-319-98515-2_5

England where there is an independent regulatory body responsible for setting service provision conditions for all operators and for monitoring operators' compliance.

Local governments can choose to delegate the management of the water public service to a private firm or directly manage it. Water governance in France is characterized by important private sector participation. Indeed, private firms manage the majority of water services, but there is still a large part of water services under direct management. Such a diversity of governance forms allows for benchmarking between direct and delegated management and raises questions about the determinants of the choice of the local government to manage or delegate. A large literature (see infra) comes with some answers about these two questions. Moreover, because contract length is typically between 8 and 20 years, switches from private management to public management are becoming more and more frequent. Some recent papers (see infra) discuss the reasons motivating governance switches. Finally, an important question is the potential scope economies as France is characterized by a myriad of—rather small—water services. A recent law[2] encourages the grouping of public services to simplify their management and to take advantage of scale economies. The expected effects of such grouping are better efficiency, increased capabilities, and more effective regulation.

This chapter presents the French institutional context and reviews some of the important results found in the literature. It is organized as follows. Section 1 introduces water regulation in France, by insisting on the various governance forms. Section 2 presents the impact of private sector participation on different measures of performance. Section 3 introduces current issues such as merging services to increase scale economies, remunicipalizations, and social tariffs. A brief conclusion follows.

1 Regulation

Governance: In-House or Contracted-Out

In France, the responsibility for public services' provision is public, but the management can either be public or private. Although some municipali-

[2] The so-called loi NOTRe (law of August 7, 2015) on the new territorial organization of the Republic.

ties manage production through direct public management and undertake all operations and investments needed for the provision of the service, the dominating organizational form is private management. In this case, the delegating authority contracts with a private operator for the management of the service. A private operator, independent of the local government, is hired to manage the service and operate facilities through one of the four different private-public arrangements. The most common is the *lease* contract in which the operator manages the service, invests in the network, and gets a financial compensation through consumer receipts. Under a *concession contract*, the operator also undertakes construction risk, as it must finance a large part of investments over the duration of the contract. This contractual agreement differs from the lease contract in that operators share risk in exchange for greater decision rights and claims on revenues. Other contracts can be chosen by the local authority such as the *gerance* contract in which it pays an operator a fixed fee or an *intermediary management contract*, that is, a gerance contract but with a small part of the operator's revenues depending on its performance. Such contracts provide few incentives to reduce costs and transfer no risks and no decision rights to the private operator. According to the Cour des Comptes (2011), the highest financial court in France, 71% of the population is covered by a private operator for water provision and 56% for water sanitation.

Contrary to other industrialized countries, there is no price-cap or rate-of-return regulation for water utilities in France as there is no national regulator. Such regulation has been replaced by a contract in case the public authority decided to contract out service provision, or a decision of the municipality board in the case of public operation. In the case of delegated management, rules have been defined to ensure that standards are respected during the operation to limit the opportunistic behavior of operators and preserve competition between firms. A first step was the introduction of the "Sapin Law" (1993) which defined a national legislative framework governing the conduct of the bidding process. Under the "Sapin Law", the institutional framework to select the private partner was the following. If the public authority chose a lease or a concession contract, it selected its partners in two steps. First, the public authority launched a classical invitation to tender which was open to all interested private water companies. Second, there was a negotiation phase between the public authority and short-listed bidders. At the end of the negotiation, the public authority chose its final partner for the duration of the contract. The selection of the private company followed the so-called *intuitu personae* principle according to which the municipality or the community set a list of criteria to select the firm considered as the best partner.[3] Following the transposition of three European directives on contract attributions (Directives 2014/23, 2014/24, and 2014/25) in 2016, the selection of the private company does not follow the *intuitu personae* principle anymore. Now the head of the public authority must select the best offer in regard to the "global economic advantage" on the basis of several "objective criteria" and guaranteeing an "effective competition". Service quality and sustainable development are also potential criteria. Figure 5.1 summarizes the different contractual options for public services in France.

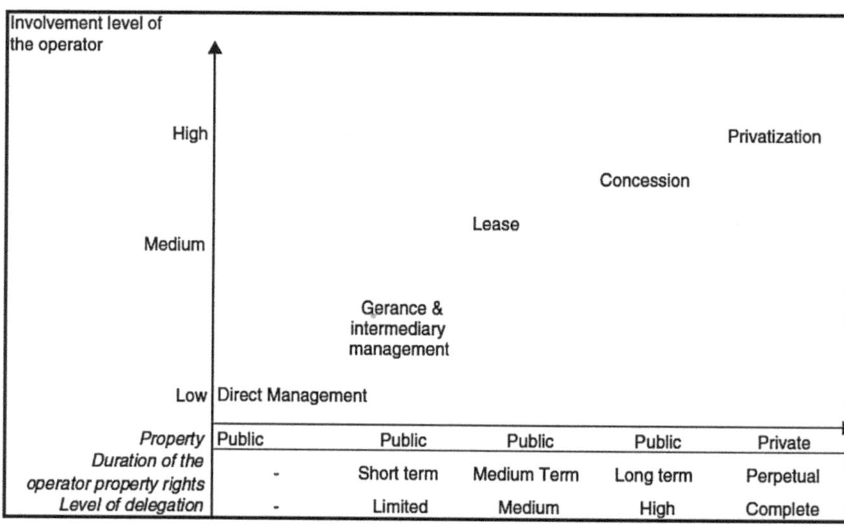

Fig. 5.1 Contractual options for public services in France (Chong et al. 2006)

[3] However, the number of bidders remained low, around 1.9 for each bidding process (Guérin-Schneider and Lorrain 2003).

Quality

Since the "Barnier Law" (1995), private operators have to produce every year an *Annual Report on Prices and Service Quality* (*Rapport prix et qualité du service* in French). These reports include yearly indicators on water public services such as detailed prices for a standard bill, leak ratio, water production, and water quality[4] among other indicators. It is thus possible to see, follow, and compare the evolution of quality and price in France. For example, water quality is relevant for more than 99% of the tests run (Porcher 2012), and the leak ratio slightly decreases every year because of the amount of investments made to prevent leaks (CGDD 2010).

Furthermore, rules have been defined to ensure that standards are respected during the operation to limit the potential opportunistic behavior of operators. These rules support water quality, duration of contracts, and information about management and provision quality. In the case of water quality, a precise definition of more than 60 verifiable quality parameters has been set by the 1992 Water Act to ensure that water services, would they be private or public, respect quality standards. Consequently, water quality is respected and is rarely below a 95% score of conformity to the standards of the microbiological analysis. Moreover, limits on duration have been implemented, and management and provision information is now required to be publicly reported. To ensure competition among operators, the "Barnier Law" (1995) limits the duration of contracts and includes an automatic renegotiation of the contract every five years. To struggle against information asymmetries, the executive power passed a decree in 2007 to force municipalities and communities to provide 14 performance indicators in the mayor's *Annual Report on Prices and Service Quality*. These performance indicators and other data about water and sewerage services have been collected from 2009 on by the *French National Agency for Water and Aquatic Environments* (*Observatoire national de l'eau et des milieux aquatiques* or ONEMA in French) to give users and citizens information about their water services. However, the data collection appears as being a failure so far as there is a lack of incentives for public authorities to fulfill the data on a regular basis.

[4]Water quality is measured as the % of compliance to the microbiological/physicochemical indicators.

Price Settings

In the case of delegated management, public authorities face the classic regulatory problem: they find themselves in an information asymmetry position and have few tools to carry out their essential tasks. However, rules have been implemented to limit opportunistic behavior by private operators. For example, in renegotiating prices, operators are constrained by the fact that in administrative contracts, all renegotiations that change the value of the initial contract by more than 5% trigger a new selection process of the private operator. Even if this safeguard is rarely used, it provides a credible power to local authorities in order to prevent opportunistic behavior from an operator.

Price setting is different whether the local community has chosen to delegate the service to a private firm or not. Under direct public management, the municipality council designs prices in order to generate revenues that allow the utility to cover its costs. European legislation (Directive 2000/60) requires the water utility budget to be balanced following the so-called full-cost recovery principle. Prices are thus set to cover operating and capital costs.[5] Administrative accounting rules are framed so that municipalities hold two separate accounts for the water utility budget. The first account is an operating budget and the second is an investment budget. Net revenues from the operating budget are automatically transferred to the investment budget in order to limit operating costs. This is usually the case if the municipality undertakes a multi-year investment program. While the "full-cost recovery" principle usually implies a zero-margin cost structure, margins are however possible, but the way they are used is highly controlled by administrative rules.

Under private management, the price structure is determined by projecting financial accounts provided by the operator over the duration of the contract. The contract includes periodic revisions of water tariffs using a price index adjusting formula. The relationship between the local

[5] There is little historical evidence of the application of this principle. However as large cities' accounts are now published every year, there is strong evidence of the application of this principle in recent years. The highest financial court in France, the Cour des Comptes (2011), notified several municipalities that their prices were too high, therefore using municipal budgets to fund non-water spendings, or too low, that is, subsidized by another municipal budget.

municipality and the firm is formalized by means of a contract that specifies a price structure, a formula of price revision, and negotiated clauses allowing for exceptional conditions. Since the bargaining power is often considered to be favorable to firms, the price structure is likely to reflect a monopolistic behavior rather than social welfare maximization.

Water Taxation

In France, negative externalities and resource protection are considered in the water tariff structure, namely, to internalize externalities, as the polluter-pays and user-pays principles are clearly stated in the law. Two taxes have been implemented to finance the six water agencies, implemented at the river basin level, for the purpose of subsidizing projects to address pollution and protect resources. The product of taxes is collected by the utilities and transferred to the basin agencies.

The per-unit rates of these taxes, set up by the basin agencies, depend on the geological characteristics of the basin. These characteristics include the origin of water and the condition of the sources, for the resource protection fee, and pollution intensity for the pollution fee. On average, the pollution fee is a 0.21 euro tax per cubic meter, while the resource protection rate is a 0.52 euro tax per cubic meter (Porcher 2017b). The level of taxes is computed using a simple formula at the municipal level: inhabitants are added to seasonal visitors (weighted by a coefficient of 0.4) and multiplied by an agglomeration coefficient going from 0.5 to 1.4 depending on the size of the city. A predetermined cost of pollution per capita which is differentiated from city to city is then multiplied by the artificially computed number of inhabitants. The result is the corresponding amount of pollution charges that is to be paid by the municipality to the water agency. It is then divided by the billed units to determine the level of tax per unit. The total amount of water taxes is equivalent to 2 billion euros per year. To the best of our knowledge, France has been the first country to implement such water taxes (in the 1960s), but similar taxes exist in the Netherlands (since the 1970s), Germany (1981), and Denmark (1997). These taxes are also designed as incentives to consume less water, especially in water-stressed regions.

Datasets

An important feature of water governance is the availability of datasets enabling comparisons between water public services. There are two main datasets in France. As mentioned above, there is no national regulator in France. There is however the abovementioned *National Observatory of Water and Aquatic Environments* (ONEMA in French) which gathers data on the different regulatory indicators of water in France since 2009. The dataset, named SISPEA, is publicly available. Data on water services characteristics can be found, such as the governance form, the length of the network, and the regulatory indicators (e.g. the rate of complaints, the index of resource protection, etc.). A complete list is reported in Appendix Table 5.1. It is in principle compulsory for local governments to fulfill every year the values of the different indicators.

The data collection of the ONEMA follows the work that was performed since the beginning of the 1990s by the *French Institute for the Environment* (IFEN in French). IFEN collected data every three years on a representative subsample of 5000 French municipalities. Datasets were available for 1998, 2001, 2004, and 2008, and some early datasets were collected in the beginning of the 1990s. The IFEN dataset is the dataset used in most of the studies reviewed in the following sections. Note that the IFEN dataset is a city-level dataset, while the SISPEA dataset is a service-level dataset. A service can be managed by a single city or by a group of cities. IFEN dataset is thus disaggregated at the city level when a service includes several cities.

Overall Distribution of Regulatory Functions

The above sections on governance forms, price setting, quality indicators, water taxes, and collection of data show that France is characterized by a rather complex organization of regulatory functions and powers (Salvetti and Canneva 2016). For example, quality standards are supervised by the Health Ministry, the incentives for the efficient use of water resources are managed by water agencies, tariff regulation or governance forms are managed by the service (see Table 5.2 in Appendix). The same applies for regulatory powers, where a various set of local and national actors audit, fine, or enforce compliance with regulation (see Table 5.3 in Appendix). Overall, the regulation functions and powers are very fragmented in the French water sector. Numerous stakeholders are involved, at different scales and in different regulation areas. Nevertheless, this multi-level governance is observed in many OECD countries (OECD 2015) and results from the conception of water sector regulation as being domain specific. For example, the Environment Ministry sets up the water treatment standards in order to respect the environment, while the Health Ministry is in charge of setting the compulsory quality standards for drinking water. A myriad of actors might however reveal a coordination problem, especially in a context of highly decentralized water services as in France. The issue of grouping or merging local governments to enforce them becomes then central in order to rationalize the number of providers and to make regulation more effective.

2 Private Sector Participation and Performance

Private Sector Participation and Price

Many papers study the relationship between private sector participation and price in France. Early studies by Ménard and Saussier (2003), Glachant and Miessner (2003), and Boyer and Garcia (2004) show that municipalities with complex services tend to use private management rather than public management. Carpentier et al. (2006) use treatment effects on a large dataset of French municipalities for 1998 and find that private management copy with harder operating environments. Accounting for the complexity of services explains a large part of the price difference between public and private operators. When there is complexity, municipalities thus prefer to pay higher prices rather than directly manage water services. The average difference, ceteris paribus, is rather small (5% per cubic meter in 1998), and the difference is not statistically significant for large municipalities.

Chong et al. (2006) use a dataset of 5000 French municipalities in 2001 and find ceteris paribus an 11-euro premium of private management relative to the direct public management on baseline bills of 120 cubic meter consumption per year. This 11-euro premium is lower to what simple descriptive comparison would find (30 euros on average). They also find that local governments are keener to contract out the management of water public services if they are more technically difficult to provide. A recent paper by Chong et al. (2015) studies the impact of contracting out on price using a large panel of 3700 municipalities and shows that, over the 1998–2008 period, contracting out increases price. However, the impact narrows or disappears when one considers cities with more than 10,000 inhabitants, as they have probably more capabilities to negotiate contracts with private operators.

There are several reasons that can be used to explain the gap between public and private performance. A first reason is that margins are expected to be higher in private management, in comparison with public management. Such a margin could create efficiency costs, because marginal prices would be set higher than marginal costs. A study of margins in France (Porcher 2014) on a subset of utilities in 2008 shows that margins seem to be higher in services directly managed, around 15 cents of euros per cubic meter, than in services managed by a private operator—around 12 cents of euros per cubic meter. The remaining difference in price between public and private management cannot be related to differences in margins. The price gap between public and private operators might best be explained by some differences in transaction costs, less competition or less effort to decrease costs and thus prices by private operators, especially in small services, or by differences in debts.

A second argument is that debt of the public service can be used to fund investment instead of using the "full-cost recovery" principle. Indeed, in France, public services such as water, waste, and public transport have their own budget ("supplementary budget" or *budget annexe* in French) which is appended to the budget of the municipality. Such a supplementary budget means that all the costs of the water industry must be covered by the price paid by consumers. This suggests that if the price does not cover the costs, the supplementary budget is in deficit, creating a debt for the public service, which cannot be refunded by increasing local taxes or using the surpluses of some other supplementary budgets (such as sanitation or transportation). Directly managed public services are typically expected to have lower prices than contracted-out services but higher debt of the supplementary budget because city councils are reluctant to increase prices in the short term. On the contrary, under *lease* contracts, private firms would tend to have higher prices than directly managed public services but lower debt of the supplementary budget. A recent study by Porcher (2017a) uses an original dataset of 116 water utilities in 2009 serving more than 9 million inhabitants in France. The study shows that debt per customer is, respectively, on average 319 euros and 133 euros under public and private management. In Porcher (2017a), debts of the supplementary budget of water are assumed to be refundable immediately, in five years or ten years under a 2% interest rate. Prices are then recomputed to consider debt refunding using the different assumptions. Under an immediate debt refunding hypothesis, there is an average difference of 150.81 euros for an annual bill between public and private management. The difference decreases to 5.57 euros for an annual bill in favor of private management under a five-year debt refunding hypothesis, and it becomes 13 euros in favor of public management under a ten-year debt refunding hypothesis. It thus shows that by using a period of debt refunding longer than the standard contracts with private operators, municipalities succeed in decreasing price in the long run despite important levels of debt.

A final argument, explained in Chong et al. (2006) and Chong et al. (2015), is that transaction costs or capabilities might explain the performance gap between public and private management. Indeed, as the gap narrows with the size of the city, even controlling for scale economies, it seems that the level of transaction costs or capabilities differs depending on the size of the city. Most probably, the lack of information and measurement about transaction costs or capabilities in these studies makes these interpretations prospective; moreover, a good measure of transaction costs should be compared with a good counterfactual allowing the computation of bureaucratic costs.

Private Sector Participation and Cost Efficiency

In countries like Germany or the UK, the availability of data on costs and revenues enables researchers to work on better measures of performance, such as cost efficiency, that is, the level of costs related to the scale characteristics of the service, such as population, billed units, and length of the network. In France, such data is not directly available but can be retrieved in some of the mayor's *Annual Report on Prices and Service Quality*. Le Lannier and Porcher (2014) collected information on costs and revenues in these reports to provide a benchmarking of big water services in France.

Using a dataset made of 177 water services serving more than 20,000 inhabitants in 2009, the authors compare the efficiency of private and public management. To measure efficiency, the authors use a mixture of data envelopment analysis (DEA) and stochastic frontier analysis (SFA) to obtain efficiency scores ranging between 0 (the less efficient) and 1 (the more efficient). An interesting contribution of the article is that the methodology allows controlling for the heterogeneity in the environment of the water services. Some environmental variables relative to the service itself, such as the type of water (ground or underground) or treatment (simple or complex), or to the geographic area—touristic area, density, and so on—are added to the model. The reader can thus check the evolution of the efficiency scores for each water services, before and after integrating the environmental variables. The results show that water services under private management are on average more complex to manage. Accounting for environmental variables increases efficiency by 0.1 under private management, while it only lifts up efficiency by 0.059 for public management. However, even after having taken environment variables and statistical noise into account, private management remains on average less efficient than public management. Public management has an efficiency score of 0.883 against 0.823 for private management. As a summary, even if the technical efficiency gap is narrowing after correcting for structural differences, it remains significantly positive. This gap partly results from a widespread technical efficiency of water services under private management.

Private Sector Participation and Quality

A few papers use quality as a performance indicator for at least two reasons. First, apart from water quality and the leak ratio, most of the regulatory quality indicators are either not collected (this is the case in the IFEN dataset) or not well fulfilled (this is the case in the SISPEA dataset). A good measure of service quality such as the rate of complaints, to capture the feeling of the user, is not available. Second, water quality is largely uniformed in France, with small disparities between services and governance forms.

There are however three studies which could be mentioned on water quality in France, even if their focus is not quality per se. The first one is Ménard and Saussier (2000) that links governance choice with quality parameters, using a dataset of 2000 observations for 1995. Their measure of quality is a dummy variable equal to 1 if a water service has been identified as failing to meet quality parameters at least once in the year and 0 otherwise. They show that there are no significant differences between public and private management. Another paper by Porcher (2012) uses a dataset of more than 2200 French municipalities observed between 1998 and 2008. The paper shows that private management is associated with higher water quality, measured as the percentage of tests passing the required microbiological standards. The impact is nevertheless rather low, around 1 point of percentage.

Finally, another way to approach quality is to focus on the leak ratio rather than water quality. The leak ratio can be used as a proxy of the care of the manager to reduce water losses. It is measured as the ratio of losses and production. In a recent paper, Chabrost et al. (2018) compute the overleak for French services, in order to see if this variable is a determinant of remunicipalizations. The overleak is the difference between the leak ratio under a given governance model and the predicted leak ratio if the service had the other governance model. For example, for services directly managed, the overleak variable is computed as the ratio of the difference between the observed leak ratio in a given period and the econometrically predicted leak ratio that would exist if the service was managed by a private company in numerator and the observed leak ratio in denominator. For directly managed services, the overleak is positive, that is, the computations show that direct management is associated with higher leak ratios.

To conclude, the literature shows that private sector participation is associated with higher water quality and better network performance, that is, lower leak ratios.

3 Some Other Issues: Scale and Scope Economies, Remunicipalizations, and Social Tariffs?

Scale and Scope Economies

The so-called loi NOTRe (2015), focused on territorial reform, is trying to enforce municipalities and inter-municipal bodies to group together to reach a more relevant scale for water provision and management. The expected result is that the overall water policy framework will be rationalized in order to ensure better regulation, efficiency, effectiveness, and accountability of water policy and water utilities. There are however several papers computing the potential scale economies of reducing the number of water services. Each water service would then increase in production together with an increase in the number of customers, reducing the average variable cost of production. Garcia and Thomas (2001) show that municipalities have an interest in merging and forming a single water utility up to a certain scale. Communities of two to five municipalities tend to have the higher degree of scale economies. Creating a water district seems to be more profitable for local communities with high population density.

Some other pieces of research compute the scope economies of bundling the management of water and sanitation services in the hands of a single operator. Chong et al. (2013) use the IFEN dataset from 1998 to 2004 and show that local governments regularly select the same governance mode (and operators) for water and sanitation services. Their results show that using the same operator for both water production and sanitation leads to a significant price reduction for consumers, between 8 and 11 euros for an annual standard bill per household. Scope economies thus seem to exist and could significantly improve the efficiency of water services.

Remunicipalizations

In a report for a NGO, Kishimoto et al. (2015) compute that nearly 50% of remunicipalizations worldwide take place in France. While delegated management for water services has been used for a long time, it is interesting to investigate what the main drivers of remunicipalizations are. Indeed, there were several cases of remunicipalizations in France, including Paris in 2009, while the option of remunicipalization has been discussed in major cities such as Bordeaux, Marseille, or Nice. In a recent paper, Chabrost et al. (2018) discuss the main drivers of remunicipalizations. The authors believe that the main determinants might be dissatisfaction with private management performances. A good proxy for underperformance of water services can be computed as excessive prices or underinvestment. Besides these efficiency concerns, other matters might be part of the explanation, especially political concerns such as ideology, willingness to be reelected by politicians, or mimetic behaviors between local governments. The intuition for the latter is that small local authorities might not have the capabilities to benchmark their performance relative to other local governments. As a result, the number of observed remunicipalizations around might be an incentive to remunicipalize itself.

The authors compute two different variables to measure performance. The first variable, *overprice*, is the ratio of the difference between the price in a given service and the price that would be observed if the service was run in the alternative governance form and the price in the service. For example, for a water service directly managed, overprice is the ratio of the difference between the current price and the price that would be observed under private management and the current price. The numerator is a "predicted price" using an endogenous regression switching model. In our case, a positive overprice means that switching from public to private management would decrease the price. On the contrary, a negative overprice means that current price is lower than what it would be under private management.

The second variable, overleak, is similarly computed, using leak ratios instead of prices. A positive overleak means that the observed leak ratio is higher than the predicted leak ratio under the other governance form. In this case, a switch to the other governance form could decrease the leak ratio.

Chabrost et al. (2018) link overprice and overleak to the probability of remunicipalizing. They find that overprice and overleak are influencing positively the probability of remunicipalization. The results show that cities paying more and experiencing more leaks under private management than what they would have had under public management are more likely to remunicipalize. The finding suggests that the observed remunicipalizations can be explained by the municipalities' willingness to increase water services efficiency.

Social Tariffs

In France, the so-called Brottes Law (2013) promotes the experimentation of discriminative pricing based on income, for example, the implementation of social tariffs for the poor. As mentioned above, in France, water tariffs have two parts, a fixed part, covering fixed costs, and a variable part, covering the variable costs of production. The "Brottes Law" (2013) allows local governments to experiment multi-tiered pricing based on the level of income. Mayol and Porcher (2018) use a unique dataset from Dunkerque, a city in the North of France, to study the impact of tariff change on consumption and to compute consumers' price elasticities.

The case study of the city of Dunkerque is particularly interesting. The city has implemented a three-part tariff and a social tariff. The three-part tariff works in the following way: a "vital consumption" tier below 75 cubic meter per household per year, a "useful consumption" tier between 75 and 200 cubic meters per household per year, and a "comfort" tier above 200 cubic meters. The marginal price, exclusive of taxes, is, respectively, 0.84, 1.56, and 2.07 euros per cubic meter. The social tariff consists in a large rebate on the first tier for consumers eligible to the universal health insurance, a social benefit for the poor in France. Consumers eligible to the universal health insurance would pay 0.32 euro per cubic meter for the first tier. Mayol and Porcher (2018) show that the change in the tariff structure increases consumption for consumers in the first tier and largely decreases consumption in the second and the third tier. Interestingly, the results show also a tendency of consumers to bunch around kink points. The authors do not discuss the efficiency costs of such measures, which are probably high, as consumer elasticities are negative and significantly different from 0, while marginal prices differ from the marginal cost of production.

Social tariffs are an interesting policy instrument to study because they raise the question of equity and efficiency. While social tariffs seem to clearly favor equity in use, they also distort prices and create efficiency costs. An alternative social policy to social tariffs is to set marginal price equal to marginal cost and to use water coupons for the poor funded by a small increase in fixed-part tariffs for the other consumers or to use fixed-part tariff exemptions for the poor (Porcher 2014). The struggle against water poverty is an interesting issue which demands further research.

4 Conclusion

This chapter gave a thorough outline of water regulation in France. Water regulation in France is highly decentralized, with different governance forms observed for water and sanitation. The use of private firms to manage water and sanitation is rather common. Contracting out increases price and quality overall. The questions of scale and scope economies, remunicipalizations, and social tariffs have been hot topics in the recent years. However, the overall expected impact of these three phenomena on the sector's efficiency is still to be assessed and discussed.

A final question that remains open is whether the French model of water regulation provides better results than the British model of water regulation. It seems that sunshine regulation can be informative for local governments and give them more capabilities in negotiating contracts with private operators. Nevertheless, the implementation of a single centralized regulator is a disregarded option at the moment, as French local governments are keen on keeping their prerogatives in managing water and sanitation public services. Moreover, the current quality of the SISPEA dataset does not allow doing a proper benchmarking of a representative set of water services. Efforts should be made in order to collect data of good quality to be able to increase the quality of water services' benchmarking.

Appendix

Table 5.1 Variables of the datasets

Variable	Definition	Datasets
Population	Population connected to the water service	SISPEA, IFEN
Price of water for a standard bill or a cubic meter	Price, inclusive of all taxes, for a standard bill of 120 cubic meters; price divided per 120 cubic meters	SISPEA, IFEN
Tax-exclusive price and taxes	Price, exclusive of taxes, and decomposition of taxes in the tax-inclusive price	IFEN
Water quality	% of tests passing the compliance to the microbiological/physicochemical indicators	SISPEA
Index of knowledge of the network	Index from 0 to 120, based on a marking system	SISPEA
Network performance	Ratio of consumed units to production	SISPEA, IFEN
Average rate of network renewal	Average rate of network renewal—measure as the percentage of new pipes on all pipes—on the last five years	SISPEA, IFEN
Amount of debt waivers and transfers to solidarity funds	Amount of debt waivers and transfers to solidarity funds in euros per cubic meter	SISPEA
Unexpected rate of water interruptions	Number of water interruptions for 1000 customers	SISPEA
Duration of debt extinguishment	Number of years to refund the debt of the water service	SISPEA
Rate of unpaid bills	Ratio of unpaid bills to total bills	SISPEA
Rate of complaints	Complaints for 1000 customers	SISPEA

Table 5.2 Distribution of regulation functions in the French water sector

Regulation functions	French organization bearing the function	Level
Quality standards for wastewater treatment	Environment Ministry	National
Quality standards for drinking water	Health Ministry	National
Supervising utilities' financing activities	Accounting chambers	Regional
Supervision of contracts with private sectors	Service (municipal or inter-municipal body)	Local
Carrying management audits on utilities	Service (municipal or inter-municipal body)/accounting court	Local/national
Promotion of innovative technologies	Water agencies	River basin
Defining public services obligations	Central government	National
Promoting demand management	Central government/service (municipal or inter-municipal body)	National/local
Incentive for efficient use of water resources	Water agencies/central government	River basin/ national
Defining technical and service standards	Central government	National
Customer engagement	Service (municipal or inter-municipal body)	Local
Incentive for efficient investment	Water agencies	River basin
Consumer protection and dispute resolution	Rights defender	National
Uniform systems of accounts	Finance Ministry	National
Information and data gathering	ONEMA	National
Analyzing utilities' investments/business plans	Service (municipal or inter-municipal body)	Local
Monitoring of service delivery performance	ONEMA	National
Tariff regulation	Service (municipal or inter-municipal body)	Local

Adapted from Salvetti and Canneva (2016)

Table 5.3 Distribution of regulator powers in the French water sector

Regulator powers	French organization bearing the power	Level
Compulsory collection of information	ONEMA	National
Mediate to resolve dispute	Administrative courts	Local
Enforce compliance with regulation	Water police for discharge/regional health agencies for drinking water quality standards	Local/ regional
Investigate cases of breaches	Administrative courts	Local
Issue guidelines and code of conduct	Central government/ONEMA	National
Impose fines/financial sanctions	Administrative courts, criminal courts, local authorities	Local
Audit water utilities	Accounting chambers	Regional
Review or approve contract terms	Service (municipal or inter-municipal body)/local representatives of central government	Local
Control profits and dividends	Accounting chambers	Regional
Veto utilities investment plans	Service (municipal or inter-municipal body)	Local
Impose or ban a particular technology	Health Ministry	National

Adapted from Salvetti and Canneva (2016)

References

Boyer, M., & Garcia, S. (2004). *Régulation et mode de gestion: une étude économétrique sur les prix et la performance dans le secteur de l'eau potable*. Working Paper 2004s-60, Cirano.

Carpentier, A., Nauges, C., Reynaud, A., & Thomas, A. (2006). Effets de la délégation sur le prix de l'eau potable en France: Une analyse à partir de la littérature sur les "effets de traitement". *Economie & Prévision, 174*(3), 1–19.

CGDD (Commissariat Général au Développement Durable). (2010). Services d'eau et d'assainissement: une inflexion des tendances ?, Le point sur n° 67, 2010.

Chabrost, M., Porcher, S., & Saussier, S. (2018). *Still Water Runs Deep: Remunicipalizations and Privatizations, or When Local Governments Disrupt the Status Quo*. Working Paper.

Chong, E., Huet, F., Saussier, S., & Steiner, F. (2006). Public-Private Partnerships and Prices: Evidence from Water Distribution in France. *Review of Industrial Organization, 29*, 149–169.

Chong, E., Desrieux, C., & Saussier, S. (2013). Putting All One's Eggs in One Basket: Relational Contracts and the Management of Local Public Services. *Journal of Economic Behavior and Organization, 89*, 167–186.

Chong, E., Saussier, S., & Silverman, B. (2015). Water Under the Bridge: Determinants of Franchise Renewal in Water Provision. *Journal of Law, Economics and Organization, 31*(1), 3–39.

Cour des Comptes. (2011). *Les services publics d'eau et d'assainissement: des évolutions encourageantes*. Report, Cour des Comptes.

Garcia, S., & Thomas, A. (2001). The Structure of Municipal Water Supply Costs: Application to a Panel of French Local Communities. *Journal of Productivity Analysis, 16*(1), 5–29.

Glachant, M., & Miessner, F. (2003). *Analyse économétrique de l'effet du mode d'organisation sur le prix des services de l'eau et de l'assainissement dans le bassin Seine-Normandie*. Working Paper, Cerna.

Guérin-Schneider, L., & Lorrain, D. (2003). Les relations puissance publique-firmes dans le secteur de l'eau et de l'assainissement. *Flux, 52–53*, 35–54.

Kishimoto, S., Lobina, E., & Petitjean, O. (2015). *Our Public Water Future: The Global Experience with Remunicipalization*. TNI Book Ed.

Le Lannier, A., & Porcher, S. (2014). Efficiency in the Public and Private French Water Utilities: Prospects for Benchmarking. *Applied Economics, 46*(5), 556–572.

Mayol, A., & Porcher, S. (2018). Tarifs discriminants et monopoles de l'eau potable: une analyse empirique de la réaction des consommateurs face aux distorsions du signal prix. *Revue Economique*, forthcoming.

Ménard, C., & Saussier, S. (2000). Contractual Choice and Performance: the Case of Water Supply in France. *Revue d'Economie Industrielle, 92*(1), 385–404.

Ménard, C., & Saussier, S. (2003). La délégation de service public, un mode organisationnel efficace ? Le cas de la distribution de l'eau en France. *Economie Publique, 12*, 99–129.

OECD. (2015). OECD Principles on Water Governance. https://www.oecd.org/cfe/regional-policy/OECD-Principles-on-Water-Governance.pdf.

Porcher, S. (2012). *Do Markets Reduce Prices?* EPPP Chair Working paper 2012-07.

Porcher, S. (2014). Efficiency and Equity in Two-Part Tariffs: The Case of Residential Water Rates. *Applied Economics, 46*(5), 539–555.

Porcher, S. (2017a). The 'Hidden Costs' of Water Provision: New Evidence from the Relationship Between Contracting-Out and Price in French Water Public Services. *Utilities Policy, 488*, 166–175.

Porcher, S. (2017b). *Water Taxes and Consumer Behavior in France*. Working Paper.

Salvetti, M., & Canneva, G. (2016). Water Sector Regulation in France: A Complex Multi-model and Multi-Level Regulatory Framework. In A. Asquer, F. Becchis, & D. Russolillo (Eds.), *The Political Economy of Local Regulation*. London: Palgrave Macmillan.

6

Germany
The German Drinking Water Sector

Astrid Cullmann, Julia Rechlitz, and Caroline Stiel

1 Introduction

The German drinking water sector has a long and stable history. Contrary to other sectors of general interest, such as electricity, gas, telecommunications, and postal services, it has not experienced major reorganization induced by liberalization or technological change. It is organized in local monopolies with about 6000 small-sized water utilities, the majority owned by the local municipality. Current challenges

A. Cullmann (✉)
German Institute for Economic Research (DIW Berlin), Berlin, Germany

Faculty of Economics and Management, Technische Universität Berlin, Berlin, Germany
e-mail: acullmann@diw.de

J. Rechlitz • C. Stiel
German Institute for Economic Research (DIW Berlin), Berlin, Germany
e-mail: JRechlitz@diw.de; CStiel@diw.de

S. Porcher, S. Saussier (eds.), *Facing the Challenges of Water Governance*, Palgrave Studies in Water Governance: Policy and Practice,
https://doi.org/10.1007/978-3-319-98515-2_6

emerge from changing demand patterns linked to demographic trends, as well as from rising costs of water purification due to pollution from agricultural land use.

2 Sector Description

Germany is a country rich in water resources. It disposes of 188 billion m^3 raw water sources, of which 25 billion m^3, that is, 13%, were used for water supply in 2013 (UBA 2017b). Out of this, 20 billion m^3 were used for agricultural and industrial production, and 5 billion m^3 for drinking water supply. Since 2004, Germany has been continuously decreasing its raw water usage and remained below the threshold of 20% of total raw water resources, whose transgression would be treated as a sign of water distress. The majority of water is collected from groundwater sources (61%, Destatis 2015b).

The provision with drinking water is a core duty of German municipalities (§ 28.2 of the German constitution (*Grundgesetz*)), which decide on how to organize local water supply. They either provide the service themselves through municipality-owned utilities or confer local provision to a water company from the private sector. The majority of water suppliers in Germany are municipally-owned, while private sector participation is low. Municipally-owned firms supply 87% of the drinking water (VKU 2017). As a result, water provision in Germany is very fragmented with a total of 5948 firms in 2013 (Destatis 2015b). Most water utilities are located in the federal states of Baden-Württemberg and Bavaria (see Table 6.1).

3 Main Issues at Stake

To continue providing high-quality and sustainable drinking water at affordable prices, German drinking water companies must address two main challenges. First, the water sector is strongly affected by changes in consumption patterns caused by demographic factors, which are society aging, rural depopulation, and urbanization. Second, by-products from farming contaminate raw water sources and raise the costs of water treatment in regions

Table 6.1 Water supply by region in 2013

Region	*N*	Water supplied to end-consumers in m³
Baden-Württemberg	1325	557,460
Bavaria	2266	730,869
Berlin	1	189,100
Brandenburg	93	110,968
Bremen	3	35,799
Hamburg	1	95,262
Hesse	448	310,379
Mecklenburg-West Pomerania	52	85,802
Lower Saxony	262	461,312
North Rhine-Westphalia	538	1,068,403
Rhineland-Palatinate	210	214,679
Saarland	47	50,761
Saxony	95	180,960
Saxony-Anhalt	66	113,296
Schleswig-Holstein	448	170,280
Thuringia	82	91,425
Total Germany	**5948**	**4,466,755**

Source: Destatis 2015b

Note: *N* number of water utilities. Connection rate—share of inhabitants connected to public water networks

with intensive agriculture. In order to comply with European environmental regulation, water suppliers have to make significant investments in the near future, while the allocation of costs is a priori unclear.

Demographic Change

The water supply sector is a capital-intensive industry. The firm-specific cost structure is characterized by large shares of fixed cost, which is necessary to operate the network reliably and ensure high-quality provision of drinking water. Londong et al. (2010) assess the share of fixed costs from 70% to 80% of firm's total cost. Moreover, most facilities have a long service life. Both of these factors mean the water supply infrastructure is inflexible and cannot adapt easily to demand changes. Besides a decrease in water consumption per capita, mostly due to a frugal water use of end-devices, the change in the population structure within a water network presents a major challenge for water supply companies (Siedentop et al. 2006a, b).

The decline in population and, hence, the decrease of consumers connected to the water supply network come along with two main issues: First, it becomes more difficult for the water utilities to ensure a high-quality water supply because of less consumption, which causes stagnation zones in the mains with the risk of sedimentation and microbiological contamination. Second, fewer consumers may lead to a change of prices and tariffs. Firm's fixed costs are distributed among less people and can lead to increasing prices per capita. Depending on the pattern of the demographic change, this might affect regions to a different degree. Hence, the demographic change in Germany belongs to one of the main issues at stake.

It is characterized by two main developments: the moving patterns of the population and the aging of society. Both lead to a simultaneous growth and decline sometimes even within the region (BBSR 2017). Today, the majority (47.4%) of the population lives in cities. Only 30.7% lives in larger villages and 21.9% in rural areas, respectively. The moving patterns of German population have intensified the urbanization over the last 20 years (BBSR 2017). While in East Germany about 25% of all rural, peripheral areas lost more than 10% of their population between 1995 and 2005, this trend even intensified until 2015, when 50% of the rural, peripheral areas recorded a decline larger than 10% (BBSR 2017). This trend can also be observed in West Germany, albeit in weakened form. Table 6.2 illustrates this moving behavior over the last 25 years.

The decline of population, especially in the rural areas, is amplified by a general decrease of the population size in Germany, which is mostly due to the aging process. Since 1990, the age structure of the German society has changed: While in 1990 the median age of the population was 37, it increased to 45 in 2013 (Destatis 2015a). Moreover, the fertility rate declined in the same period. Both lead to a trend of a shrinking population, which is also expected to continue in the future (Destatis 2015a).

Table 6.2 Local population development

			Share of the municipalities with a population decline more than 10%	
			1995 to 2005	2005 to 2015
West Germany	Major cities	Central	1.5	0.0
	Mid-sized cities	Central	0.2	0.0
		Peripheral	2.8	0.0
	Big towns	Central	0.4	0.6
		Peripheral	2.5	1.5
	Small town	Central	0.2	0.6
		Peripheral	0.9	3.7
	Rural communities	Central	4.1	4.9
		Peripheral	9.7	12.3
	Total	Central	1.1	1.3
		Peripheral	6.2	8.2
East Germany	Major cities	Central	40.0	0.0
	Mid-sized cities	Central	36.4	12.1
		Peripheral	75.4	29.5
	Big towns	Central	14.0	19.3
		Peripheral	49.4	46.0
	Small town	Central	17.1	27.1
		Peripheral	37.9	51.4
	Rural communities	Central	10.3	25.6
		Peripheral	24.2	53.6
	Total	Central	19.1	21.1
		Peripheral	35.0	50.2

Source: BBSR (2017)

Pollution Caused by Nitrates

Condition of Groundwater Resources

Since 1991 the protection of water against pollution caused by nitrates from agricultural sources is on the European political agenda (EC Nitrates Directive (91/676/EEC)). The Directive 2006/118/EG "on the protection of groundwater against pollution and deterioration" determines a uniform groundwater quality standard issuing a threshold of maximum 50 mg/l nitrate throughout Europe. If the threshold is exceeded, measures must be taken to reduce the nitrate concentration in the groundwater.

In addition, the directive stipulates that in the event of increasing nitrate concentrations over time, preventive measures must be initiated as soon as three-quarters of the threshold (37.5 mg/l) are reached.

Nitrate is a composite of fertilizers and other agrochemical products used in agriculture, which trickles down into the groundwater. As a consequence, groundwater resources in regions with intensive farming and livestock breeding are particularly vulnerable to nitrate pollution.

Figure 6.1 illustrates chemical groundwater quality across Germany. It is based upon data from the EUA monitoring network, which includes 706 nitrate measuring points in areas with agricultural land use. The dark grey areas cover regions with bad groundwater quality, that is, nitrate threshold values are exceeded, while light grey areas denote regions where groundwater is in a good condition.

Areas suffering from bad groundwater quality are spread throughout Germany. However, regional clusters of measuring points are recognizable and coincide with areas of intensive agricultural land use, such as Lower Saxony in the northeastern parts of Germany. More specifically, for the reporting period (2012 to 2014), 28% of the German measuring points of the EU nitrate monitoring network had measured concentrations of more than 50 mg/l, and 23% of the measuring points found concentrations between 25 mg/l and 50 mg/l (Table 6.3, first row). Despite first attempts to reduce nitrate pollution in critical regions, hardly any improvement is observed when groundwater quality is compared to the previous monitoring period 2008 to 2011 (Table 6.3, second row). For that reason, the European Commission started an infringement procedure against Germany in 2014 and referred Germany to the Court of Justice of the EU in 2016 (EC 2016).

Implications for the Water Sector

Nitrate pollution of groundwater resources constitutes a significant challenge for German drinking water companies, since nitrate removal during water treatment is expensive. A recent study commissioned by the German Environment Agency estimates costs for nitrate removal at 0.55

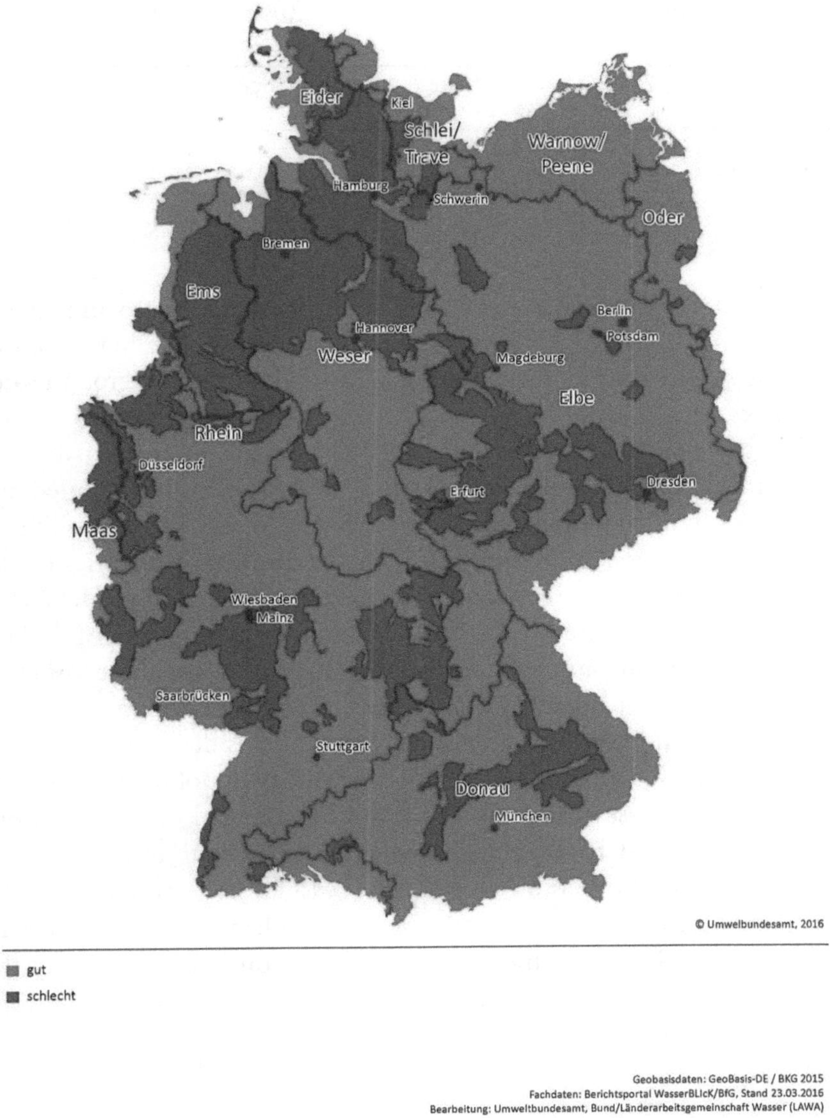

Fig. 6.1 Chemical groundwater quality in Germany 2016. Source: Umweltbundesamt (2017a)

Table 6.3 Distribution of nitrate concentration in EUA network in 2008–2011 and 2012–2014

	≤25 mg/l %	>25–40 mg/l %	>40–50 mg/l %	>50 mg/l %	*N*
2012–2014	49.3	14.2	8.5	28.0	697
2008–2011	48.3	15.0	8.7	28.0	692

Source: BMUB/BMEL (2017)
N Number of EUA nitrate measuring points

to 0.76 EUR/m³ (Oelmann et al. 2017).[1] This would lead to an increase in the water bill from 32 to 45% per year for an average household of four persons living in regions with >50 mg/l nitrate in groundwater resources.

For this reason, currently discussed policies aim for preventive measures, including deepening wells, shifting them, or blending polluted raw waters with uncontaminated water streams. While the feasibility of blending raw water sources depends on the chemical properties of the different sources, it is the least costly measure. Oelmann et al. (2017) consider different case studies and calculate total costs of 5.1 billion EUR, resulting in imputed costs of 21,000 EUR/a for the considered water supply companies. Costs for relocation and deepening are significantly larger and vary by region. Their exact level is determined by local topography, soil conditions, and the availability of unpolluted groundwater sources in those sediments. Estimates for imputed costs range from 36,000 to 55,000 EUR/a (Oelmann et al. 2017). In addition to the investment costs for new dwells come the costs for new connection to the existing pipeline network and the dismantlement of old dwells. Furthermore, deepening the dwell merely postpones the problem to the future, since eventually also deeper groundwater resources will be contaminated through seepage.

[1] The main processes for nitrate removal include biological denitrification, reverse osmosis, and ion exchange (CARIX). According to the German Environment Agency, in 2017 there were three water utilities which applied biological denitrification in plants built explicitly for this purpose (Wasserwerk Broichhof, Wasserwerk Stadtwerke Aschaffenburg, Wasserwerk Frankfurt Flughafen). Furthermore, an unknown number of water plants remove nitrate jointly with other composites such as sulfate in processes of desalination.

An alternative strategy consists of cooperation with the agricultural sector and in fostering the reduction of agrochemical product usage. Several water supply companies established information campaigns for farmers, pay farmers for reduced fertilizer usage, or purchase agricultural land in vulnerable areas. The purchased farm land is subsequently let for rent to farmers who comply with certain ecological standards (e.g., organic farming).

Common to all strategies is the allocation of costs to water supply companies, that is, eventually to end-consumers of drinking water, instead of the agricultural sector, which violates the costs-by-cause principle. This issue is not yet addressed by national environmental regulation.

4 Governance and Regulation

General Regulatory Environment

German cartel law fully exempts the water sector from general competition law and explicitly allows regional authorities to grant monopoly rights for all steps of the value chain to a single water supplier (*Gesetz gegen Wettbewerbsbeschränkungen* GWB §31.1.2). In contrast to the electricity and gas sector, there is no obligation for unbundling generation and retail from networks operation, with the result that the majority of monopolists are vertically integrated. They usually serve all steps of the value chain from water extraction to retailing to end-consumers, even though some firms specialize in bulk water supply, with purified water being traded locally between neighboring municipalities.[2] In summary, water supply is traditionally organized in local monopolies in the borders of the municipal area.

[2] In some cases, water is extracted far away from consumption points and transported over long distances (*Fernwasserversorgung*). The city of Munich, for instance, is served with water from the Alps, and water extracted in the Harz Mountains is transported to the cities of Hannover, Bremen, and Göttingen. Further long-distance routes exist in the *land* Baden-Württemberg (*Bodenseewasserversorgung*), Saxony (*Landestalsperrenverwaltung*), and Thüringen (*Thüringer Fernwasserversorgung*) (Bundeskartellamt 2016).

Water supply is subject to strict environmental and hygienic regulation. Following the EU water directive (2000/60/EC), national and regional laws specify quality and quantity standards for ground- and surface water, thereby regulating raw water usage. For raw water extraction, 13 out of 16 German states charge water suppliers with a variable fee ranging from 0.002 to 0.31 EUR/m³ (BDEW 2015). In the majority of the cases, revenues from extraction fees are earmarked for investments into environmental protection. The decree *TrinkwV* from 2001 stipulates health standards for the provision with drinking water and regulates maximum permissible values for harmful constituents.

Firm Governance and Ownership Structure

Municipalities can choose from a variety of legal structures for organizing water supply. Drinking water supply can be provided by a subunit of the local administration (*Regiebetrieb*, *Eigenbetrieb*) or conferred to a legally independent firm under public law (*Anstalt öffentlichen Rechts*) or private law (GmbH, AG). Cooperation among municipalities is also frequent, formalized through a legal association of different municipalities with the declared goal of joint water provision (*Zweckverband*). Furthermore, municipalities can cooperate with the private sector, either through building joint ventures (water utilities in mixed ownership), through selling the water networks to a private firm (full privatization), or through contracting on single steps of the value chain (*fremde Betriebsführung*).

Each type of organization implies different degrees of municipal influence, as well as rules of taxation and accounting. Organizations under public law are specifically tailored to economic activities of the state, which allows them to apply a distinct tariff scheme exempted from general cartel law (see Sect. 3.2), to enact compulsory connection to the main services (*Anschluss- und Benutzungszwang*) or to employ public servants (Oberndörfer et al. 2011). Furthermore, organizations under public law are exempted from corporate taxation if they do not intend to make profits. Managers' autonomy in organizations under public law is usually limited, and the local government can influence day-to-day decision-making.

Utilities organized under private law, on the other hand, resemble in their governance structure firms of the private sector. Managers are legally independent in day-to-day decision-making, and municipal influence is exerted through advisory boards (*Gesellschafterversammlung*) with a focus on general firm strategies. Private labor law and standard taxation rules apply.

There is no official statistics on ownership structure in German water supply. The industry association *Bundesverband der Energie- und Wasserwirtschaft (BDEW)* represents 1558 German water utilities and regularly publishes figures on water supply. In 2012, 35% of its members were organized under private law, and 65% under public law (BDEW 2015). Since only fully state-owned firms can organize under public law, this implies a lower bound of at least 65% state-owned firms in German water supply. Stiel (2017) analyzed 1170 municipally-owned water utilities based on microdata from the Federal Statistical Office and found that 89% of them opt for an organization under public law, among which 40% participate in municipal cooperation (*Zweckverband*) and 11% are private-law companies. Half of the latter are in mixed ownership with private minorities.

Tariff Regulation

Regulation of water tariffs depends on the legal form of the water company. Notably, differences exist between utilities organized under private sector law and those under public law. While fully private water companies and those in mixed ownership must be organized under private law, fully state-owned utilities can choose between private and public law. Legal structures under public law imply specific rules tailored to the public sector (e.g., in terms of taxation, ownership influence, etc.) and provide the utilities with the possibility to charge fees instead of prices for water consumption. Fees and prices are subject to two distinct regulatory frameworks.

Ex Ante Regulation

Each federal state has a law called *Kommunalabgabengesetz*, which stipulates guidelines for the calculation of water *fees*. While the exact content might differ, general rules include cost recovery (neither long-term surplus nor losses), due proportionality to the services provided, and equality (no arbitrary discrimination of consumers) (BDEW 2015). The level of water *fees* must be approved by the local parliament of the municipality and is published as official decree *Gebührenverordnung*. The calculation of water *prices* is, a priori, not regulated ex ante, and firms can set prices without formal approval of a public authority. However, a judgment from the Federal Court of Justice from the year 2015 clarifies that state-owned firms must follow similar guidelines for the calculation of water prices since the principles of equality, proportionality, and cost recovery apply to any financial conduct of the state (decision no. VIII ZR 106/14, cited in Bundeskartellamt 2016).

Ex Post Regulation

German cartel law *Gesetz gegen Wettbewerbsbeschränkungen* includes a specific section on water supply, which allows the regional and federal cartel offices to open an investigation if they suspect a water utility of charging unreasonably high tariffs (§31–31b GWB). However, a revision of the cartel law in 2011 specified that investigations may only be opened with respect to prices and do not apply to fees (§185.1 GWB). During the investigation, the water utility must either prove (1) that price differences compared to other water companies are justified by (exogenously given) production conditions or (2) that revenues do not reasonably exceed costs. The first procedure is commonly used, and the latter has not yet been applied, mostly due to difficulties in defining appropriate cost levels by the cartel offices (Bundeskartellamt 2016). If investigations were to be successful, orders for price cuts can be issued with respect to future price schemes. However, there are no retroactive sanctions. For this reason, in the past the cartel offices often applied an alternative routine and resorted to general competition law following §19 GWB (Bundeskartellamt 2016). While the burden of proof is reversed and now lies at the cartel offices, orders for price cuts can be retroactive.

Examination of fees is more limited. Each county disposes of an inspecting authority, which surveils the activities of subordinate municipalities

(*Kommunalaufsicht*). The mandate, however, is restricted to control of legality, that is, the authority can only investigate whether the level of fees is in line with existing (municipal) law and, in particular, whether the principles of equivalence, cost recovery, and proportionality are respected. It usually does not involve cost efficiency analyses or comparisons across firms, partially due to lack of expertise and financial resources. Rather, reported costs are taken as given.

Tariffs

Table 6.4 gives an overview on average water tariffs for a typical household in 2013 by region and illustrates price differences of up to 79% between

Table 6.4 Water tariffs across Germany 2013

Region	Variable charge per m³ in EUR	Fixed charge per year in EUR	Total costs for consumption of 80 m³ per year in EUR
Baden-Württemberg	2.01	33.14	206.18
Bavaria	1.45	45.19	193.94
Berlin	2.17	17.58	161.19
Brandenburg	1.52	79.62	201.22
Bremen	1.98	28.62	187.02
Hamburg	1.72	62.92	200.52
Hesse	1.98	33.86	187.02
Mecklenburg-West Pomerania	1.61	87.57	216.37
Lower Saxony	1.23	61.74	160.14
North Rhine-Westphalia	1.62	118.90	248.50
Rhineland-Palatinate	1.70	77.42	213.42
Saarland	1.87	96.99	246.59
Saxony	1.92	114.21	267.81
Saxony-Anhalt	1.64	101.84	233.04
Schleswig-Holstein	1.42	47.17	160.77
Thuringia	2.00	126.07	286.07
Germany	**1.69**	**70.98**	**206.18**

Source: Destatis (2014)

Note: EUR in 2013 values. Gross tariffs collected at municipal level and averaged per region

regions. There are substantial differences in water tariffs across Germany. These differences cannot be fully explained by exogenous production conditions such as topography, land use, population density, or climatic conditions (Monopolkommission 2010, 2014; Bundeskartellamt 2016).

The federal and regional cartel offices opened several investigations, the most prominent being the case against the water supplier of the city of Berlin (*BWB*). While they successfully enforced tariff cuts in some cases (e.g., 20% for *Mainova*, 18% for *BWB*, 15% for *Stadtwerke Mainz*), their mandate is limited to the prosecution of excessive water *prices*. Furthermore, firms can avoid the implementation of sanctions once they reorganize their legal status under public law.[3] Consequently, both the Federal Cartel Office and the Federal Network Agency remain very critical of the coexistence of two distinct regulatory frameworks and argue in favor of a uniform regulatory scheme, where the existing rules in cartel law on water prices extend to water fees. The latest revision of cartel law in 2013 did not react to this concern but instead anchored the dual framework explicitly in law (§185.1 GWB).

5 Empirical Studies on German Water Sector

The water supply industry in Germany faces many challenges such as an increasing nitrate concentration in groundwater and demographic change. Despite the necessary adjustments to these challenges, it is important that water supply utilities operate their business technically efficient, minimize their cost, and provide high-quality services. Hence, a firm-specific analysis of the production process and the cost structure is necessary to examine whether these objectives are met. Otherwise, this shows potential for improvements. This section focuses on empirical studies, which evaluate firm performance that is based on efficiency analysis and regulatory benchmarking. While earlier studies mainly focus on the comparison of the firms' efficiencies, the latest approaches

[3] According to the Federal Cartel Office, such a behavior could be observed in at least four cases, which further weakened the power of investigation (Bundeskartellamt 2016).

have become more advanced by also analyzing the effect of corporatization as well as economies of scale and scope. Since firm-specific data on German water supply utilities is rather scarce due to data protection reasons, the current state of research in this field still leaves many open questions for further analysis.

General Overview

Hanusch and Cantner (1991) conducted one of the first studies that analyzes the relative performance of German water utilities based on 13 water utilities. By setting up a pure production model, they apply the corrected ordinary least squares (COLS) estimator to estimate firm-specific efficiency scores. With a mean efficiency score of 0.963, they detect on average only a small potential for efficiency gains in firm's production process. Zschille and Walter (2012) use a data development analysis (DEA) to assess the technical efficiency of water supply companies. They combine this approach with a bootstrapped truncated regression model[4] in order to account for the firm's operating environment, which otherwise might influence efficiency results. They consider, among other things, water leakage, share of groundwater usage, and population density. Based on a sample of 373 observations in 2006, they find rather low efficiency scores which vary widely across firms. Especially small firms need to improve their production performance.

Scale and Scope Economies

Since German water supply industry is fragmented, special attention is given to the analysis of economies of scale and scope. In a cross-sectional analysis of 47 companies in 2000 and 2001, Sauer (2005a, b) determines potential economies of scale of water utilities in rural areas. Using a generalized McFadden cost function, he finds that the optimal firm size is substantially larger than the ones observed. Hence, especially smaller companies can benefit from utilizing these economies of scale. Drawing

[4] This model is based on an approach by Simar and Wilson (2007).

on a larger data set of 364 water utilities in whole Germany in 2006, Zschille (2015) deepens the analysis: He investigates the gains of a horizontal integration of adjacent water utilities. The analysis rests on conditional DEA which also considers different, firm-specific operational environments. He finds that the potential efficiency gain of merging is small compared to the possible increase of firm-specific technical efficiency by its own. The economies of scale of the water supply industry are further analyzed by Zschille (2014). The estimation of a production model using DEA and conditional DEA provides a greater insight into the scaling properties and delivers non-parametric measures of the sector-specific scale elasticity. The results show that the majority of the observed firms operate under non-decreasing returns to scale, while only large firms exhibit constant or non-increasing returns to scale. Still, the optimal firm size lies above the median firm size in the used data set. These findings are confirmed by the estimated scale elasticities indicating some potential benefits from adapting the size, but they are rather small.

Zschille (2016a) is the first one to take a detailed cost structure of German water utilities into consideration. He estimates a cost function and measures economies of scale (EOS). The analysis is based on a panel data set provided by Research Data Centers of the Federal Statistical Office and the statistical offices of the German states covering the years 2004, 2007, and 2010.[5] He shows that the costs differ depending on the river basin at which the firm is located. The total costs are negatively affected by the population density of the area served as well as the fact that the water utility serves only one municipality. Again, economies of scale are identified just for some companies, which are rather small. Table 6.5 summarizes the estimated EOS. Note that EOS below 1 indicates the corresponding firm is exposed to diseconomies of scale. Firms with EOS greater than 1 can benefit from economies of scale.

[5] For a detailed description of the used data set, see Zschille (2016b).

Table 6.5 Estimated economies of scale (EOS)

	Min.	25%	Median	Mean	75%	Max.
EOS	0.816	0.956	0.987	0.988	1.019	1.139
Number of observations with:						
EOS < 1	407					
EOS = 1	0					
EOS > 1	258					

Source: Zschille (2016a)

Note: Estimated EOS are obtained by estimating a translog cost function for 655 firms.

Governance

The impact of different governance types is addressed in the empirical literature as well. Haug (2007) examines the impact of the type of municipal influence on the production performance of public German water utilities. Firstly, a production model is again estimated using DEA; afterward the effect of the type of municipal control is estimated relying on OLS and Tobit regression. Based on a sample of 34 East-German water utilities, he finds that reduced municipal influence affects the firm-specific efficiency scores significantly in a positive way, while the effect of other factors such as population density or the diversification of supply is negligibly small. Ruester and Zschille (2010) investigate the effect of different governance types under different public and private ownership shares on retail prices of water utilities. Based on a data set of 765 observations from the year 2003 and a two-stage Heckman model, they find that retail prices increase with private sector participation. The most comprehensive study on the impact of governance in the water sector is a recent analysis by Stiel (2017). She analyzes a panel of 1170 municipality-owned water supply companies observed over the years 2003 to 2014 and investigates how total factor productivity is affected by different governance types and reorganization. Total factor productivity is estimated in a structural production framework using a control function approach. Results show that the productivity level does not change significantly between firms with a governance type under public or private law (see Table 6.6). Furthermore, companies with private sector participation do not perform on a higher productivity level than those with public ownership.

Table 6.6 Influence of governance on total factor productivity for German water suppliers (2003–2014)

Governance type				
Corporatized under private law		Organized under public law		
Fully municipality-owned	Private sector minority (<50% of the shares)	Fully municipality-owned		Private sector minority (<50% of the shares)
		Single municipal owner	Intermunicipal cooperation	Intermunicipal cooperation
−0.008 (0.004)	0.019 (0.017)	–	−0.001 (0.002)	−0.001 (0.002)

Source: Stiel (2017)

Note: Result from regressing total factor productivity of 1170 water utilities for 2003–2014 ($N = 7578$) on governance, firm size, and population density. Reference category: organization under public law with a single municipal owner (*Eigenbetrieb*). White standard errors in parentheses

In summary, the empirical papers examine different aspects of the German water supply industry: They evaluate the firm performance, search for potential benefits associated with different governance types and find evidence for potential gains of economies of scale and scope. Since their analyses differ in the applied method and data sets, some deviation in the results might be explained.

6 Conclusions

The German drinking water sector has not experienced major reorganization induced by liberalization or technological change. Current challenges in the drinking water sector emerge in particular from demographic changes as well as nitrate pollution by the agricultural sector leading to rising costs. These costs are allocated to the water supply companies violating the costs-by-cause principle, an issue not yet addressed by national environmental regulation.

In Germany, municipalities can choose from a variety of legal structures for organizing water supply (under private law, public law, with private sector participation or without private sector participation), implying different degrees of municipal influence. Empirical results however show that productivity of the water supply firms does not depend on the different governance types. With regard to the water tariffs, we note substantial differences in water tariffs across Germany. As these differences cannot be fully explained by exogenous production conditions, the cartel offices opened several investigations against water suppliers and successfully enforced tariff cuts in some cases.

References

BBSR. (2017). *Raumordnungsbericht 2017*. Bonn: Bundesamt für Bauwesen und Raumordnung.

BDEW. (2015). *Profile of the German Water Sector*. Bonn: wvgw Wirtschafts- und Verlagsgesellschaft.

BMUB/BMEL. (2017). *Nitratbericht 2016. Gemeinsamer Bericht der Bundesministerien für Umwelt, Naturschutz, Bau und Reaktorsicherheit sowie für Ernährung und Landwirtschaft*. Bonn: Bundesministerium für Umwelt, Naturschutz, Bau und Reaktorsicherheit.

Bundeskartellamt. (2016). *Bericht über die großstädtische Trinkwasserversorgung in Deutschland*. Bonn: Bundeskartellamt.

Destatis. (2014). 1000 Liter Trinkwasser kosteten 2013 im Durchschnitt 1,69 Euro. Press Release 14/110.

Destatis. (2015a). *Bevölkerung Deutschlands bis 2060. 13. koordinierte Bevölkerungsvorausberechnung*. Bonn: Statistisches Bundesamt.

Destatis. (2015b). *Öffentliche Wasserversorgung und öffentliche Abwasserentsorgung*. Öffentliche Wasserversorgung. Fachserie 19 Reihe 2.1.1. Wiesbaden.

EC. (2016). Water: Commission Refers Germany to the Court of Justice of the EU Over Water Pollution Caused by Nitrates. EU Commission Press Release. IP 16/1453.

Hanusch, H., & Cantner, U. (1991). Produktion öffentlicher Leistungen—Effizienz und technischer Fortschritt. *Jahrbücher für Nationalökonomie und Statistik, 208*(4), 369–384.

Haug, P. (2007). *Local Government Control and Efficiency of the Water Industry: An Empirical Analysis of Water Suppliers in East Germany*. Halle Institute for Economic Research, Discussion Paper Number 3, February.

Londong, J., Alexeeva-Steininger, J., Welch Guerra, M., Meyer, D., Sauer, S., Steinbrück, S., Schröder, M., Kobel, C., Steinke, M., & Müllerstedt, N. (2010). Auswirkungen des demografischen Wandels auf die technische Infrastruktur von Abfallentsorgung, Wasserversorgung und Abwasserbeseitigung in ländlichen Regionen in den neuen Bundesländern Kurzexpertise im Auftrag des Bundesministeriums des Inneren. Weimar.

Monopolkommission. (2010). Mehr Wettbewerb, weniger Ausnahmen. Achtzehntes Hauptgutachten der Monopolkommission gemäß §44 Abs. 1 GWB 2008/2009. Bonn.

Monopolkommission. (2014). Eine Wettbewerbsordnung für die Finanzmärkte. Zwanzigstes Hauptgutachten der Monopolkommission gemäß §44 Abs. 1 GWB 2012/2013. Bonn.

Oberndörfer, M., Insam, A., Schmitz, M., Fritsche, N., & Blaschke, U. (2011). Kriterien für die Rechtsformwahl von öffentlichen Unternehmen. *Public Governance. Zeitschrift für öffentliches Management, 2011*, 12–15.

Oelmann, M., Czichy, C., Scheele, U., Zaun, S., Dördelmann, O., Harms, E., Penning, M., Kaupe, M., Bergmann, A., & Steenpaß, C. (2017). *Quantifizierung der landwirtschaftlich verursachten Kosten zur Sicherung der Trinkwasserbereitstellung*. Study Commissioned by the German Environment Agency.

Ruester, S., & Zschille, M. (2010). The Impact of Governance Structure on Firm Performance: An Application to the German Water Distribution Sector. *Utilities Policy, 18*(3), 154–162.

Sauer, J. F. (2005a). *The Economics and Efficiency of Water Supply Infrastructure*. Berlin: Logos Verlag.

Sauer, J. F. (2005b). Economies of Scale and Firm Size Optimum in Rural Water Supply. *Water Resources Research, 41*(11), 1–13.

Siedentop, S., Schiller, G., & Koziol, M. (2006a). *Infrastrukturkostenrechnung in der Regionalplanung*. Werkstatt: Praxis Heft 43. Bonn.

Siedentop, S., Schiller, G., & Koziol, M. (2006b). *Siedlungsentwicklung und Infrastrukturfolgekosten—Bilanzierung und Strategieentwicklung*. Study Commissioned by the Federal Ministry of Transport, Construction, and Infrastructure Development and the Federal Office for Building and Regional Planning.

Simar, L., & Wilson, P. W. (2007). Estimation and Inference in Two-Stage, Semiparametric Models of Production Processes. *Journal of Econometrics, 136*(1), 31–64.

Stiel, C. (2017). *Modern Public Enterprises: Organisational Innovation and Productivity*. DIW Berlin—German Institute for Economic Research, Discussion Paper 1713.

Umweltbundesamt. (2017a). *Factsheet: Wieviel zahlen Trinkwasserkunden für die Überdüngung?* Dessau.

Umweltbundesamt. (2017b). Wasserressourcen und ihre Nutzung. Retrieved from http://www.umweltbundesamt.de/daten/wasser/wasserressourcen-ihre-Nutzung. Accessed online 2018-02-09.

Verband Kommunaler Unternehmen. (2017). Zahlen, Daten, Fakten 2017. Verband Kommunaler Unternehmen e.V. Berlin.

Zschille, M. (2014). Nonparametric Measures of Returns to Scale: An Application to German Water Supply. *Empirical Economics, 47*, 1–25.

Zschille, M. (2015). Consolidating the Water Industry: An Analysis of the Potential Gains from Horizontal Integration in a Conditional Efficiency Framework. *Journal of Productivity Analysis, 44*(1), 97–114.

Zschille, M. (2016a). *Cost Structure and Economies of Scale in German Water Supply*. DIW Berlin—German Institute for Economic Research, Discussion Paper 1576.

Zschille, M. (2016b). *Official Data on German Water Supply (Statistik über die öffentliche Wasserversorgung) 2004, 2007, 2010*. DIW Berlin—German Institute for Economic Research, Data Documentation 84.

Zschille, M., & Walter, M. (2012). The Performance of German Water Utilities: A (Semi)-Parametric Analysis. *Applied Economics, 44*(29), 3749–3764.

7

UK
Strong and Weak Lock-in of Water Governance Outcomes in England

Emanuele Lobina

1 Introduction

This chapter looks at the governance of privatised water supply and sewerage in England, with particular reference to the workings of economic institutions and their implications for sustainable water development. The chapter aims at taking stock of nearly 30 years of water and sewerage privatisation—an experiment that arguably remains unparalleled worldwide for its depth and breadth—in view of its relevance for theory and policy. This offers an opportunity to reflect on the validity of those perspectives of Austrian economics that have influenced the institutional design of privatisation in England (Littlechild 1978) and that favour private over public service provision (Sawyer 2009). This stocktaking exercise is relevant for policymakers who are considering the lessons of the emerging remunicipalisation trend (Kishimoto et al. 2015), while pressures to adopt public-private partnerships (PPPs) and other forms of

E. Lobina (✉)
PSIRU, Business Faculty, University of Greenwich, London, UK
e-mail: e.lobina@gre.ac.uk

S. Porcher, S. Saussier (eds.), *Facing the Challenges of Water Governance*, Palgrave Studies in Water Governance: Policy and Practice,
https://doi.org/10.1007/978-3-319-98515-2_7

privatisation remain unabated (Iossa and Saussier 2018)—not least for British policymakers at a time when water privatisation in England is highly controversial.

Focussing on England rather than the whole UK warrants an explanation. On the one hand, neglecting the experience of water service reform in Wales, Scotland, and Northern Ireland makes it impossible to fully capture the diversity of water governance in Great Britain—from a privatised utility turned into a not-for-profit company limited by guarantee in Wales to a public utility responsible for managing operations and relying on the wholesale adoption of PPPs for delivering infrastructure investment in Scotland and a public utility providing fully subsidised services to domestic users in Northern Ireland (Tinson and Kenway 2013; Lobina and Terhorst 2005; Northern Ireland Water 2017). On the other hand, narrowing the attention to England enables an in-depth analysis of a textbook case of outright privatisation whose influence of national and international water policies is second only to the French model of privatisation by delegation. More precisely, the virtually perpetual transfer of managerial control and asset ownership to the private sector is not the most emulated aspect of the English model of water privatisation—even Chile's pervasive privatisation programme consists in the award of concessions to public-private joint ventures, in some cases for an unlimited duration and in others for a fixed-term duration (Lobina and Hall 2007). Aspects of the English model that have resulted in more significant levels of policy diffusion include the establishment of vertically integrated monopolies providing water and sewerage services at river basin level (Lobina and Hall 2008b) and, most notably, its economic regulatory institutions. For example, Italy—as well as most Latin American countries—has taken inspiration from England's Ofwat (Office of Water Services, currently known as the Water Services Regulation Authority) to embrace price-cap regulation (Lobina 2005a; Foster 2005). In this light, close consideration of the English experience appears legitimate.

The remainder of the chapter is structured as follows. Section 2 offers a brief historical overview of water governance in the two centuries preceding the 1989 privatisation. Section 3 illustrates the governance framework of privatised water and sewerage services and its evolution from 1989 to present. Section 4 sketches a framework for the investigation of the duality of governance regimes and organisational efficiency[1] outcomes and proceeds to look at the tensions between different dimensions of sustainability arising from the governance of privatised water and sewerage. Section 5 discusses the implications of findings for the public versus private debate, inclusive of the Williamsonian transaction cost economics and Willig's industrial economics. The chapter ends with concluding remarks in section 6.

2 A Historical Overview of Water Governance: 1789–1989

Before 1800 the majority of urban water services in England and Wales were provided by the private sector. The period from 1840 to 1900 saw a sustained process of municipalisation—that is, the expansion of municipal ownership whether by takeover of private companies or establishment of greenfield municipal utilities. This process was a response to the private sector's reluctance to develop service provision, due to the lack of profitability in expanding access and supplying water of good quality, and was motivated by the ability of municipal enterprises to tap cheap investment finance and prioritise service development over commercial considerations. The pace of municipalisation accelerated after the national cholera epidemics of 1832 and 1849 and became particularly intense between 1861 and 1881, when the share of towns served by municipally owned undertakings went from 40.8% to 80.2%—a share that reached 90.1% in 1901. Most water services were municipalised by World War I; the exception was represented by 28 private companies (the so-called statutory water companies) that in 1974 supplied water to 25% of the population (Hassan 1998; Juuti and Katko 2005).

[1] For the purposes of this chapter, I look at two components of organisational efficiency: productive efficiency and distributive efficiency. By productive efficiency, I mean the generation of value through the production of water services irrespective of whether and how this value is redistributed to stakeholders external to the service provider. Some call this internal efficiency; see, for example, Vickers and Yarrow (1988). By distributive efficiency, I refer to the redistribution of value to stakeholders external to the service provider irrespective of how this value has been generated. Some call this distributive equity; see, for example, Mookherjee (2006).

By 1970, only about 3% of the rural population remained unserved by networked systems (Hassan 1998): virtual universal coverage was achieved thanks to the overwhelming contribution of the public sector, both operationally and financially (Hall and Lobina 2009). For all its social and economic importance, universalisation of access to water is an unsung achievement of the public sector. The difficulties encountered by the municipal undertakings in tackling water pollution—due to a combination of planning difficulties and politicised management, a failure to recover costs through charges, and organisational fragmentation—appear to have received greater attention (Hassan 1998). The achievements and difficulties of municipal providers characterise the so-called Keynesian period when public ownership and subsidised charges were part of a governmental policy pursuing security of supply and social equity objectives (e.g. in light of the "ability-to-pay" principle), as a way of supporting both public health and economic growth (Bakker 2001, 2003). Conversely, wastewater treatment was neglected as this was not the express responsibility of any local authority and the situation was compounded by years of underinvestment in sewerage infrastructure, which was not seen as strategically important as water supply, particularly during World War II and its economic aftermath. This led to growing levels of water pollution and public disquiet (Hassan 1998; Jordan et al. 1977).

By the early 1970s, the British government was reconsidering its policy approach, and this led to the restructuring of the water industry with the 1973 Water Act. The restructuring did not involve the 28 "statutory water companies" and saw the nationalisation (without payment of compensation to local authorities) and amalgamation of 157 water undertakings, 1393 sanitary authorities, and 29 river authorities into 10 governmentally owned and financed regional water authorities (RWAs) (9 of which in England and 1 in Wales) responsible for water supply, sewerage, wastewater treatment, and water resource management at river basin level. The RWAs were to apply full cost recovery, with the introduction of direct billing and the removal of subsidies and cross-subsidies, and deliver a sustained investment programme to meet the quality requirements of EC directives—for higher standards for the quality of drinking water, cleanliness of beaches, and wastewater treatment—while facing severe fiscal constraints in accessing governmental funding. In fact, the RWAs were subject to strong pressures to cut their expenditure and contribute to reduce the public sector borrowing requirement at a time when the UK was under IMF conditionality (Hassan 1998; Vickers and Yarrow 1988; Hall and Lobina 2008). The restructuring induced a gradual change in organisational culture with the introduction of managerialism and a move towards basing payment—according to the "benefit principle"—on the costs imposed by individual consumers (Bakker 2001; Jordan et al. 1977). The RWAs became more driven by productive efficiency due to targets to reduce operating and capital expenditure, while industrial reorganisation on a regional basis was designed to achieve economies of scale and scope. However, regulatory constraints on price increases meant that productive efficiencies were achieved at the cost of service degradation and underinvestment. In turn, this led to customer dissatisfaction (Hassan 1998; Vickers and Yarrow 1988). The new business culture, leaner organisational structure, and negative public opinion of the RWAs meant that the water industry was ready for privatisation.

In 1979, Margaret Thatcher became prime minister with an ideological agenda to promote popular capitalism. In addition to neoliberal ideology, what made water privatisation attractive to the British government was the prospect of financing the expenditure required to meet EU law requirements while reducing the fiscal pressure on public finances. Before the publication of governmental proposals to privatise the RWAs, the 1983 Water Act completed the nationalisation process by severing all remaining links between local authorities and RWAs, as all appointments were to be made by government. The general public lost the right to attend the meetings of the RWAs. Instead, consumer consultative committees were created in each region (Hassan 1998; de la Motte 2005). The Act also promoted the commercialisation of the RWAs—by enabling them to access private capital markets—thus paving the way for the privatisation of the water industry (Ofwat and Defra 2006). In 1986, the government published a consultative White Paper on water privatisation, which emphasised that privatisation and regulation would result in greater efficiency and that profit would represent an effective incentive for the private companies to achieve efficiency (Ogden 1997). It also published a report by "Austrian" economist Stephen Littlechild, commissioned by the Department of the Environment. This was revised in 1988, and the two reports (Littlechild 1986, 1988) came to inform the regulatory framework of privatisation. The 1989 Water Act provided for the privatisation of the RWAs—stripped of their water resource management and environmental regulatory duties that were transferred to a new National Rivers Authority, thus retaining responsibility for providing water supply, sewerage, and wastewater treatment services. The RWAs were privatised in November 1989 when their shares were floated on the stock exchange (Hassan 1998; Vickers and Yarrow 1988; Ofwat and Defra 2006). To ensure the success of the flotation and make it an attractive investment proposition, the government wrote off all the debt of the RWAs and provided a cash injection (termed "green dowry") to the balance sheet of the privatised companies—public subsidies totalling £6.5 billion to which was added a further £7.7 billion in tax relief on the companies' profits (Hall and Lobina 2008).

3 Governance Framework: Overview and Evolution, 1989 to Present

In November 1989, the privatised water and sewerage regional monopolies were renamed water and sewerage companies (WaSCs) and began operating alongside the statutory private water-only companies (WOCs). The WaSCs were monopolies licenced by the government—licences were awarded initially for 25 years and could be terminated with a minimum notification period of 10 years. They were protected from takeover for the first five years by a governmental "golden share" after which takeovers—but not mergers—have been allowed. Three new regulators were created. Drinking water quality was monitored by the Drinking Water Inspectorate. The National Rivers Authority was responsible for monitoring river and environmental pollution until 1996 when this was succeeded by the Environment Agency. Economic regulation was carried out by Ofwat. Customer representation was the responsibility of ten regional Customer Service Committees initially established as an integral part of Ofwat; since 2005 this responsibility sits with the independent Consumer Council for Water (de la Motte 2005; Ofwat and Defra 2006).

Ofwat is a non-ministerial government department—independent of government and accountable to Parliament—tasked with reviewing the private companies' investment plans and, on the basis of the plans deemed to be justifiable, setting price limits for the following five years. This regulatory mechanism, known as price-cap regulation, revolves around the formula RPI + K, which determines the extent to which prices can be adjusted on a yearly basis to inflation, financial requirements, and past performance against operational targets. The system is premised on the expectation that the regulated companies will have an incentive to be always more efficient as they are enabled to retain the additional profits thus generated. At the subsequent price review, Ofwat is then in a position to incorporate productivity gains in setting new performance targets so that efficiency gains are passed on to consumers. Because these targets are comparatively assessed and the performance of the best companies is used as a yardstick in determining the K of the other companies, the system is said to contain an element of "yardstick competition". Ofwat's primary duties include the protection of consumers' interests—for example, through the promotion of competition—and ensuring the companies' financeability. The latter implies securing a rate of return on capital that is sufficient to induce lenders and shareholders to, respectively, make loans to and hold shares in the companies (Ofwat and Defra 2006; de la Motte 2005; Ménard 2017; Bayliss 2017; Ogden 1997).

Ofwat is a powerful and resourceful regulator that in the financial year 2016–2017 had 174 permanent employees and a budget of over £25 million (Ofwat 2017). In addition to running the five-yearly price reviews and having the power to issue fines for breach of customer service standards, Ofwat may amend the companies' licences and initiate policy reviews. It thus wields significant influence on shaping the regulatory framework, and, ultimately, its hard and soft power derives from the centrality of its role in ensuring the success of privatisation. There is a convergence of interests between the economic regulator and the regulated companies as their very existence depends on privatisation, and this explains a symbiotic relationship that is strengthened by a "revolving door" process between the industry and Ofwat (Bayliss 2017). Since 1989, British politics have been dominated by a "Westminster Consensus" on water privatisation. In fact, government has either been led by a

Conservative Party that unquestionably embraced neoliberalism or, from 1997 to 2010, by a Labour Party that, having abandoned its commitment to the common ownership of the means of production, had come to accept water privatisation. The dominance and possible imminent demise of this "Westminster Consensus" are important factors to understand the evolution of water service governance to date. Few salient developments may serve as background to the next sections.

With the election of a new Labour government in 1997, the regulatory environment became more challenging for the private companies. First, the government imposed a windfall tax on utility company profits that recouped most of the £7.7 billion in tax relief from which the water companies' shareholders had benefited at privatisation. Then, the government pressured Ofwat to impose a much more stringent price cap in the 1999 review which required 12% cuts in prices. Finally, the 1999 Water Act banned disconnection of users for non-payment, as well as the use of prepayment meters. The ban, which remains in force to date, was in response to the public health hazards associated with cut-offs. Following privatisation the number of households disconnected had in fact tripled in the first five years, with 18,636 households disconnected in 1994. The combined result of these regulatory measures was to squeeze the profitability of the water industry. The rate of return on capital was halved, from 12% to 6%. The response of many companies to this was to withdraw equity capital as far as possible and instead use debt to finance the great majority of operations. In Wales, the regional utility was taken over by a consortium of US energy companies, who transferred all the assets, liabilities, and statutory functions to Welsh Water, a not-for-profit company limited by guarantee and financed entirely by debt (Lobina and Hall 2001, 2008a; Ofwat 2016).

In 2002, Ofwat made the companies' licences virtually perpetual, thus contradicting its mandate to promote competition. Prior to then, the 25-year licences would be automatically extended unless the government gave companies at least 10 years' notice before terminating the licences. In 2002, Ofwat extended the minimum notice period of termination to 25 years to avoid that the approach of 2004, when a notice would have to be given to terminate any licence in 2014, could drive up the costs of raising finance. Ofwat did so by publishing a consultation document containing the proposal to amend the licences of all companies when Parliament was suspended for the summer holidays and could not discuss the proposal. By so doing, Ofwat served the companies' short-term interest in reducing the costs of accessing finance while giving them a long-term guarantee against the threat of losing their monopoly rights (Lobina and Hall 2008a). Conversely, in 2008, Ofwat acted to promote competition in the market—that is, retail competition in ancillary services like billing and metering—by instigating a review alongside a separate review launched by the UK government. Retail competition is highly unusual in water supply and wastewater because in these highly capital-intensive services the majority of costs borne by consumers are derived from the cost of capital so that the greatest potential for savings lies with reducing the cost of capital rather than operating costs (Lobina and Hall 2008b).

Governance has evolved not only in response to changes in the regulatory framework but also in utility ownership. These changes have increasingly occurred since the expiry of the governmental golden share in WaSCs and, by 2001, saw the emergence of a mixed pattern of domestic and international ownership for both WaSCs and WOCs. If, for example, multinationals Suez, Enron, and RWE owned one WaSC each, Vivendi (now known as Veolia) was the owner of six WOCs. Only occasionally did former foreign owners give way to domestic owners, as in the case of Welsh Water discussed above (Lobina and Hall 2001). By 2012, the landscape had changed dramatically with the growing importance of private equity funds, financial groups, and Asian multinationals. Only three of the ten WaSCs remained listed on the stock exchange, the other four including Thames Water were owned by private equity funds and financial groups, one was a not-for-profit company limited by guarantee, and the remaining two were owned by Asian multinationals. Private equity funds also owned the majority of WOCs after Suez and Veolia had sold their stakes in seven of these companies. The increasing prominence of financialisation in the sector creates a number of regulatory challenges partly due to the reduced transparency that comes with delisting companies from the stock exchange and partly due to the practice, common among private equity funds, of buying up firms and turning them around to extract a quick profit (Hall and Lobina 2012; Bayliss 2017; Tinson and Kenway 2013).

4 Privatisation and Sustainability

In setting out its overall approach to sustainability, Ofwat (2010) identifies five principles: (1) a safe and reliable water and sewerage service for consumers that minimises the impacts on the environment now and in the future; (2) consumers that continue to get a fair deal and receive a level of service that consistently meets their needs; (3) financially robust sectors that are able to meet consumers' needs at a fair cost, into the future; (4) companies that remain accountable to their consumers; and (5) using the best available information to support decision-making. Ofwat's contributions to the debate on the merits of water and sewerage privatisation in England have rarely been framed in terms of comparative advantage to enhance sustainability. Establishing whether this is due to the nature of the public versus private debate or other factors is something that remains outside the scope of this chapter. Yet my contention is that discussing public and private sector performance relative to the advancement of sustainable water development—or other non-individualistic social welfare functions[2]—allows for injecting a healthy dose of realism in the debate. For commentators who were to choose this approach would be compelled to interrogate the whys and hows of the complex interdependencies and tensions between the different dimensions of sustainability through space and time.

Ofwat and the companies have argued that two measures of the success of privatisation are the amount of investment delivered by the WaSCs and the efficiency increases since then. For example, in 2013 Ofwat claimed that £111 billion had been invested since privatisation, equivalent to roughly double pre-privatisation levels (Ofwat 2013). In this respect, it should be noted that the biggest factor driving increases in capital expenditure was EU law on higher standards for the quality of drinking water, cleanliness of beaches, and wastewater treatment, so that the required investment levels would have been delivered irrespective of privatisation.[3] Furthermore, maintaining the growth rates in capital

[2] On non-individualistic social welfare functions, see Desmarais-Tremblay (2016).

[3] The underlying assumption here is that the UK government's commitment to respecting EU law would not depend on the public or private ownership of water utilities.

expenditure achieved under public ownership between 1985 and 1989—at an average annual growth rate of 8%—would have allowed for far greater investments than the £55 billion delivered by the privatised companies from 1990 to 2004 (Hall and Lobina 2008).

As to the second argument, the industry association Water UK commissioned consultancy Frontier Economics to quantify the productivity gains achieved by the nine English WaSCs since privatisation. The report, commissioned in response to the Labour Party's pledge to renationalise the water industry if returned to power (Roberts 2017), finds that total factor productivity growth between 1992/1993 and 2016/2017 has increased by 64% on a quality-adjusted basis and 27% without quality adjustment (Wright and Huggins 2017). First, it should be noted that the study by Frontier Economics is concerned with productivity growth under privatisation and does not assess the relative efficiency of public and private water operations. Second, Dieter Helm (2018) questions the relevance of comparing the efficiency of today's private companies with that of public enterprises in the pre-IT world of the 1970s and 1980s, when manual labour accounted for a much higher share of total costs. In that sense, interestingly, Saal and Parker (2000) find no evidence of increased efficiency under privatisation when comparing public sector performance between 1985 and 1989 with private sector performance between 1990 and 1995. Furthermore, Hall and Lobina's (2008) elaboration of Saal and Parker (2001) and Saal and Reid's (2004) data shows that there is no evidence of total factor productivity growth after privatisation when considering the 1985–1999 period. Finally, for the period 1985–2000, Saal et al. (2007) find that "while technical change improved after privatization, productivity growth did not improve".

In the absence of conclusive evidence on the relative efficiency of public and private ownership, it has become fashionable to argue that what matters for sectoral performance is regulation, not ownership—as if it was possible, in real-world governance, to isolate ownership and property rights from regulation or other formal and informal institutions. One way of avoiding the fallacies associated with this form of blackboard economics is to treat the complexity of both governance and sustainable water development seriously, in recognition of the fact that the latter is the normative objective of the former. On the one hand, governance can be seen as the multi-actor interaction informing the process and outcome of institutional change—of which, water service reform and regulation are part and parcel—whereby interaction is dictated by combinations of actors' interests, beliefs, and power—and institutions comprise formal and informal rules, norms, and customs (Lobina 2013).

On the other hand, sustainable water development can be seen as the viable reproduction of a utility's operations and capital investment—which supports the viable reproduction of a community and the reproduction of interconnected financial, economic, technical, social, environmental, and political regimes—consistently with collective expectations (Lobina 2013). Because the policy process and policy outcome of water service reform are intertwined and path dependent, different governance systems may produce a variety of organisational efficiency outcomes subject to lock-in, understood as a temporary rather than permanent condition. In turn, the more or less temporary lock-in of organisational efficiency, categorised as strong and weak lock-in of public efficiency/inefficiency and private efficiency/inefficiency, will affect the reproduction of governance regimes (Lobina 2017a, c). Hence, a key criterion for comparative institutional analysis is the directionality of collective agency, resulting from the multi-actor interaction that informs the process and outcome of institutional change in the sector. This framework enables evaluating the sustainability implications of privatised water supply and sewerage in England.

Financial Sustainability

Private ownership, full cost recovery, and price-cap regulation have resulted in the financeability or financial viability of the sector. This reflects the fact that guaranteeing financeability is prominent among Ofwat's primary duties. A prima facie case for the claim that privatisation has guaranteed financial sustainability rests on the aggregate value of investments delivered since privatisation—estimated at £126 billion—that have largely been funded by debt (NAO 2015). In this view, the 40% tariff increases above inflation that have occurred since 1989 (NAO 2015) and the £18.1 billion in profits paid to the shareholders of the nine English WaSCs in the ten years from 2007 to 2016 (Bayliss and Hall 2017) constitute a fair price to pay for the industry's liquidity.

However, short-term liquidity does not necessarily correspond to long-term financial sustainability, and concerns have been raised in regard to financial resilience, or the financial capacity to withstand internal and external financial shocks. For example, it has been noted that the £18.1 billion in dividends paid by the nine English WaSCs between 2007 and 2016 represent over 96% of the £18.8 billion in post-tax profits recorded by the companies in the same period. Using almost all profits to finance dividend payment has contributed to an increasingly higher gearing, the ratio between debt and equity, and led in turn to the growing cost of servicing the debt. Aggregate net finance costs for the period 2007–2016 thus averaged £1.2 billion. These challenges to achieving financial sustainability are compounded by the practice of paying disproportionate amounts in dividends, with three WaSCs paying more in dividends than their total pre-tax profits over a period of ten years. In 2016, the aggregate amount paid by all WaSCs for both net interest payment and dividends was in excess of £2.9 billion and corresponded to 28% of turnover (Bayliss and Hall 2017).

The rent-seeking practices of private equity owners appear to represent more cause for concern. These practices—which are integral to financialisation—include raising debts to pay a "special dividend" to shareholders, especially in the course of acquisitions, adding acquisition debt to the regulated utility, and issuing intergroup loans to the regulated utility. "Special dividends"—so-called because they come on top of the payment of ordinary dividends—can be as high as the £717 million paid by Yorkshire Water to the shareholders who bought the company for £3 billion in 2007. Much like the practice of financing the payment of ordinary dividends through loans, and that of adding acquisition debt to the regulated utility, the payment of "special dividends" produces an upward pressure on gearing. Similarly, the practice of contracting intergroup loans means that gearing increases, while value is extracted from the regulated utility, specifically in the form of (tax-deductible) interest paid to the utility's own shareholders—for example, in 2013 Southern Water's accounts showed £67.9 million in interest on loans of £633.9 million contracted from its parent company (Bayliss 2014, 2017). Similar practices of asset sweating are in shareholders' financial interest but risk saddling future generations of consumers with unsustainable mountains of debt.

Economic Sustainability

The impact of privatisation on the reproduction and viability of the economic system depends on whether consumers receive good value for their money, whether whole asset value is maintained, and whether taxpayers benefit from adequate contributions to fiscal expenditure. As regards consumers, the House of Commons Public Accounts Committee stated in 2016 that: "Ofwat, like other economic regulators, has repeatedly overestimated the cost of finance in successive price reviews.... As a result, water companies made windfall gains of at least £1.2 billion between 2010 and 2015 from bills being higher than necessary" (House of Commons Committee of Public Accounts 2016). The practice condemned by the Public Accounts Committee—also known as "gaming"—consists in the companies' overestimating the investment required in the price review period, followed by Ofwat allowing the companies to charge higher prices as calculated on this basis. Because the real expenditure is lower, the companies can pocket the difference as increased profit. This practice has been a persistent feature of regulatory governance since privatisation: the private companies' capital underspend from 1995–96 to 2005–06 has fuelled dividends to the tune of £4.3 billion (Hall and Lobina 2008). In addition, a number of companies—including Severn Trent, Southern Water, and Thames Water—have been found responsible by the Serious Fraud Office and Ofwat for providing false regulatory data with the effect of inflating water bills. For example, in 2006 Ofwat found—after the illicit practice had been revealed by a whistleblower—that Severn Trent customers would pay £42 million more by 2009–2010 (Lobina and Hall 2008b).

Some of the techniques of value extraction that risk undermining the maintenance of whole asset value fall outside the scope of regulatory scrutiny. Bayliss and Hall (2017) note that the full amounts gained through these techniques—which include shareholder loans at higher than market rates of interest, the sale of land and property, and the sale of ownership stakes in the utility—are difficult to determine due to dense corporate structures and to limited disclosure. The sale of land and property to remunerate shareholders is reminiscent of asset stripping in Tallinn, Estonia, in the early 2000s—the private shareholders in local concessionaire Tallinna Vesi were a subsidiary of the WaSC United Utilities and of the multinational IWL—effectively in the form of special dividends funded by reducing the overcapitalisation of Tallinna Vesi's balance sheet and the large amount of idle money on Tallinna Vesi's bank account (Lobina and Hall 2003).

Thanks to a generous tax regime, the contributions made by WaSCs to general fiscal expenditure can hardly be described as adequate. "Despite the profitability of the sector, water companies pay little tax. In a ten-year review of company accounts, the combined tax charge in profit and loss accounts comes to £1.7bn on a pre-tax profit of £20.7bn—around 8%. Actual tax paid can be much lower: in 2013 for example, the shareholders extracted nearly £1bn in dividends, but paid only £1m in taxes" (Bayliss and Hall 2017: 4). Thames Water acknowledges in its Annual Report 2013/2014 that they "have not paid substantial amounts of corporation tax in recent years, principally due to the capital allowances (Thames Water) have legitimately received to incentivise (their) infrastructure investments" (Thames Water 2014: 43). Thames Water (2014) also argues that the deferment of tax payment is justified by high levels of investment in infrastructure—to the tune of £1 billion per year in the period 2010–2015—and that tax relief enables utilities to lower customers' bills. While this is true in principle, in practice WaSCs have proved able to extract profits irrespective of the price levels set by Ofwat—for example, through gaming, asset sweating, and practices falling outside the scope of the regulator—with the effect of exerting an upward pressure on future prices. Also, as discussed below, the cost-effectiveness of Thames Water's investment programme is debatable and poses challenges for achieving technical sustainability.

Technical Sustainability

Technical sustainability consists in the ability of utilities to maintain, renew, and expand infrastructure and to upgrade operational systems so as to deliver quality services in the long term.[4] Thames Water, the largest

[4] This section focusses on leakage reduction, which is an indicator of organisational efficiency adopted across the global water sector. On the persistent problem of sewage flooding in England, see Lobina and Hall (2001) and Priestley (2016). The latter finds that in 2015–2016, 4344 properties were internally flooded by sewage in England and Wales.

WaSC, appears to epitomise the challenges faced by the private water industry to achieve technical sustainability amid a culture of managed under-investment. In June 2018, Ofwat imposed a penalty of £120 million on Thames Water for failing to cut leaks (International Water Association 2018). Following a rising number of annual bursts of trunk mains—including eight high-profile bursts from October to December 2016, which repeatedly flooded homes, damaged businesses, and caused risk to life and road closures—Thames Water commissioned an independent review of the major bursts and leaks occurred in 2016, their potential causes, and Thames Water's own responses to the incidents. The review—carried out by Paul Cuttill, former Chief Operating Officer of EDF Energy Networks—found that the current rate of replacement for trunk mains was estimated at 0.19% per year and that Thames Water's replacement programme was not keeping pace with deterioration (Cuttill 2017). In Japan, where all water utilities are publicly-owned to date, the average rate of renewal of water pipelines in 2015 was 0.74% per year.[5] In Denmark, where the overwhelming majority of water utilities are municipally-owned, a survey of 61 utilities found that the average renewal rate in 2016 exceeded 1% per year.[6]

In 2017, average leakage in England was 20%, with leakage in Thames Water's service area being estimated at more than double the national average. Average leakage in England was down from the national average of 31% recorded in 1994/1995, but it had remained stable since 2013. Also in 2017, Ofwat set leakage reduction targets for the period 2015–2020 of zero for six companies, meaning that no further reduction was expected of the companies, while for three other companies the targets would allow for an increase compared to the leakage levels already achieved by the companies (Carrington 2017; Lobina and Hall 2001). The relative ambition of leakage reduction targets can be explained by the principles underpinning Ofwat's leakage policy. Since the introduction of leakage targets in

---

[5] I owe this insight to Junji Hashimoto. For details, see the Ministry of Health, Labour and Welfare of Japan's "Current trend of water administration", 12 September 2017, particularly page 5 https://www.mhlw.go.jp/file/06-Seisakujouhou-10900000-Kenkoukyoku/0000203990.pdf.

[6] I owe this insight to Niels Knudsen. For details, see the Danish Water and Wastewater Association DANVA's "Water in Figures 2017—DANVA Statistics and Benchmarking", page 24 http://www.e-pages.dk/danva/215/24.

the mid-1990s, these have been informed by economic considerations such as conformity with the so-called sustainable economic leakage level (SELL)—or the level at which it would cost more to make further reductions in leakage than to produce the water from another source (Lobina and Hall 2001, 2008b).[7] The implications of the sustainable economic leakage level for environmental sustainability are discussed below.

Social Sustainability

An important part of social sustainability is to ensure the reproduction and viability of water services while enhancing social inclusivity and expanding access to affordable water and sewerage. The profit-seeking practices of private providers—including the fraudulent misreporting of data, the gaming and asset sweating practices discussed above—exert an upward pressure on prices and represent a key determinant of an alarming increase in water poverty (Lobina and Hall 2008a). In 2014–2015, an estimated 23% of households in England were spending more than 3% of their income on water and sewerage, and an additional 11% were spending over 5% of their income on water services (Ofwat 2015). This meant that more than a third of households in England—that is, 34%—were affected by water poverty, defined as the condition of spending more than 3% of household disposable income in water and sewerage bills. Overall, in 2013 water bills represented more than 5% of average household spending for the poorest 10% of households compared to an average of around 2.3% for all households (NAO 2015). As a result, arrears in water charges are a contributing factor of destitution in the UK (Fitzpatrick et al. 2016). Finally, there is evidence that those who fail to pay their bills tend to be poorer households and that the problem of water poverty is compounded by affordability problems in other utility sectors, while bad debts have risen by 44% between 2010 and 2015 and there has been a 305% increase in the number of people seeking help with their water debts from the National Debtline (Bayliss and Hall 2018).

[7] See also Ofwat's website on Leakage: https://www.ofwat.gov.uk/households/supply-and-standards/leakage.

Figures on water poverty in England in 2014–2015 remained similar to the figures for England and Wales in 2008–2009 and 2009–2010 (Ofwat 2015; Bradshaw and Huby 2013). However, they indicated a marked increase from the 9% of water poor recorded in the period 1997–1998 to 2002–2003 and the 15% recorded prior to 1997. Not only is there a clear correlation between price increases and affordability problems. Private companies' profit-seeking tactics and regulatory complacency are co-determinants of this alarming increase in water poverty (Lobina and Hall 2008a). In that sense, Ofwat has been more part of the problem than of the solution. Irrespective of Ofwat's statutory responsibility for protecting the interests of all consumers, it has failed for over 25 years to take action to adequately address the issue of water affordability by a growing section of society. Ofwat has, in fact, interpreted its duty to protect customers as one to defend economic equity and the "benefit principle" according to which individual consumers should be charged for the costs they impose on the system, not in view of their ability to pay. Ofwat demanded that governmental agencies should intervene to alleviate the social costs of water charges, and it has even refused to allow one water company to extend their vulnerable groups scheme to customers who were not in receipt of benefits, or to non-metered customers. This contrasts with the robust action that Ofwat has taken to protect the commercial interests of the companies and ensure their ability to remain profitable—for example, with its 2002 decision to extend the advance termination notice to 25 years—and suggests that Ofwat has for too long selectively defined its own remit to exclude the protection of vulnerable consumers, either through pricing mechanisms or the promotion of policy measures. Ofwat's stance on the growing problem of water poverty can thus be described—at least until recently—as "principled inflexibility with the weak and accommodating leniency towards the powerful" (Lobina and Hall 2008a: 113).

Ofwat (2015) expects that, as a result of the 2014 price review, average water bills will fall by 5% in real terms by 2020 and is introducing incentives in the price review for companies to strengthen customer orientation and deliver affordable services that customers are willing to pay for (Bayliss and Hall 2018). This, however, is no guarantee that the problem of water poverty will be resolved in the short or in the long term. Ofwat had already imposed a 12% average price cut with the 1999 price review, but prices began rising sharply again in 2004 and in the space of only four years water poverty did reach alarming levels comparable to those recorded in 2014–2015 (Lobina and Hall 2008a). Overall, the English experience with water privatisation is one of regressive redistribution as rent is extracted from the many, including vulnerable consumers, in favour of the few—from domestic and foreign shareholders to lavishly paid managers (Bayliss and Hall 2018; Lobina 2017b).

Environmental Sustainability

Environmental sustainability requires that water services are delivered in a way that prevents or minimises harm to the environment. Sewage flooding is a persistent problem in England which often results in pollution incidents. If Shaoul (1998: 32) finds that "(t)here was little improvement in industry performance and five companies had not reached their target performance in 1995", in 2000 the Customer Service Committee for the North West of England described raw sewage flooding into homes as a misery (Lobina and Hall 2001). In 2015–2016 alone, a total of 37,434 areas were externally flooded by sewage in England and Wales. South West Water was facing a £1.7 million fine for poor performance on pollution, having been responsible for 171 sewage pollution incidents between 2015 and 2016. Offenders also included Southern Water—fined £2 million for flooding beaches in Kent with raw sewage—and other WaSCs (Priestley 2016).

Ofwat argues that SELL—whereby water companies are required to fix leaks "as long as the cost of doing so is less than the cost of not fixing the leak. The cost of not fixing a leak includes environmental damage and the cost of developing new water resources to compensate for the water lost through leaks"—gives consumers the best value for money.[8] However, conservation NGO WWF has complained that SELL meant that "it is cheaper to drain a river dry than fix a leak and is one of the reasons we have seen so many rivers dry up this spring" (Carrington 2017). In addition, the higher

[8] See Ofwat's website on leakage: https://www.ofwat.gov.uk/households/supply-and-standards/leakage.

the leakage, the heavier the environmental footprint of water service delivery owing to the impact of abstracting, treating, and transporting drinking water that is not consumed—for example, the impact associated with additional requirements in terms of electricity consumed and chemicals used. It is thus questionable that SELL gives consumers the best environmental value for money. Interestingly, Ofwat has recently expressed concern that "SELL is not driving companies to become more efficient in how they tackle leakage" (Vallely 2017). It remains to be seen however whether environmentally sustainable leakage levels can be achieved under private ownership. The international experience shows in fact that the lowest leakage levels are achieved in countries where water utilities are entirely or predominantly publicly owned and managed. In the Netherlands, where all water supply operators are publicly owned, average leakage is around 4% (Lobina and Hall 2000; Roeber 2008). In Japan, where all water supply operators are public, the average leakage level is 7.5% (Marques et al. 2011). In Germany, where public water operators serve nearly 80% of the national population, average leakage is around 7% (Wackerbauer 2009a, b).[9]

Political Sustainability

Political sustainability comprises the ability of a society to reproduce the political and societal consensus that supports the viable and sustainable reproduction of a system of water governance. Failure to reproduce that consensus typically leads to weakened trust in the potential sustainability of a system of governance, and that might usher in its demise and facilitate its replacement with an alternative system of governance (Lobina 2017c). The persistence of the private governance regime in the English water sector has been underpinned by the "Westminster Consensus" between the two major parties in British politics, but this political consensus is apparently coming to an end as the Labour Party is reconsidering its position on the merits of privatisation. The Labour Party did in fact include proposals for the public ownership of water utilities in their 2017 Manifesto (Labour Party 2017), proposals that proved particularly

[9] For a more comprehensive discussion of leakage rates and international comparisons, see Hall and Lobina (2008).

popular at the 2017 general election and that have galvanised a resurgence of activism for the renationalisation of water utilities.[10] All this is taking place against the backdrop of a public opinion that has increasingly favoured a return to the public ownership of water: if an opinion poll found in 2006 that 56% of respondents would welcome the eventual renationalisation of water services, this percentage has grown to 71% in 2012 and to 83% in a 2017 poll (Lobina 2018b).

Whether and when public ownership will replace the current private governance regime depend among other factors on the ability of the advocacy coalition backing water renationalisation to take advantage of political opportunity structures to achieve their intended mobilisation outcome and on the responses of the advocacy coalition backing water privatisation (Lobina et al. 2011; Lobina 2017c). In that sense, Ofwat has shifted emphasis in recent years, particularly since late 2017, and appears to be moving away from its traditional sympathy for the privatised companies. Ofwat has in fact signalled its intention to pay greater attention to the needs of consumers in the upcoming price review and to intervene to curb excessive returns by demanding the companies to share part of their gains with the end consumer. Bayliss and Hall (2018) cast doubt on the feasibility of these regulatory proposals as credit rating agency Moody's has reacted by assigning a negative outlook to three of the nine English WaSCs—even if Moody's itself had described the amount of proposed gain sharing as modest—after having given a negative outlook to five WaSCs as a result of the growing pressure to meet customers' needs. They also point to the conflict, highlighted by the recent proposals, between Ofwat's duty to guarantee the financeability of the companies and its remit to protect the interests of consumers:

> (t)hese latest measures do little to challenge the real concerns in the sector of financial engineering and lack of transparency, beyond the gain-sharing and a superficial requirement for firms to justify payments of dividends and directors' pay in (the 2019 price review). Yet the developments outlined above show that when the regulator does impose even slightly stricter rules on companies, their credit ratings fall. The regulator is thus caught in an impossible bind. (Bayliss and Hall 2018: 5)

[10] See https://weownit.org.uk/act-now/lets-bring-water-public-ownership.

5 Discussion of Findings and Emerging Issues

Before discussing the findings on the strong and weak lock-in of water governance outcomes in England, it is helpful to recall the theories that influenced the adoption of privatisation in 1989. Sawyer (2009) identifies Austrian economics and neoclassical economics as influential theories that favour private over public service provision. The Austrian tradition posits that profits are an important inducement in the attainment of productive efficiency and that the market failure resulting from natural monopoly can be remedied by the introduction of competition. For its part, neoclassical economics proposes that it is possible to remedy the market failure resulting from natural monopoly by resorting to regulation. Also, neoclassical welfare economics remains woefully silent on issues of distribution. These and other perspectives—an example of which is Oliver Williamson's transaction cost economics—tend to embrace ideas of organisational efficiency that, like productive efficiency, are premised more on the extraction than the distribution of value (Sawyer 2009; Lobina 2018a).

Lock-in of Water Governance Outcomes Under Public Ownership in England

Water governance under public ownership has not been without problems, but it is important to recognise its central role in the universalisation of access to water in the second post-war period—what can be described as a triumph of distributive efficiency. Together with factors ranging from planning difficulties to politicised management and organisational fragmentation, macroeconomic considerations derived first from domestic industrial policy and then from international conditionality have all contributed to difficulties with achieving productive efficiency and generating the economic value necessary to achieve financeability and other forms of sustainability. In turn, failure to invest in infrastructure led to growing levels of environmental pollution that were instrumental in the deterioration of collective trust in public ownership and negatively affected distributive efficiency towards future generations.

Still, despite these difficulties, it is under public ownership that the water industry was restructured and attained levels of productive efficiency that the privatised companies struggled to match in the first period of privatisation. The lock-in of public inefficiency was therefore weak as, once collective agency indicated a clear institutional trajectory, public ownership showed its ability to achieve distributive efficiency (as far as access is concerned) and to move from a position of productive inefficiency to one of productive efficiency.

Two theoretical implications can be derived. First, the relatively strong lock-in of distributive efficiency and the weak lock-in of productive inefficiency under public ownership appear to question Oliver Williamson's (2000) assessment of the public sector as the organisational mode of last resort, to choose "when all else fails".[11] In fact, distributional issues—including distributive efficiency—are central to water service governance and, more broadly, the organisation of public services (Dagdeviren and Robertson 2016; Sawyer 2009).[12] Second, the findings lend support to Lobina's (2013) treatment of Lorrain's paradox, or the paradox of multiple agency. Lorrain's paradox can be illustrated as follows: (a) Lorrain (1997) argues that the plurality of objectives pursued by the public sector, including social justice, goes to the detriment of productive efficiency and finds in favour of private over public ownership; (b) Lobina (2013) observes that the private sector's profit maximisation imperative represents a strong incentive to extract value to the detriment of distributive efficiency, whereas the public sector—being less constrained by high-powered incentives to extract net gains—can more easily align productive and distributive efficiency; and, (c) Lobina (2013) concludes that the public sector has a comparative advantage to promote sustainable water development in all its dimensions.

[11] Dagdeviren and Robertson (2016) also find that Williamson's (2000) assessment of the public sector as the organisational mode of last resort does not reflect the empirical reality of the water sector.

[12] In this vein, Lobina (2018a) argues that the alignment of productive and distributive efficiency is of fundamental importance for enhancing sustainable water development and the human right to water.

Lock-in of Water Governance Outcomes Under Private Ownership in England

Water governance after the 1989 privatisation[13] has favoured the extraction of economic value with a view to guaranteeing short-term financeability and shareholder remuneration. This emphasis has been to the detriment of long-term financeability and economic, technical, social, and political sustainability. The English experience with water privatisation is therefore characterised by the strong lock-in of productive (internal) efficiency and distributive inefficiency, together with highly regressive outcomes and cost shifting to future generations. More precisely, the strong lock-in of private productive efficiency and distributive inefficiency can be seen as the result of the alignment of (1) private operators' high-powered incentives to extract net gains, (2) private operators' ability to pursue commercial objectives (an ability which results from the joint effect of asymmetric power and favourable institutions), and (3) the convergence of the interests of like-minded and powerful actors such as the private companies, Ofwat and credit rating agencies.

The upshot is threefold. First, the findings lend support to Lobina's (2013) treatment of Willig's paradox, which evokes an iron law of private monopolistic interest. Willig's paradox can be illustrated as follows: (1) Willig (1994) argues that insulation from self-serving political interference results in the superior efficiency of regulated private enterprises over public undertakings. This insulation arises from the fact that the institutional framework supporting privatisation guarantees profitability in order to attract private investment; (2) Lobina (2013) observes that the private sector is capable of taking advantage of asymmetric power and institutions favouring its commercial interests—drawing on a repertoire of tactics that include adventurous bids and systematic renegotiation—while public decision makers face high costs to steer private agents away from an undesired course of action; and, therefore, (3) Lobina (2013) concludes thus: "it is the institutional framework expected to promote efficiency by shielding private operators from non-commercial demands

[13] Similar considerations can however be made on the privatisation period prior to the municipalisation process of the late nineteenth and early twentieth centuries.

(Willig 1994) that allows private operators to abuse of their monopoly power" (Lobina 2013: 123). Second, considering the importance of distributive efficiency for sustainable water development, the strong lock-in of distributive inefficiency under private ownership further shows the implausibility of characterising the public sector as the organisational mode of last resort for the provision of water services. Third, once it becomes apparent that the Williamsonian assumption—that the public sector is the organisational mode of last resort, to choose when all else fails—does not hold, similar assumptions of private sector superiority entertained by Austrian, neoclassical, and industrial economics are equally disproved.

International Significance of the Findings

Despite the institutional specificities that characterise private water governance in England, it would be a mistake to dismiss the strong lock-in of private distributive inefficiency as attributable to the exceptionalism of the English case. Indeed, looking at the experience with over 40 contracts in more than 20 countries—covering diverse contractual typologies such as concessions, leases, contracts awarded to a private operator or a mixed-economy enterprise, and different institutional settings across developed, transition, and developing countries—Lobina (2005b, 2013) finds evidence of systematic private interest-seeking irrespective of the type of contractual and regulatory arrangement adopted. While the interest-seeking tactics deployed by the private sector vary in function of the local institutional setting, the aim of these tactics remains the extraction of net gains. So, for example, the use of adventurous bids and systematic renegotiation under French-style privatisation by delegation is arguably equivalent—in terms of the pursuit of monopolistic gain—to the practice of gaming observed in England. Interestingly, Williamson (1976) identifies the role of sympathetic regulators as one of the contributing factors of what can be effectively described as the strong lock-in of private (distributive) inefficiency under franchise arrangements. There seems to be something fundamental that explains behavioural similarities between private water operators in different institutional contexts. In addition to the private sector's capability to take advantage of asymmetric power and formal institutions to extract net gains in what is a natural monopoly environment (Lobina 2013), I suggest that such fundamental explanation is that all private firms pursue profit maximisation as their raison d'être (Friedman 1953; Vickers 1985). All this questions the exceptionalism of the governance outcomes of water privatisation in England.

6 Conclusions

Taking stock of nearly 30 years of water and sewerage privatisation in England and placing the analysis in a historical and international context, this chapter has looked at the governance outcomes of privatisation and their implications for sustainable water development. Attention has been paid to how water governance has resulted in different organisational efficiency outcomes and, in turn, affected systemic reproduction and viability across the different regimes of water sustainability: financial, economic, technical, social, environmental, and political.

Water governance after the 1989 privatisation has favoured the extraction of economic value with a view to guaranteeing short-term financeability and shareholder remuneration. This emphasis has been to the detriment of long-term financeability and other areas of sustainability. The English experience with water privatisation is characterised by the strong lock-in of productive (internal) efficiency and distributive inefficiency, together with highly regressive outcomes and cost shifting to future generations. This strong lock-in can be explained by the alignment of private operators' high-powered incentives to extract net gains, their ability to pursue commercial objectives in their institutional environment, and the convergence of interests between like-minded and powerful actors. Hence, the rigidity with which shareholders and financial interests are protected at the expense of societal and environmental interests makes privatisation inimical to sustainable water development.

Lessons can be drawn that are relevant to both theory and policy. First, it is important to unpack the notion of organisational efficiency and to carefully consider how productive and distributive efficiency impact on sustainable water development under different forms of ownership and governance.

Second, once the comparative institutional analysis of public and private water operations is informed by ambitious social welfare functions like sustainable water development, it becomes apparent that the assumption that the public sector is an intrinsically inferior organisational mode does not hold. Third and finally, the international experience with water privatisation and a variety of PPPs suggests that, as the interest-seeking tactics deployed by the private sector vary in function of the local institutional setting, the governance problems discussed in this chapter—which are ultimately due to the unsustainability of private sector interests, in that the directionality of collective agency resulting from the governance of private water and sewerage services goes counter to collective expectations of development—have broader relevance beyond the English case.

Acknowledgements I would like to thank Stéphane Saussier, as well as Kate Bayliss, Olivier Favereau, David Hall, Junji Hashimoto, Niels Knudsen and Malcolm Sawyer, for helpful comments and suggestions. All remaining errors are mine. The financial support of the Independent Social Research Foundation is gratefully acknowledged (http://www.isrf.org/about/fellows-and-projects/emanuele-lobina).

References

Bakker, K. J. (2001). Paying for Water: Water Pricing and Equity in England and Wales. *Transactions of the Institute of British Geographers, 26*(2), 143–164.

Bakker, K. J. (2003). *An Uncooperative Commodity: Privatizing Water in England and Wales*. Oxford and New York: Oxford University Press.

Bayliss, K. (2014). *Case Study: The Financialisation of Water in England and Wales*. FESSUD Working Paper Series No 52 ISSN 2052-8035. Retrieved from http://fessud.eu/wp-content/uploads/2015/03/Case-study-the-financialisation-of-Water-in-England-and-Wales-Bayliss-working-paper-REVISED_annexes-working-paper-52.pdf.

Bayliss, K. (2017). Material Cultures of Water Financialisation in England and Wales. *New Political Economy, 22*(4), 383–397.

Bayliss, K., & Hall, D. (2017). *Bringing Water into Public Ownership: Costs and Benefits*. PSIRU Reports, May. Retrieved from http://gala.gre.ac.uk/17277/3/17277%20HALL_Bringing_Water_into_Public_Ownership_2017.pdf.

Bayliss, K., & Hall, D. (2018). Written Evidence Submitted by Dr Kate Bayliss, SOAS, University of London and University of Leeds and Professor David Hall, PSIRU, University of Greenwich. Written evidence submitted to the Regulation of the Water Industry Inquiry. Environment, Food and Rural Affairs Committee. Retrieved from http://data.parliament.uk/writtenevidence/committeeevidence.svc/evidencedocument/environment-food-and-rural-affairs-committee/regulation-of-the-water-industry/written/84472.html.

Bradshaw, J., & Huby, M. (2013). Water Poverty in England and Wales. *Journal of Poverty and Social Justice, 21*(2), 137–148. Retrieved from http://dro.dur.ac.uk/13131/1/13131.pdf?DDD34+mrnv91+mvrl45+d700tmt.

Carrington, D. (2017). Water Companies Losing Vast Amounts Through Leakage, as Drought Fears Rise. *The Guardian*, May 11. Retrieved from https://www.theguardian.com/environment/2017/may/11/water-companies-losing-vast-amounts-through-leakage-raising-drought-fears.

Cuttill, P. (2017). *Thames Water Trunk Mains Forensic Review*. Final Findings Report, Commissioned by Thames Water, March 24. Retrieved from https://www.thameswater.co.uk/trunkmainsreview.

Dagdeviren, H., & Robertson, S. A. (2016). A Critical Assessment of Transaction Cost Theory and Governance of Public Services with Special Reference to Water and Sanitation. *Cambridge Journal of Economics, 40*(6), 1707–1724.

de la Motte, R. (2005). D10m: WaterTime National Context Report–UK. *WaterTime Deliverable D10m*. Watertime Project. Retrieved from http://www.watertime.net/docs/WP1/NCR/D10m_UK.doc.

Desmarais-Tremblay, M. (2016). The Normative Problem of Merit Goods in Perspective. *Forum for Social Economics*, 1–29. https://doi.org/10.1080/07360932.2016.1196593.

Fitzpatrick, S., Bramley, G., Sosenko, F., Blenkinsopp, J., Johnsen, S., Littlewood, M., Netto, G., & Watts, B. (2016). *Destitution in the UK*. York: The Joseph Rowntree Foundation. Retrieved from https://www.jrf.org.uk/file/49076/download?token=UAmOk9PN&filetype=full-report.

Foster, V. (2005). *Ten Years of Water Service Reform in Latin America: Toward an Anglo-French Model*. Water Supply and Sanitation Sector Board Discussion Paper Series No. 3. Washington, DC: World Bank. Retrieved from http://documents.worldbank.org/curated/en/962281468269682959/Ten-years-of-water-service-reform-in-Latin-America-toward-an-Anglo-French-model.

Friedman, M. (1953). *The Methodology of Positive Economics, Essays in Positive Economics*. Chicago: Chicago University Press.

Hall, D., & Lobina, E. (2008). *From a Private Past to a Public Future?—The Problems of Water in England and Wales*. PSIRU Reports, Commissioned by the GMB Union, November 2007. Retrieved from http://www.psiru.org/sites/default/files/2008-02-W-UK.doc.

Hall, D., & Lobina, E. (2009). Water Privatization. In P. Arestis & M. Sawyer (Eds.), *Critical Essays on the Privatization Experience, International Papers in Political Economy Series* (pp. 75–120). Basingstoke and New York: Palgrave Macmillan.

Hall, D., & Lobina, E. (2012). *Water Companies and Trends in Europe 2012*. PSIRU Report Commissioned by EPSU, August. Retrieved from https://www.epsu.org/sites/default/files/article/files/2012_Water_companies-EWCS.pdf.

Hassan, J. (1998). *A History of Water in Modern England and Wales*. Manchester: Manchester University Press.

Helm, D. (2018). Water Boarding, Cross Regulation Network Paper: 9, February 14. Retrieved from http://www.dieterhelm.co.uk/assets/secure/documents/Water-Boarding-14.02.18.pdf.

House of Commons Committee of Public Accounts. (2016). *Economic Regulation of the Water Sector*. Report, Together with Formal Minutes Relating to the Report. Fifteenth Report of Session 2015–16. HC 505. London: The Stationery Office, 13 January. Retrieved from https://www.publications.parliament.uk/pa/cm201516/cmselect/cmpubacc/505/505.pdf.

Hutton, W. (2018). We Can Undo Privatisation. And It Won't Cost Us a Penny. *The Guardian*, January 9. Retrieved from https://www.theguardian.com/commentisfree/2018/jan/09/nationalise-rail-gas-water-privately-owned.

International Water Association. (2018). Thames Water Hit with £120 Million Penalty over Leakage, The Source, 8 June 2018. Retrieved from https://www.thesourcemagazine.org/thames-water-hit-with-120-million-penalty-over-leakages.

Iossa, E., & Saussier, S. (2018). Public Private Partnerships in Europe for Building and Managing Public Infrastructures: An Economic Perspective. *Annals of Public and Cooperative Economics, 89*(1), 25–48.

Jordan, A. G., Richardson, J. J., & Kimber, R. H. (1977). The Origins of the Water Act of 1973. *Public Administration, 55*(3), 317–334.

Juuti, P. S., & Katko, T. S. (Eds.). (2005). *Water, Time and European Cities. History Matters for the Futures*. Watertime Project. Retrieved from http://www.watertime.net/docs/WP3/WTEC.pdf.

Kishimoto, S., Lobina, E., & Petitjean, O. (Eds.). (2015). *Our Public Water Future: The Global Experience with Remunicipalisation*. Amsterdam, London,

Paris, Cape Town and Brussels: Transnational Institute, Public Services International Research Unit, Multinationals Observatory, Municipal Services Project and the European Federation of Public Service Unions. Retrieved from http://www.psiru.org/sites/default/files/2015-04-W-OurPublicWater-FutureFINAL.pdf.

Labour Party. (2017). For the Many Not the Few—The Labour Party Manifesto 2017. Retrieved from https://labour.org.uk/wp-content/uploads/2017/10/labour-manifesto-2017.pdf.

Littlechild, S. C. (1978). *The Fallacy of the Mixed Economy: An Austrian Critique of Conventional Mainstream Economics and of British Economic Policy*. Transatlantic Arts.

Littlechild, S. C. (1986). *Economic Regulation of Privatised Water Authorities: A Report Submitted to the Department of the Environment*. HM Stationery Office.

Littlechild, S. C. (1988). Economic Regulation of Privatised Water Authorities and Some Further Reflections. *Oxford Review of Economic Policy, 4*(2), 40–68.

Lobina, E. (2005a). WaterTime National Context Report—Italy, *WaterTime Deliverable* D10f, March 4. Research Project on "Decision Making in Water Systems in European Cities" (WATERTIME), European Commission, 5th Framework Programme, 2002–2005. Contract No. EVK4-2002-0095. Retrieved from http://www.watertime.net/docs/WP1/NCR/D10f_Italy.doc.

Lobina, E. (2005b). Problems with Private Water Concessions: A Review of Experiences and Analysis of Dynamics. *International Journal of Water Resources Development, 21*(1), 55–87.

Lobina, E. (2013). Remediable Institutional Alignment and Water Service Reform: Beyond Rational Choice. *International Journal of Water Governance, 1*(1/2), 109–132.

Lobina, E. (2017a). A Better Future for Water: What Role for Theory? In R. Kiddey (Ed.), *Today's Future—Challenges & Opportunities Across the Social Sciences* (Issue XIII, pp. 16–21). ISRF Bulletin, June. Retrieved from https://issuu.com/isrf/docs/isrf_bulletin_issue_xiii.

Lobina, E. (2017b). Categorical Errors on Water Renationalisation. *PSIRU Briefs*, June 3. Retrieved from http://gala.gre.ac.uk/17380/7/17380%20LOBINA_Categorical_Errors_on_Water_Renationalisation_2017.pdf.

Lobina, E. (2017c). Water Remunicipalisation: Between Pendulum Swings and Paradigm Advocacy. In S. Bell, A. Allen, P. Hofmann, & T. H. Teh (Eds.), *Urban Water Trajectories* (pp. 149–161). London: Springer. Retrieved from http://gala.gre.ac.uk/14983/1/14983%20LOBINA_Water_Remunicipalisation_2017.pdf.

Lobina, E. (2018a). *For Another Idea of Efficiency in the Water Sector*. PSIRU Discussion Papers, University of Greenwich. Retrieved from http://gala.gre.ac.uk/view/authors/531.html.

Lobina, E. (2018b). *Commentary on the European Commission's "Study on Water Services in Selected Member States"*. A PSIRU Report Commissioned by the European Federation of Public Service Unions, February. Retrieved from http://gala.gre.ac.uk/19211/7/19211%20LOBINA%20European_Commission%E2%80%99s_Study_on_Water_Services_2018.pdf.

Lobina, E., & Hall, D. (2000). Public Sector Alternatives to Water Supply and Sewerage Privatization: Case Studies. *International Journal of Water Resources Development, 16*(1), 35–55.

Lobina, E., & Hall, D. (2001). *UK Water Privatisation—A Briefing*. PSIRU Reports, February. Retrieved from http://www.psiru.org/reports/2001-02-W-UK-over.doc.

Lobina, E., & Hall, D. (2003). *Problems with Private Water Concessions: A Review of Experience*. PSIRU Reports, June. Retrieved from http://www.psiru.org/reports/2003-06-W-over.doc.

Lobina, E., & Hall, D. (2007). *Water Privatisation and Restructuring in Latin America, 2007*. PSIRU Reports, September. Retrieved from http://www.psiru.org/sites/default/files/2007-09-W-Latam.doc.

Lobina, E., & Hall, D. (2008a). Water. In S. Thomas (Ed.), *Poor Choices: The Limits of Competitive Markets in the Provision of Essential Services to Low-Income Consumers* (pp. 93–122). London: Energywatch. Retrieved from http://www.psiru.org/sites/default/files/2008-09-EW-PoorChoicesWater.pdf.

Lobina, E., & Hall, D. (2008b). The Illusions of Competition in the Water Sector. A UNISON Response to the Cave Review of Competition and Innovation in Water Markets, September. Retrieved from http://www.psiru.org/sites/default/files/2008-09-W-UKcompet.doc.

Lobina, E., & Terhorst, P. (2005). D19: WaterTime Case Study—Edinburgh, UK, *WaterTime Deliverable* D19, January 29. Research Project on "Decision Making in Water Systems in European Cities" (WATERTIME), European Commission, 5th Framework Programme, 2002–2005. Contract No. EVK4-2002-0095. Retrieved from http://www.watertime.net/docs/WP2/D19_Edinburgh.doc.

Lobina, E., Terhorst, P., & Popov, V. (2011). Policy Networks and Social Resistance to Water Privatisation in Latin America. *Elsevier Procedia Social and Behavioural Sciences, 10*, 19–25. Retrieved from https://ac.els-cdn.com/S187704281100005X/1-s2.0-S187704281100005X-main.pdf?_tid=3b9a22d1-ea79-4283-b1a6-e5a69c13dcb0&acdnat=1529265352_aaeb615c02d616f76188d248de0d2cfd.

Lorrain, D. (1997). Introduction: The Expansion of the Market. In D. Lorrain & G. Stoker (Eds.), *The Privatization of Urban Services in Europe* (pp. 1–26). London: Pinter.

Marques, R. C., Berg, S. V., & Yane, S. (2011). *Performance Benchmarking Analysis of Japanese Water Utilities*. Proceedings of the Pi 2011 Conference. Retrieved from http://www.researchgate.net/publication/228749894_Performance_Benchmarking_Analysis_of_Japanese_Water_Utilities/file/79e4150eae5ddae1dc.pdf.

Ménard, C. (2017). Meso-institutions: The Variety of Regulatory Arrangements in the Water Sector. *Utilities Policy, 49*, 6–19.

Mookherjee, D. (2006). Decentralization, Hierarchies, and Incentives: A Mechanism Design Perspective. *Journal of Economic Literature, 44*(2), 367–390.

NAO (National Audit Office). (2015). HC 487 Session 2015–2016 The Economic Regulation of the Water Sector. Retrieved from https://www.nao.org.uk/wp-content/uploads/2014/07/The-economic-regulation-of-the-water-sector.pdf.

Northern Ireland Water. (2017). Scheme of Charges 2017/18. Northern Ireland Water, May 2018. Retrieved from niwater.com/sitefiles/resources/pdf/water-charges/201718/niwater2017-18schemeofcharges.pdf.

Ofwat (Water Services Regulation Authority). (2010). Delivering Sustainable Water—Ofwat's Strategy. Ofwat, March 2010. Retrieved from ofwat.gov.uk/wpcontent/uploads/2015/10/rpt_fwd_20100303ofwatstrategy.pdf.

Ofwat (Water Services Regulation Authority). (2013). Ofwat Confirms Changes to Deliver for Customers. Press Release PN 09/13, July 25. Retrieved from https://www.ofwat.gov.uk/pn-0913-ofwat-confirms-changes-to-deliver-for-customers.

Ofwat (Water Services Regulation Authority). (2015). *Affordability and Debt 2014–15*. December. Retrieved from https://www.ofwat.gov.uk/wp-content/uploads/2015/12/prs_web20151201affordability.pdf.

Ofwat (Water Services Regulation Authority). (2016). *Costs and Benefits of Introducing Competition to Residential Customers in England—Emerging Findings*. July 18. Retrieved from https://www.ofwat.gov.uk/wp-content/uploads/2016/07/pap_tec20160715rrremergingfindings.pdf.

Ofwat (Water Services Regulation Authority). (2017). *Water Services Regulation Authority (Ofwat) Annual Report and Accounts 2016–17 for the Period 1 April 2016 to 31 March 2017*. June 28. Retrieved from https://www.gov.uk/government/uploads/system/uploads/attachment_data/file/623009/ofwat-annual-report-web.pdf.

Ofwat (Water Services Regulation Authority), Defra (Department for Environment Food and Rural Affairs). (2006). *The Development of the Water Industry in England and Wales*. Ofwat and Defra Report, January 27. Retrieved from https://www.ofwat.gov.uk/wp-content/uploads/2015/11/rpt_com_devwatindust270106.pdf.

Ogden, S. G. (1997). Accounting for Organizational Performance: The Construction of the Customer in the Privatized Water Industry. *Accounting, Organizations and Society, 22*(6), 529–556.

Priestley, S. (2016). *Sewer Flooding*. Briefing Paper Number CBP7839, House of Commons Library, December 19. Retrieved from https://researchbriefings.parliament.uk/ResearchBriefing/Summary/CBP-7839.

Roberts, M. (2017). Productivity in the Water Industry. Blog, Water UK, November 28. Retrieved from https://www.water.org.uk/news-water-uk/blog/productivity-water-industry.

Roeber, B. (2008). *Institutional Developments: Amsterdam Water Supply*. Presentation at RPW Strategy Meeting, Brussels, January 7–8.

Saal, D., & Parker, D. (2000). The Impact of Privatization and Regulation on the Water and Sewerage Industry in England and Wales: A Translog Cost Function Model. *Managerial and Decision Economics, 21*(6), 253–268.

Saal, D., & Parker, D. (2001). Productivity and Price Performance in the Privatised Water and Sewage Companies of England and Wales. *Journal of Regulatory Economics*, 61–90. Retrieved from http://www.springerlink.com/content/m3j6018112134q78/.

Saal, D., & Reid, S. (2004). An Investigation into Opex Productivity Trends and Causes in the Water Industry in England & Wales—1992–93 to 2002–03. Stone and Webster Consultants, Main Report—Final May 2004. Retrieved from https://www.researchgate.net/profile/David_Saal/publication/40500702_An_investigation_into_OPEX_productivity_trends_and_causes_in_the_water_industry_in_England_and_Wales_-_1992-93_to_2002-03/links/00b7d529c767a74518000000.pdf.

Saal, D., Parker, D., & Weyman-Jones, T. (2007). Determining the Contribution of Technical Change, Efficiency Change and Scale Change to Productivity Growth in the Privatized English and Welsh Water and Sewerage Industry: 1985–2000. *Journal of Productivity Analysis, 28*, 127–139. https://doi.org/10.1007/s11123-007-0040-z.

Sawyer, M. (2009). Theoretical Approaches to Explaining and Understanding Privatisation. In M. Frangakis, C. Hermann, J. Huffschmid, & K. Lóránt (Eds.), *Privatisation Against the European Social Model* (pp. 61–76). London: Palgrave Macmillan.

Shaoul, J. (1998). *Water Clean Up and Transparency: Accountability of the Regulatory Process in the Water Industry*. University of Manchester.

Thames Water. (2014). Investing in Our Future—Thames Water Utilities Limited Annual Report and Financial Statements 2013/14. Retrieved from https://corporate.thameswater.co.uk/-/media/Site-Content/Thames-Water/Corporate/AboutUs/Investors/Archives/2013%2D%2D-2014/Thames-Water-Utilities-Limited-Annual-Report.pdf.

Tinson, A., & Kenway, P. (2013). *The Water Industry: A Case to Answer*. New Policy Institute Report Commissioned by UNISON, April 24. Retrieved from https://www.npi.org.uk/files/8213/7545/1688/Water_industry_a_case_to_answer_unison_version.pdf.

Vallely, L. (2017). Ofwat Takes "Tougher Line" on Leakage, Utility Week, July 11.

Vickers, J. (1985). Delegation and the Theory of the Firm. *The Economic Journal, 95*, 138–147.

Vickers, J., & Yarrow, G. K. (1988). *Privatization: An Economic Analysis*. Cambridge and London: MIT Press.

Wackerbauer, J. (2009a). *Public or Private Water Management: The German Experience*. Paper Presented at the Second Annual Conference on Competition and Regulation in Network Industries, Centre for European Policy Studies, Brussels, November 20.

Wackerbauer, J. (2009b). *The Water Sector in Germany*. Working Paper CIRIEC No. 2009/11, ISSN 2070-8289.

Williamson, O. E. (1976). Franchise Bidding for Natural Monopoly—In General and with Respect to CATV. *Bell Journal of Economics, 7*, 73–104.

Williamson, O. E. (2000). The New Institutional Economics: Taking Stock, Looking Ahead. *Journal of Economic Literature, 38*, 595–613.

Willig, R. D. (1994). Public Versus Regulated Private Enterprise. *Proceedings of the World Bank Annual Conference on Development Economics 1993*.

Wright, J., & Huggins, M. (2017). *Productivity Improvement in the Water and Sewerage Industry in England Since Privatisation*. Final Report for Water UK. Frontier Economics, September 29. Retrieved from https://www.dropbox.com/s/39373yt7ir364iq/Water%20UK%20Frontier%20-%20Productivity.PDF?dl=0.

8

Latin American Countries

Water and Sanitation in Latin America and the Caribbean: An Update on the State of the Sector

Salvador Bertoméu-Sánchez and Tomás Serebrisky

1 Introduction

Latin America and the Caribbean (LAC) is endowed with abundant fresh water: In 2014, it accounted for 32 percent of the world's internal renewable water resources (FAO 2016). Historically, however, access to improved

We are grateful to Antonio Estache for comments and suggestions. Any mistake or misinterpretation is our responsibility and ours only and should not be attributed to any of the institutions we are affiliated with.

S. Bertoméu-Sánchez (✉)
European Center for Advanced Research in Economics and Statistics (ECARES), Université libre de Bruxelles, Bruxelles, Belgium
e-mail: sbertome@ulb.ac.be

T. Serebrisky
Principal Economic Advisor, Department of Infrastructure, Inter-American Development Bank, Washington, DC, USA
e-mail: tserebrisky@iadb.org

S. Porcher, S. Saussier (eds.), *Facing the Challenges of Water Governance*, Palgrave Studies in Water Governance: Policy and Practice,
https://doi.org/10.1007/978-3-319-98515-2_8

water and sanitation services in the region was unequal, service quality was poor, and financing of the sector was inadequate.

To deal with such issues, starting in the early 1990s, countries adopted major reforms of their water and sanitation sectors. By the end of the decade, almost every country in the region had undertaken such sector reforms (Foster 2005). The objectives of these reforms were to extend access to water and sanitation services to 100 percent of the population and improve the efficiency and financial viability of the sector.

To achieve these objectives, most LAC countries introduced legislation focused on financing and governance. Efforts were made to attract private participation to the sector, in order to fill the financing gap. Initial aggressive efforts in some countries led to very high levels of private participation by the end of the 1990s (86 percent of total financing of the sector in Chile and 62 percent in Argentina). By the mid-2000s, however, private participation had returned to its earlier (low) levels (Foster 2005).

Many countries also created autonomous regulatory bodies to supervise the activities of the private sector, in order to ensure that quality standards were met in a cost-effective manner. These reforms went farther than the private sector participation reforms: By the end of the 1990s, more than half of the countries in the region had created a regulatory body in the sector (see Appendix Table 8.4 for a list of the regulators in each country). Incongruences between the level of government at which the sector was regulated (centrally) and managed (more locally) limited the positive impact of such reforms, however.

Thus, following the reforms that started during the early 1990s throughout LAC countries, progress in access to water and sanitation services was made, but there is evidence that such progress has been significantly constrained by affordability and governance issues. This chapter describes the key characteristics of the sector in LAC. The next section looks at sector performance (access to water and sanitation and tariffs in the water sector). The following sections describe governance (regulation and the role of private participation), briefly review the empirical literature, and summarize the chapter's main findings.

2 Sector Performance

Access to Improved Water Sources

Access to improved drinking water sources has increased since the beginning of the reform waves that started in the early 1990s. For the region as a whole, access increased from 81 percent of the population in 1990 to 95 percent in 2015 (Fig. 8.1). All but two countries (the Dominican Republic and Haiti) increased access during this period. Belize, Bolivia, El Salvador, Honduras, and Paraguay, which had very low access rates in 1990, increased access to more than 95 percent.

Progress in LAC was remarkable by world standards. So great was the improvement that in 2015, the only set of countries that had a higher access rate than LAC were developed countries, where 99 percent of the population had access to improved drinking water (Fig. 8.2).

Improvement was fueled by increases in urban access. Rural access remains low: The region performs better than Sub-Saharan Africa and Europe and Central Asia and better than the average for developing countries (84 percent vs. 83 percent), but it lags well behind developed countries, where 98 percent of the rural population has access (Fig. 8.2).

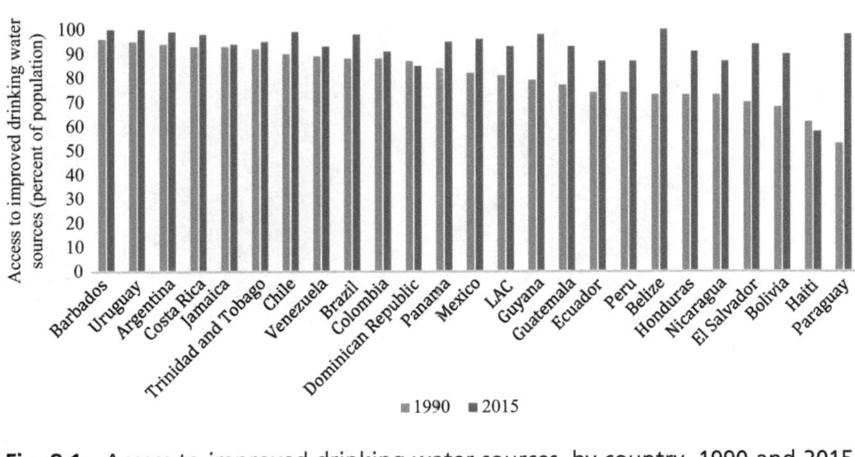

Fig. 8.1 Access to improved drinking water sources, by country, 1990 and 2015. Source: World Health Organization (2016) (accessed on July 4, 2017)

In some countries, such as Argentina, Paraguay, and Uruguay, rural rates are close to urban ones. In others, such as Bolivia, Colombia, Ecuador, Nicaragua, Peru, and Venezuela, rural access rates are below 80 percent, considerably below urban rates (Fig. 8.3). Efforts to

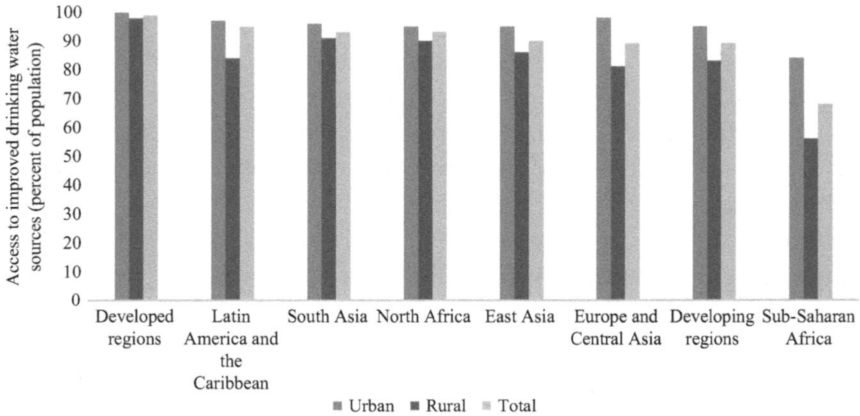

Fig. 8.2 Access to improved drinking water sources in rural and urban areas, by world region, 2015. Source: World Health Organization (2016) (accessed on July 4, 2017)

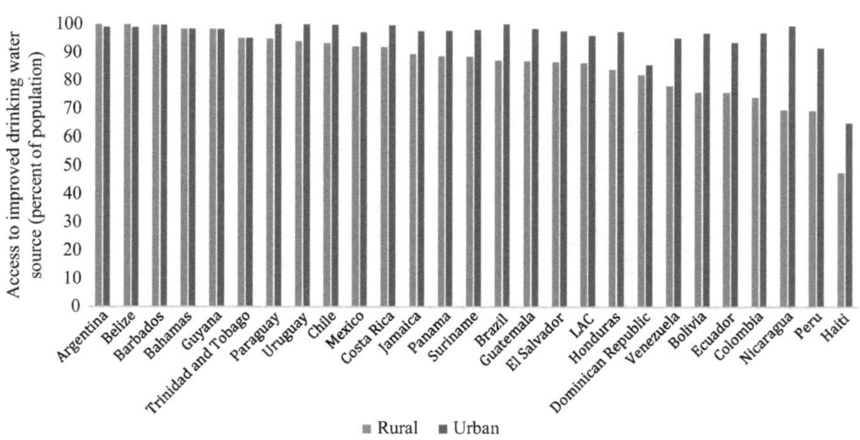

Fig. 8.3 Access to improved drinking water sources in urban and rural areas, by country, 1990 and 2015. Source: Joint Monitoring Programme of the World Health Organization and UNICEF (2016) (accessed on July 4, 2017)

improve access have focused on urban areas, possibly because of their high growth rates in terms of population and increasing demand for water services.[1]

Access to Improved Sanitation Services

Access to sanitation improved in LAC between 1990 and 2015, but the average figure reached just 81 percent (Fig. 8.4). Some countries made good progress: Between 1990 and 2015, access increased from 28 percent to 50 percent in Bolivia, from 48 percent to 83 in Honduras, and from 57 percent to 85 percent in Ecuador. However, and despite important improvements, some countries were in 2015 at worryingly low levels, such as Bolivia, Guatemala, Haiti, and Nicaragua.

Access to improved sanitation services exceeded 90 percent in urban areas of most countries in the region in 2015 (Fig. 8.5). It was much

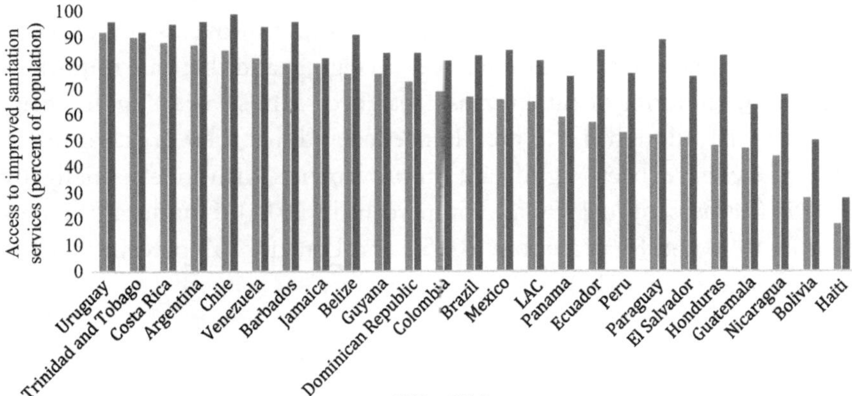

Fig. 8.4 Access to improved sanitation, by country, 1990 and 2015. Source: Joint Monitoring Programme of the World Health Organization and UNICEF (2016) (accessed on July 4, 2017)

[1] LAC is the most urbanized region in the world, having increased its urbanization rate from 73 percent in 1995 to 80 percent in 2014 (Arroyo et al. 2015).

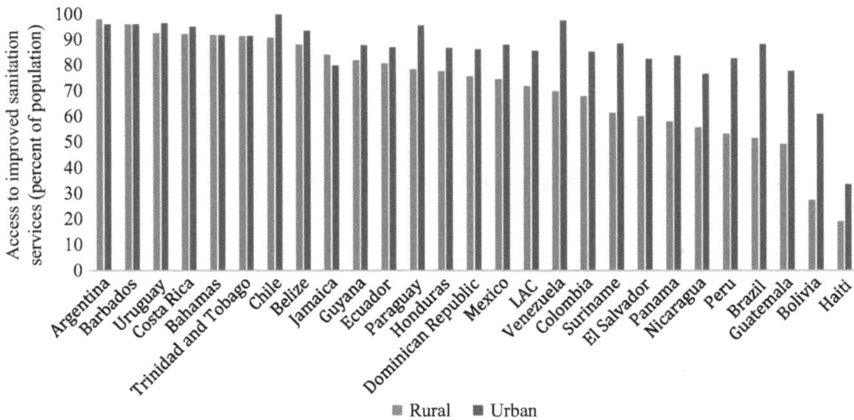

Fig. 8.5 Access to improved sanitation services in urban and rural areas, by country, 2015. Source: Joint Monitoring Programme of the World Health Organization and UNICEF (2016) (accessed on July 4, 2017)

lower in rural areas: In some countries (Guatemala, Bolivia, and Haiti), less than half the rural population had access to improved sanitation.

Overall, access to sanitation services is lagging, and there are important inequalities in terms of rural versus urban access. It is worth highlighting, however, that though efforts are still needed and that there is a clear urban bias, there have been significant improvements, driven essentially by Bolivia (28 percent in 1990 and 50 percent in 2015), Honduras (48 percent in 1990 and 83 percent in 2015), and Ecuador (57 percent in 1990 and 85 percent in 2015).

Water Tariffs

Water tariffs play a key role in improving the efficiency of water use. On the demand size, they determine the price water users pay for water, thus having social equity consequences. On the supply side, they determine the extent to which water utilities can recover costs, thus having financial sustainability consequences.

Increasing block tariffs (IBTs) have traditionally been the choice of water regulators and policymakers in developing countries, because they supposedly protect the poor. In such an IBT setup, consumption is structured in blocks, with the first block corresponding to the lowest level of consumption and the last block corresponding to the highest level of consumption. Water users are charged a fixed unit price for the units consumed in the first block, a higher price for units consumed in the second block, and so on. In terms of policy, three decisions must be taken when setting up an IBT structure: (1) the number of blocks, (2) the volume of water associated with each block, and (3) the unit price of water associated with each block.

IBTs were long thought to be efficient, pro-poor, thus attending to equity concerns, and environmentally advantageous for the following reasons:

- Wealthier households and industrial firms cross-subsidize poor households, promoting equity.
- Higher prices associated with higher blocks of consumption discourage irresponsible water use, supporting environmental objectives.
- Block charges are consistent with marginal cost pricing, which promotes economic efficiency.

But doubts have been raised about their effectiveness in developing countries. Boland and Whittington (1998) conclude that they increase inefficiency, inequity, complexity, opacity, and instability and create forecasting difficulties. They claim that simpler structures—such as two-part tariffs, which charge a fixed price per month plus a variable amount based on the volume consumed—can achieve better results.

IBTs are the most common water tariff structure in LAC. Capitals[2] in all countries except Chile, Colombia, and the Dominican Republic, which have two-part water tariffs, and Haiti, which has a simple linear tariff, use IBTs (Fig. 8.6).[3] However, the design of IBT structures

[2] For some countries, due to unavailability of data on its capital cities, the analysis has been done on the most important city, in economic and population terms.

[3] The Bahamas has not been included in the figures due to their considerably higher tariffs with respect to the rest of the region, creating graphical distortions.

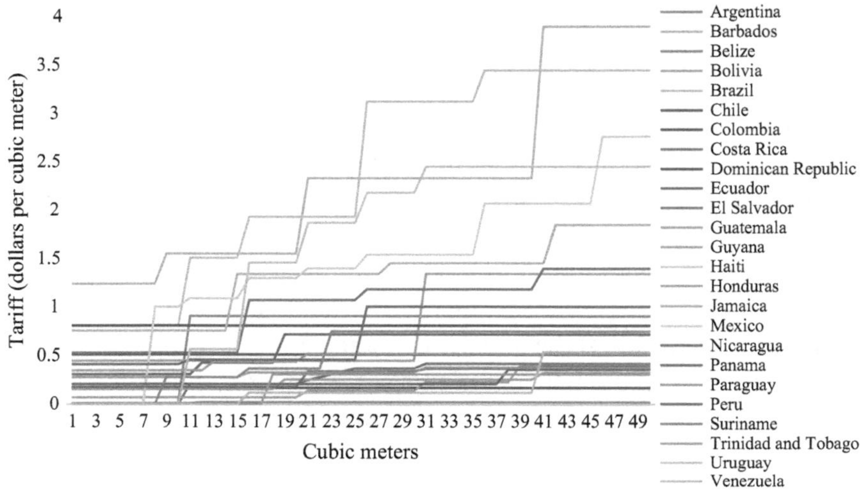

Fig. 8.6 Water tariffs in selected capitals, 2016. Sources: Data from the International Benchmarking Network for Water and Sanitation Utilities (IBNET) and Global Water Intelligence (2016)

varies significantly across countries in terms of level of fixed and variable charges and number and width of blocks.

These differences in the level of fixed and variable charges and the number and width of blocks imply that monthly water bills vary widely across countries (Fig. 8.7). For low levels of monthly consumption (3 cubic meters), the difference between the lowest (Honduras) and highest (the Bahamas and Uruguay) bill is about $12. At the highest level of consumption (50 cubic meters), this difference grows to $230. Excluding the outlier in the sample (the Bahamas), significant differences in water bills start appearing at consumption levels of 15 cubic meters. The average bill for consumption of 15 cubic meters a month, considered to be the minimum amount of water required at the household level to satisfy basic needs, is $8.50 in LAC (see Appendix Fig. 8.19),[4] slightly less than

[4]At 15 cubic meters of consumption, the lowest bill is in Honduras ($1.12), and the highest is in the Bahamas ($53.87).

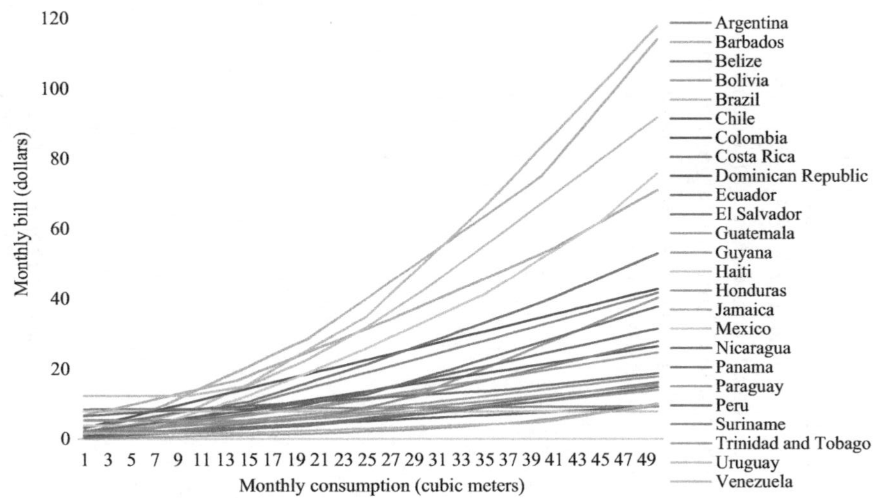

Fig. 8.7 Monthly water bill in selected capitals, 2016. Sources: Data from the International Benchmarking Network for Water and Sanitation Utilities (IBNET) and Global Water Intelligence (2016)

in other developing regions, such as Sub-Saharan Africa ($9.90) or the Middle East and North Africa ($10.40). LAC households are charged less, on average, than these two developing regions' households for the minimum level of consumption to cover for basic needs. But prices vary widely across countries. In most countries in LAC, the unit price is slightly higher for higher levels of consumption, pointing toward a tariff structure penalizing high levels of consumption.

Another indicator that reveals the heterogeneity in water prices in LAC is the average price per cubic meter (Fig. 8.8). In most LAC capitals, the average price per cubic meter increases as a function of the consumption level, albeit not steeply. Some countries charge the same rate per cubic meter, regardless of the level of consumption. No country in LAC charges less per cubic meter for higher levels of consumption.[5]

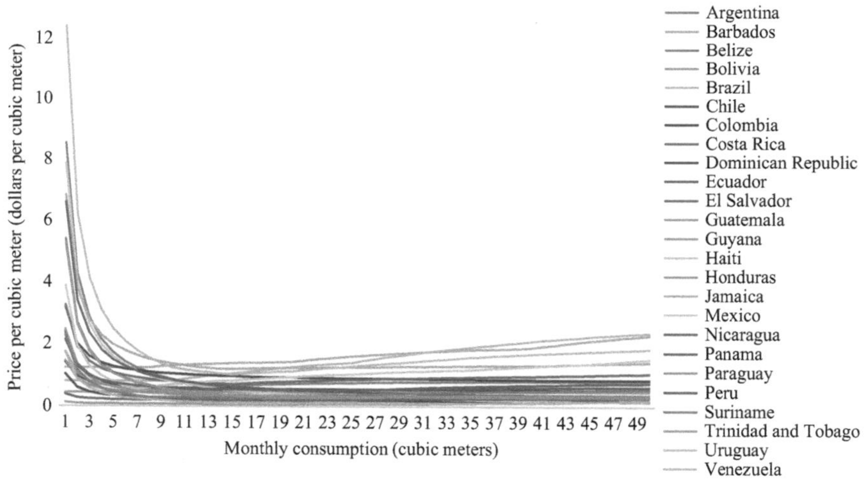

Fig. 8.8 Average price of water in selected capitals, 2016. Sources: Data from the International Benchmarking Network for Water and Sanitation Utilities (IBNET) and Global Water Intelligence (2016)

[5] In some countries outside the region, such as Spain, the average price of water falls as consumption rises.

Unreliable Access to Water

Urban access to water increased markedly after 1990, but the share of the population with continuous access actually declined slightly between 2008 and 2012, according to survey of selected cities in Argentina, Bolivia, Brazil, Colombia, Ecuador, Mexico, Panama, Peru, Uruguay, and Venezuela, falling from 87.9 percent to 86.5 percent (CAF 2014). Moreover, among the share of the population that does not have a continuous access to water services (13.5 percent in 2012), there was a significant increase between 2008 and 2012 in the share of users who had access less than once every 15 days, from 38.0 percent to 44.4 percent (Fig. 8.9). On the positive side, among people who lacked continuous water service, there was an increase in the share of people with access a few hours every day (from 44.6 percent to 56.3 percent) and a significant decrease in the share of users with access only every two or three days (from 14.0 percent to 7.4 percent).

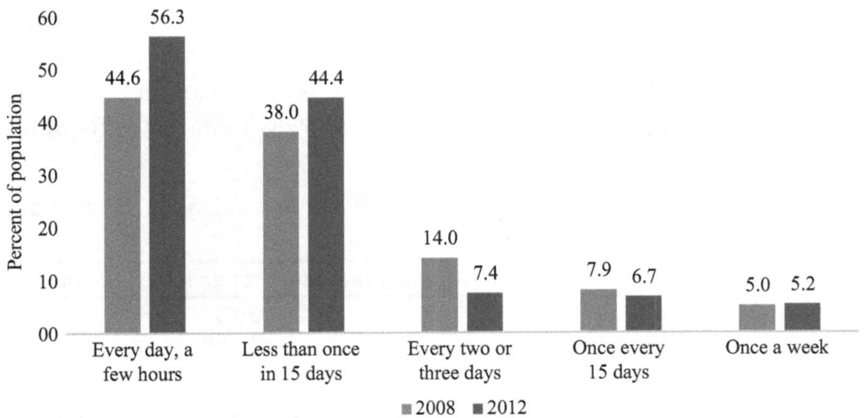

Fig. 8.9 Urban access to water in Latin America and the Caribbean, 2008 and 2012. Source: CAF (2014). Note: Survey conducted by the Banco de Desarrollo de América Latina in cities in Argentina, Bolivia, Brazil, Colombia, Ecuador, Mexico, Panama, Peru, Uruguay, and Venezuela

Health Consequences of Unsafe Water and Sanitation

Limited and unreliable access has negative effects on public health. The mortality rate attributed to unsafe water and sanitation is much lower in LAC (3.15 per 100,000 people) than in other developing regions (Table 8.1). It is significantly higher than in developed regions, such as Europe (0.65 per 100,000 people) and North America (0.6 per 100,000 people), however (see Appendix Fig. 8.20 for national mortality rates in LAC).

Another measure of the burden imposed by inadequate water and sanitation services is disability-adjusted life years (DALYs).[6] In 2012 LAC countries collectively lost 980,000 DALYs. The highest absolute figures were in Haiti, Brazil, Mexico, Guatemala, Bolivia, and Colombia (Fig. 8.10). On a population-adjusted basis, the heaviest burdens were in Haiti, Bolivia, Guatemala, and Honduras.

The share of Unaccounted-for water is also high in LAC. In 2012, 39 percent of total water supply was unaccounted for (it leaked or did not make it to its destination for some other reason) (Fig. 8.11). In Costa Rica, Panama, or Uruguay, the figure exceeded 50 percent.

Untreated wastewater is pumped into rivers and oceans, with severe consequences for both the environment and health. Even in urban areas, on average only 54 percent of wastewater was collected between 2008 and 2012 (the figure ranged from 10 percent in the Dominican Republic

Table 8.1 Mortality rates attributed to unsafe water and sanitation, by developing region, 2012

Region	Mortality rate (deaths per 100,000 people)
Sub-Saharan Africa	39.5
Middle East and North Africa	11.58
Southeast Asia	7.73
Latin America and the Caribbean	3.15

Source: World Health Organization (2016) (accessed on July 4, 2017)

[6] One DALY can be thought of as representing one lost year of healthy life. The sum of DALYs across the population can be thought of as a measure of the gap between the current health status and an ideal health situation, in which the entire population survives to old age, free of disease and disability.

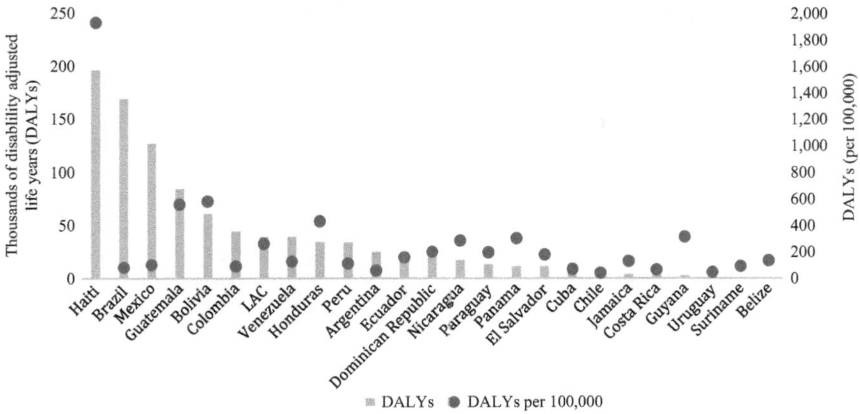

Fig. 8.10 Disability-adjusted life years (DALYs) lost as a result of inadequate access to water and sanitation, by country, 2012. Source: World Health Organization (2016) (accessed on July 4, 2017)

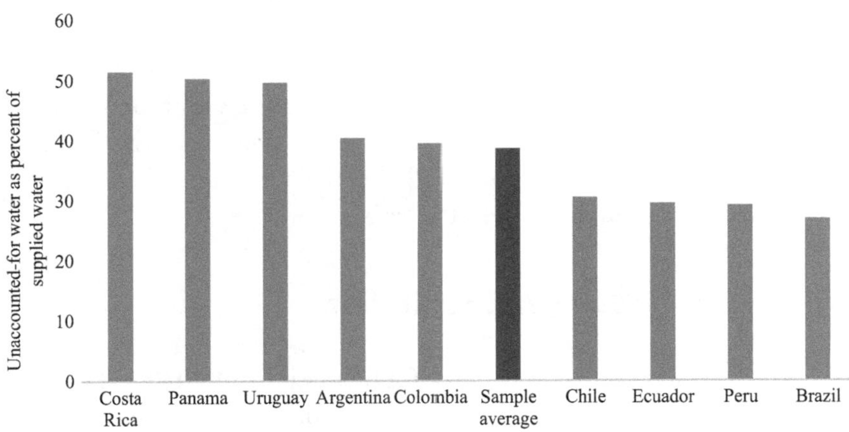

Fig. 8.11 Unaccounted-for water as percent of supplied water in capitals of selected countries, 2012. Source: Asociación de entes reguladores de agua potable y saneamiento de las Américas (ADERASA) (cited in Lentini 2015)

to 96 percent in Chile) (Fig. 8.12). Worse yet, only 26 percent of urban wastewater was treated (the figure ranged from 2 percent in the Dominican Republic to 69 percent in Chile). Data are not available for rural areas, but the problems are much worse there.

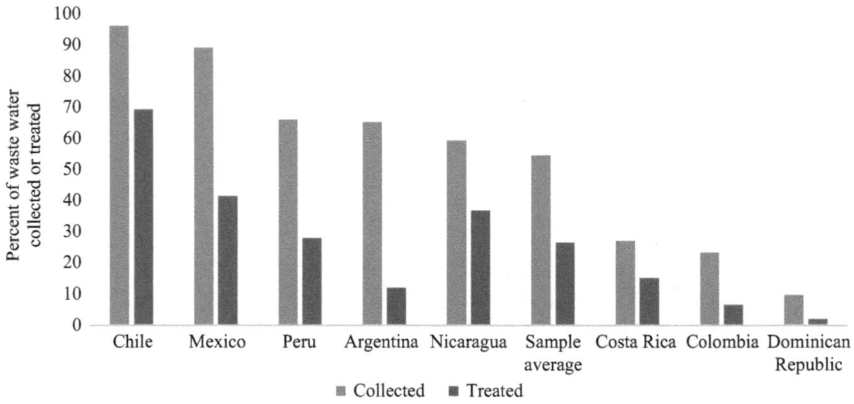

Fig. 8.12 Share of wastewater collected and treated in capitals of selected countries. Source: AQUASTAT database of the Food and Agriculture Organization (accessed on July 31, 2017). Note: Data are averages for 2008–2012

3 Sector Financing and Governance

Performance of the water and sanitation sector is closely related to its financing and governance. This section examines the evolution of public and private financing of the sector over the past decade, the financial performance of sector operators, and the characteristics of the sector's governance on a country-by-country basis.

Investment in Water and Sanitation

LAC needs to spend 0.30 percent of GDP a year until 2030 to meet the water and sanitation needs in order to maintain actual access rates in urban areas and improve access rates in rural areas by closing on the coverage and quality gaps (Arroyo et al. 2015). Average annual investment in the water and sanitation sector over this period ranges from 0.19 to 0.38 percent of GDP, with an average of 0.28 percent of GDP (Fig. 8.13). Almost all of this spending was public.

Moreover, new infrastructure put in place to close on the coverage and quality gaps directly imply additional maintenance and operating costs. Thus, the necessary spending in water and sanitation should be above

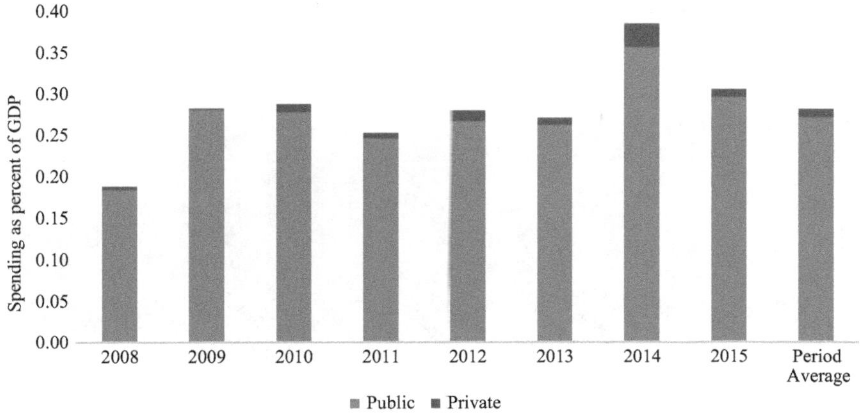

Fig. 8.13 Average spending on water and sanitation in Latin America and the Caribbean as percent of GDP, 2008–2015. Source: Infralatam (accessed on July 19, 2017) and World Bank PPI database (accessed on May 12, 2017). Note: Data are for 2008–2015

0.30 percent of GDP, and an effort is then needed to increase the sector's financing, whether the source is public or private, with respect to recent years in LAC.

Spending varied widely across countries (Fig. 8.14). It was lowest in Chile (0.086 percent of GDP) and highest in Peru (0.65 percent of GDP).

The relationship between spending and access is not clear. El Salvador, Guatemala, Honduras, and Panama spent considerably less than average and had below-average access rates for both water and sanitation and suboptimal service quality. But other countries, such as Argentina, Chile, Costa Rica, Mexico, and Paraguay, spent significantly less than the sample average and had high rates of access to both water and sanitation, and in yet others (including Bolivia, Colombia, Nicaragua, and Peru), access to water and sanitation was considerably weak, but spending was well above average, which could indicate that the necessary steps are being taken to close on the coverage and quality gaps.

The private sector has an important role to play in LAC, particularly given the constraints on government budgets. After declining considerably between 2000 and 2011, it increased in 2012–2015 (Fig. 8.15),

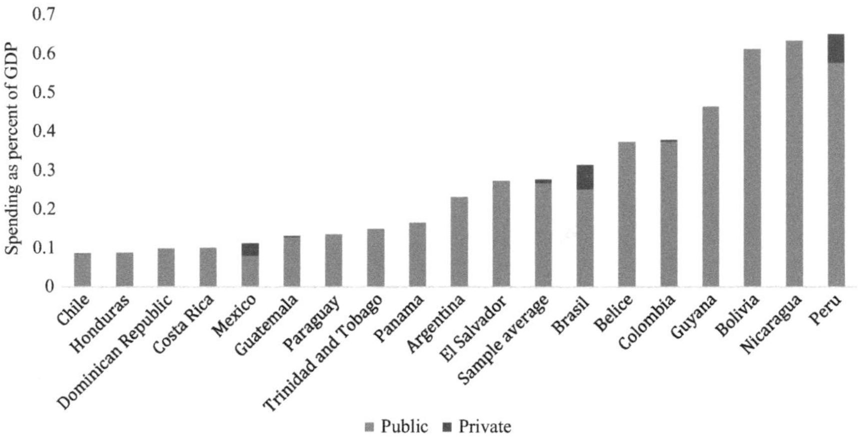

Fig. 8.14 Average spending on water and sanitation as percent of GDP in selected countries. Source: Infralatam (accessed on July 19, 2017) and World Bank PPI database (accessed on May 12, 2017). Note: Data are for 2008–2015

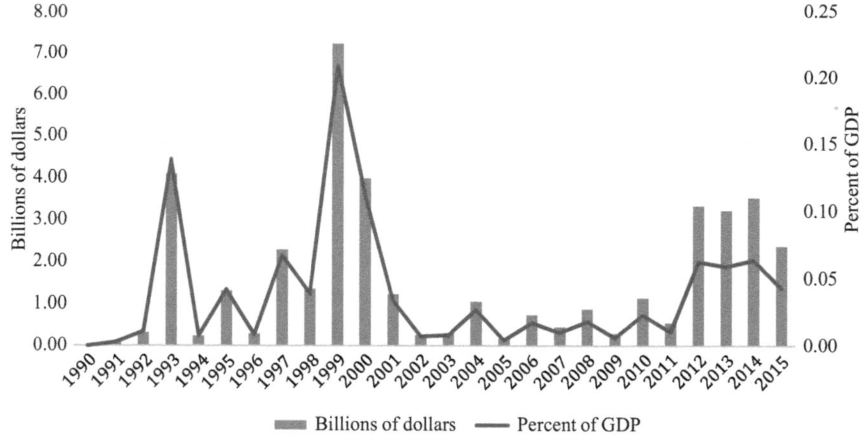

Fig. 8.15 Investment in public-private partnerships in water and sanitation in Latin America and the Caribbean, 1990–2015. Source: World Bank PPI database (accessed on May 12, 2017)

making LAC the leading region in the world for private participation in the sector (Fig. 8.16). More than three-quarters of commitments were concentrated in a few countries, however (including Argentina, Brazil, Chile, and Mexico). Efforts are thus needed in the region to increase

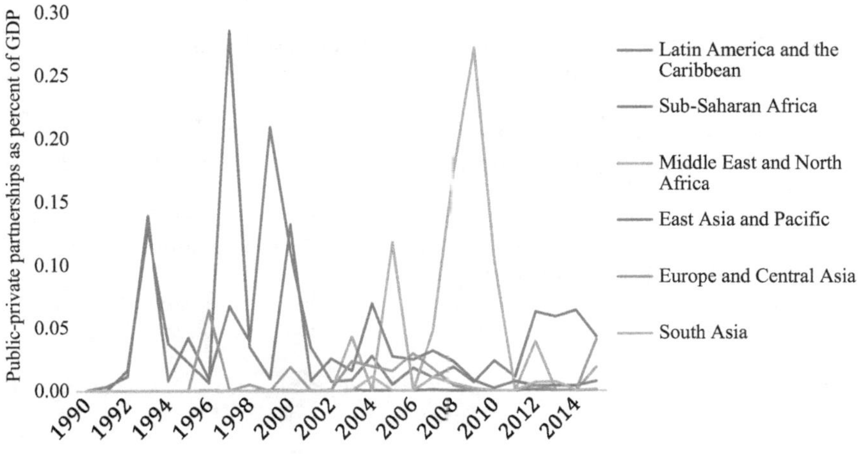

Fig. 8.16 Investment in public-private partnership in water and sanitation, by region, 1990–2015. Source: World Bank PPI database (accessed on December 5, 2017)

private participation throughout the region, not only to reach the necessary investment levels in the sector but also to release pressure on government spending.

The nature of private sector involvement in the sector has changed since 1990. During the 1990s, 42 percent of projects were brownfield, 21 percent were greenfield, 21 percent were divestitures, and 16 percent were management and lease contracts (Fig. 8.17) (see Box 8.1 for a description of these types of contracts). From the early 2000s until the beginning of the financial and economic crisis in 2008, brownfield projects accounted for 73 percent of all commitments. In 2009–2016, 68 percent of projects were brownfield projects, 28 percent were greenfield, and 4 percent were management and lease contracts. The private sector thus appears to be increasingly attracted to contract types entailing new investments and increased power of decision. Divestitures and management and lease contracts seem to have lost their appeal, as they do not imply new investments and have a higher dependence to the public sector.

Throughout the last decade, financing in the sector has not reached the necessary levels to fill in the coverage and quality gaps, and efforts should be made in that sense during the next 15 years; in that sense, not only the

Box 8.1 Summary of Characterization of Contract Types (The World Bank PPI Database)

1. **Management and lease contracts:** Take-over of the management of a public asset by a private firm for a fixed duration (the ownership and investment decisions remain with public sector)
2. **Greenfield projects:** Construction and operation of a new facility by a private operator or a partnership between private and public actors for a duration specified in the project contract
3. **Brownfield projects:** Take-over of the operations, improvement, expansion, and/or rehabilitation of an existing asset by a private firm or a public-private partnership
4. **Divestitures:** Full or partial transfer of ownership of a public asset to a private firm through a direct sale or a public offering

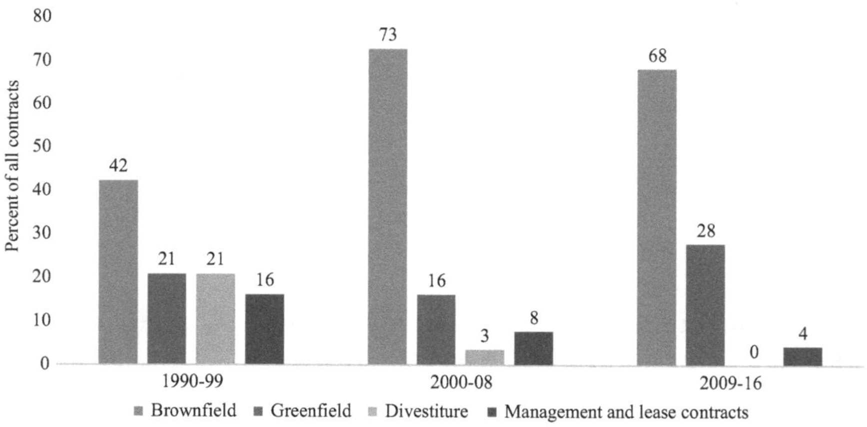

Fig. 8.17 Distribution of public-private partnership contracts in water and sanitation in Latin America and the Caribbean by type, 1990–2016. Source: World Bank PPI database (accessed on July 31, 2017)

public sector should increase the spending in water and sanitation, but efforts should be made to increase attractiveness of the sector to private participation, as current spending does not totally cover necessary financing in terms of new infrastructure but consequently neither in terms of new maintenance and operating costs.

Financial Performance of Water Operators

Insufficient financial resources prevent operators from expanding the network and increasing the coverage rate and quality of service, thus having negative consequences in terms of efficiency and equity. Excessive dependence on the government creates financing uncertainty and opens the door to political interference and clientelism (Ducci and García Merino 2013).

Two financial measures—the operating cost coverage ratio and the leverage rate—reveal the financial performance of water operators. The operating cost coverage ratio is the ratio of operating revenue over operating costs (excluding depreciation). The leverage rate (total liabilities over net worth) is an indicator of indebtedness.

Information on these indicators is scarce in LAC. The very limited data available suggest that very few countries have water utilities with adequate operating cost coverage ratios.

An operating cost coverage ratio of 1.30 allows a private operator to cover depreciation costs, taxes, and net financial results. A ratio above 1.30 allows it to invest in new infrastructure to improve coverage or for better quality. A ratio below 1.30 implies the need to find external financing sources, generally government funding (Ducci and García Merino 2013).

Operating cost coverage ratios vary widely across countries (Fig. 8.18).[7] Chile, Colombia, and Costa Rica have operating cost coverage ratios that are well above 1.30. Argentina and Panama do not cover their operating costs with their operating revenues. On average, operating revenues cover operating costs, but the average ratio (1.17) is below the recommended level of 1.30.

Ducci and García Merino (2013) recommend that leverage not exceed 100 percent of net worth. The sample average leverage rate is 57 percent (see Fig. 8.18). Some countries have much higher averages, however. Brazil and Chile exceed 100 percent, but these high ratios reflect the easier access to financial markets that these countries' operators have. Other countries are well below the sample average, such as Costa Rica, Uruguay, and Panama. These low leverage rates show a significant room

[7] There appears to be little variability within countries, although the sample is small (see Table 8.5).

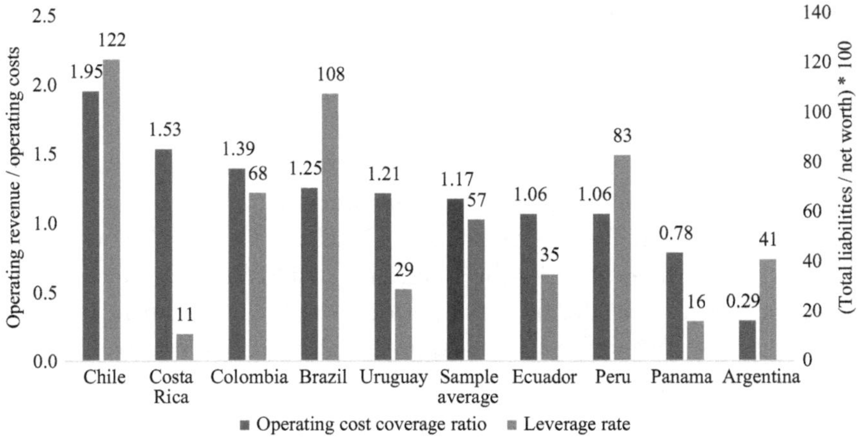

Fig. 8.18 Financial performance of water operators in capitals of selected countries, 2012. Source: Asociación de entes reguladores de agua potable y saneamiento de las Américas (ADERASA) (cited in Lentini 2015)

for maneuver in terms of accessing financial markets, but to do so, they should complement this room with enough attractiveness to investors.

The financial performance of LAC capital cities' water operators shows a significant variability between the region's countries. On the one hand, some countries show operating cost coverage ratios considerably high, implying a higher financial capability that will allow them not only to cover their operating and financial costs but also to invest to improve the service quality and coverage. Other countries find themselves in the opposite situation, not being able to cover their operating costs. Likewise, these countries appear to have considerably low leverage rates, which could indicate that their dependence on government financing is rather high, as they are not appealing enough to financial markets. Indeed, if operating revenues do not manage to cover operating costs, consequently, returns on investment should be relatively low, thus unattractive to potential investors.

Regulation of the Water and Sanitation Sector

To improve the governance of the water and sanitation sector and enhance its attractiveness to the private sector, many countries in LAC decentralized the provision of water and sanitation services in the

1990s and created independent regulatory agencies.[8] By the mid-2000s, however, service provision was decentralized, but regulation was carried out at the central level. This model created conflicts that endangered the well-functioning of the sector and limited the positive effects of reforms. It also made the sector unattractive to the private sector (Foster 2005).

Today, about 75 percent of LAC countries have independent agencies regulating their water sectors, and about 70 percent have private participation in the water and sanitation sector (Table 8.2). About a quarter of countries with independent regulatory agencies do not have private sector participation, and about 20 percent of countries with private participation do not have regulatory agencies.

Very few LAC countries have subnational independent regulatory agencies which, without having analyzed the decentralization level of water management (see below), would imply that no solution has been brought to the central regulation versus decentralized water management conflict described by Foster in 2005. It is thus worth analyzing whether centralization or decentralization levels of the water and sanitation sector regulation and its management coincide or, as was the case more than a decade ago, diverge, creating conflicts and affecting negatively the well-functioning of the sector.

Of the 19 countries in LAC that have independent regulatory agencies in the water sector, 10 have the same level of centralization in the regulation and management of the sector (Table 8.3). These countries have kept the decision-making of both areas at the central level. In the other nine countries, regulation of the sector is carried out at a less decentralized level (typically the central level) and management of the sector at a highly decentralized level (typically the local level).

Incongruences thus remain between the level of decentralization of sector regulation and management. They could be one factor behind the limited quality of service, the poor financial performance of water utilities, and the very low coverage rates in rural areas.

---

[8] The term *independence* does not imply complete independence from the political power but rather that financial, administrative, and operating autonomy limits political interference.

Table 8.2 Presence of independent regulatory agency and private participation in water and sanitation sector, 2017

Item	Independent regulatory agency	Public-private partnership (PPP)	Independent regulatory agency and PPP	PPP but no independent regulatory agency	Independent regulatory agency but no PPP	No PPP and no independent regulatory agency	Subnational independent regulatory agency	Multisector independent regulatory agency
All countries								
Number	19	18	13	5	6	2	3	7
Share (percent)	73	69	50	19	23	8	12	27
Sample size	26	26	26	26	26	26	26	26
Developing countries								
Number	17	17	13	4	4	2	3	5
Share (percent)	74	74	57	17	17	9	13	22
Sample size	23	23	23	23	23	23	23	23
Developed countries								
Number	2	1	0	1	2	0	0	2
Share (percent)	67	33	0	33	67	0	0	67
Sample size	3	3	3	3	3	3	3	3

Sources: Bertoméu-Sánchez et al. (2017) and data collected by the authors

Table 8.3 Level of government handling regulation and management of the water sector in countries in Latin America and the Caribbean with independent regulatory agencies, 2017

Country	Regulation	Management
Argentina	Regional	Local
Barbados	Central	Central
Belize	Central	Central
Bolivia	Central	Local
Brazil	State	Local
Chile	Central	Regional
Colombia	Central	Local
Costa Rica	Central	Central
Dominican Republic	Central	Central
Ecuador	Canton	Local
El Salvador	Central	Central
Honduras	Central	Local
Jamaica	Central	Central
Mexico	Central	Local
Nicaragua	Central	Central
Panama	Central	Central
Paraguay	Central	Central
Peru	Central	Local
Uruguay	Central	Central

Sources: Herrera and Post (2014) and data collected by authors

4 Brief Review of the Empirical Literature

Recent studies of LAC have confirmed the positive effects of higher-quality infrastructure on the economy and living standards documented in Aschauer's seminal 1989 paper. Lanau (2017), for instance, shows that better infrastructure raises growth and investment and that improved infrastructure could yield substantial economic benefits in the region.

Empirical studies focus on four areas:

- Performance indicators
- The effects of privatization on sector performance
- Governance and policy aspects of the sector
- Tariff schemes and their linkages to performance indicators and governance characteristics

Performance Indicators

Performance of the water and sanitation sector is traditionally studied through its effects on the economy, public health, and the environment. In their 2017 report, Fay et al. describe that water access has improved significantly over the last years but that sanitation coverage is an increasingly urgent challenge; likewise, inefficient public spending is one of the causes behind the challenges in the sector. Gamper-Rabindran et al. (2010) find strong positive correlations between the introduction of piped water and reductions in the infant mortality rate in Brazil. Schady (2015) reviews the literature and finds that the stronger evidence on the relationship between access to water and sanitation infrastructure and child outcomes in LAC is for extensions of coverage and is limited to child mortality; it does not cover other health aspects, such as morbidity, nutritional status, or development. Grafton et al. (2011) analyze a ten-country household survey. They find that concerns about the environment have a significant effect on some self-reported water-saving behaviors. Molinos-Senante and Donoso (2016) propose a water rate for Chile that creates incentives to improve water use sustainability.

Effects of Privatization

In their study of Brazil, Barbosa et al. (2016) argue that privatization is a consistent manner of ensuring accessibility to water services, as the private sector provides with the financing that the public sector does not have the capacity to cover. Clarke et al. (2009) find that following the introduction of private sector participation in Latin America, the share of households connected to piped water and sewerage rose significantly and the introduction of the private sector did not have any negative effects on the poor. Galliani et al. (2005) show that child mortality in Argentina fell by about eight percent in areas that introduced the private sector in their water services; the reduction was largest in the poorest areas. In their study of all infrastructure sectors in nine Latin American countries, Sirtaine et al. (2005) find that the financial returns of private investment were modest and that returns to many concessions were below the cost of capital. Andrés et al. (2013) find that the performance of water utilities in LAC is not highly correlated with the type of ownership (public or private).

Effects of Governance

Bardhan and Mookherjee (2006) show that, for developing countries and infrastructure in general, the effect of switching from centralization to decentralization depends on the financing mechanism of local governments. Herrera and Post (2014) find that decentralization does not necessarily stop political interference in the water and sanitation sector. Barde (2017) concludes that access rates in Brazil increased considerably between 2000 and 2010 in areas in which local governments were responsible for service. Barbosa et al. (2016) show that utilities operate best at the metropolitan level.

Bertoméu-Sánchez et al. (2017) find that globally, though having an independent regulatory agency has positive effects on the functioning of the sector, such an agency is not a necessary or sufficient condition for attracting private sector participation, although in LAC having an independent regulatory agency increases the odds of attracting private financing to the sector. Estache et al. (2001) for Latin America and Andrés et al. (2013) conclude that although private participation has been crucial in increasing sector performance, it must be accompanied by transparent and accountable regulation, which significantly improves performance and makes the poor better off as a result. Ferro et al. (2011) highlight that regulation can have positive effects on efficiency as long as information is available, quality standards are well specified, and other conditions are met.

Tariffs

Zetland and Gasson (2013) show that globally, higher water tariffs are correlated with lower per capita consumption, lower water availability, higher demand, and lower risk of shortage; at the national level, higher tariffs are also correlated with higher GDP and better governance. Jiménez et al. (2016) show that in Santo Domingo, in the Dominican Republic, regardless of the price they pay, consumers' satisfaction and willingness to pay are positively related to the quality of water service. Ferro and Lentini (2013) find that financial sustainability is strongly correlated with access to improved water rates.

Nauges and Whittington (2017) and Whittington et al. (2015) show that IBTs perform poorly in targeting subsidies to the poor. These tariff schemes also introduce price distortions that create economic efficiency losses, though welfare losses are relatively low. Barde and Lehmann (2014) show that means-tested tariffs distribute more income to the poor than IBTs but that the share of poor consumers benefiting from water subsidies is lower than it is with means-tested tariffs. Molinos-Senante and Donoso (2016) propose a water rate that incentivizes water use sustainability and equity among consumers.

5 Concluding Remarks

Several important findings emerge from this study:

- Thanks to the reforms undertaken in the 1990s, access to improved water is now almost universal in LAC (95 percent). Efforts are still needed in the sanitation sector and in rural areas in both sectors.
- IBTs, which have been strongly criticized in some of the literature and are the main type of tariff used in the region, do not penalize high consumption levels sufficiently. The average bill for 15 cubic meters of water (the minimum needed to cover basic needs) is lower in LAC than in the Middle East and North Africa or Sub-Saharan Africa.
- Service quality remains weak. The continuity of service has declined in recent years, and in urban areas (for which data are available), almost 40 percent of supplied water is wasted (the situation is worse in rural areas, although data are not available). Poor service quality affects the economy and both human health and the environment.
- The financial health and sustainability of water utilities are weak. Access to capital markets is difficult for most utilities in the region.
- Most countries have independent regulatory agencies in the sector. The literature suggests that they improve the provision of water and sanitation services but do not necessarily increase private sector participation, which remains a challenge.
- The divergence between the level of decentralization of regulation and management persists in almost half the countries in the region, limiting the scope for improving sector performance.

Appendix

Table 8.4 Regulators of the water and sanitation sector in Latin America and the Caribbean, by country

Country	Organization	Member of ADERASA	Separate regulatory body	Year of establishment	Level of regulation
Argentina	Asociación de Entes Reguladores de Agua y Saneamiento de la República Argentina	Yes	Yes	1991	Province
Bahamas, the	Water and Sewerage Corporation	No	No	1976	National
Barbados	Fair Trading Commission	No	Yes	2001 (1955)	National
Belize	Public Utilities Commission	Yes	Yes	1999	National
Bolivia	Autoridad de Fiscalización y Control Social de Agua y Saneamiento Básico	Yes	Yes	2000	National
Brazil	Associação Brasileira de Agências de Regulação	Yes	Yes	1999	State
Chile	Superintendencia de Servicios Sanitarios	Yes	Yes	1990	National
Colombia	Comisión de Regulación de Agua Potable y Saneamiento Básico	Yes	Yes	1994	National
	Superintendencia de Servicios Públicos Domiciliarios	No	Yes	1991	National
Costa Rica	Autoridad Reguladora de los Servicios Públicos de Costa Rica	Yes	Yes	1999	National
Dominican Republic	Instituto Nacional de Aguas Potables y Alcantarillados	Yes	Yes	1962	National (25 of 31 provinces)
Ecuador	Empresa Municipal de Agua Potable y Alcantarillado de Guayaquil	Yes	Yes	2001	Canton
	Agencia de Regulación y Control de Agua	No	Yes	2014	National
El Salvador	Administración Nacional de Acueductos y Alcantarillados	Yes	Yes	1961	National

(*continued*)

Table 8.4 (continued)

Country	Organization	Member of ADERASA	Separate regulatory body	Year of establishment	Level of regulation
Guatemala	Ministry of Public Health and Social Assistance	No	No	1944	National
	Ministry of Environment and Natural Resources	No	No	2000	National
Guyana, CR	Public Utilities Commission	No	No	1990	National
Haiti	National Directorate for Water Supply and Sanitation (Ministry of Public Works)	No	No	–	National
Honduras	Ente Regulador de Servicios de Agua Potable y Saneamiento	Yes	Yes	2003	National
Jamaica	The Office of Utilities Regulation	No	Yes	1995	National
Mexico	Asociación Nacional de Empresas de Agua y Saneamiento de México	Yes	No	1992	National
Nicaragua	Instituto Nicaragüense de Acueductos y Alcantarillado Sanitario	Yes	Yes	1979	National
Panama	Autoridad Nacional de los Servicios Públicos	Yes	Yes	1996	National
Paraguay	Ente Regulador de Servicios Sanitarios	Yes	Yes	2000	National
Peru	Superintendencia Nacional de Servicios de Saneamiento	Yes	Yes	1992	National
Suriname	Ministry of Natural Resources	No	No	1991	National
Trinidad and Tobago	Regulated Industries Commission	No	Yes	1998	National
Uruguay	Unidad Reguladora de Servicios Energía y Agua	Yes	Yes	2002	National
Venezuela	Ministry of Popular Power for Ecosocialism and Water	No	No	2015	National

Asociación de Entes Reguladores de Agua y Saneamiento de Las Américas

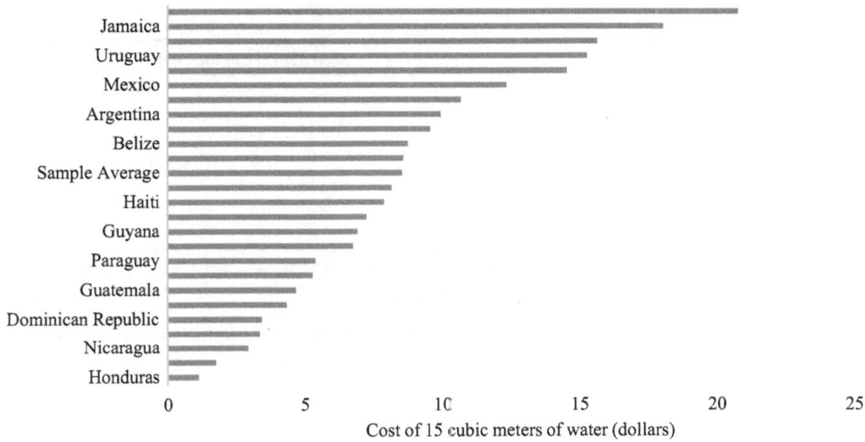

Fig. 8.19 Cost of water in selected capitals, 2016. Sources: Data from the International Benchmarking Network for Water and Sanitation Utilities (IBNET) and Global Water Intelligence (2016)

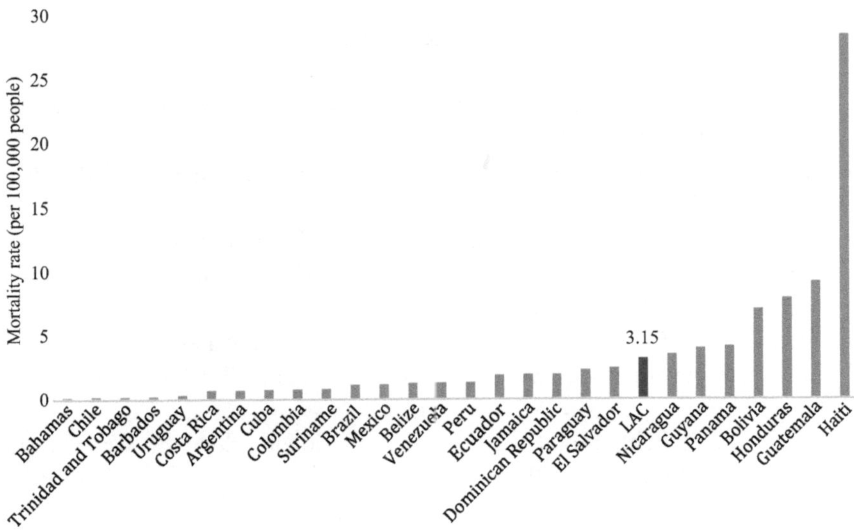

Fig. 8.20 Mortality rates attributed to unsafe water and sanitation, by country 2012. Source: World Health Organization (2016) (accessed on July 4, 2017)

Table 8.5 Operating cost coverage ratios in selected countries

Country	Average	Standard deviation
Argentina	0.86	0.36
Brazil	1.43	0.17
Chile	1.54	0.22
Colombia	1.24	0.17
Ecuador	1.27	0.30
Peru	0.86	0.18

Source: Asociación de entes reguladores de agua potable y saneamiento de las Américas (ADERASA) (cited in Lentini 2015)

Note: Table includes countries for which data are available on more than one operator

References

Andrés, L. A., Schwartz, J., & Guasch, J. L. (2013). *Uncovering the Drivers of Utility Performance: Lessons from Latin America and the Caribbean on the Role of the Private Sector, Regulation, and Governance in the Power, Water, and Telecommunication Sectors*. Washington, DC: World Bank.

Arroyo, V., Ballestero, M., & Mejía, M. (2015). *Inseguridad económica del agua en latinoamérica: De la abundancia a la inseguridad*. Caracas: Corporación Andina de Fomento (CAF).

Aschauer, D. (1989). Is Public Expenditure Productive? *Journal of Monetary Economics, 23*(2), 177–200.

Barbosa, A., De Lima, S. C., & Brusca, I. (2016). Governance and Efficiency in Brazilian Water Utilities: A Dynamic Analysis in the Process of Universal Access. *Utilities Policy, 43*, 82–96.

Barde, J. A. (2017). What Determines Access to Piped Water in Rural Areas? Evidence from Small-Case Supply Systems in Rural Brazil. *World Development, 95*, 88–110.

Barde, J. A., & Lehmann, P. (2014). Distributional Effects of Water Tariff Reforms: An Empirical Study from Lima, Peru. *Water Resources and Economics, 6*, 30–57.

Bardhan, P., & Mookherjee, D. (2006). Decentralization and Accountability in Infrastructure Delivery in Developing Countries. *Economic Journal, 116*, 101–127.

Bertoméu-Sánchez, S., Camós, D., & Estache, A. (2017). *Do Private Water Utility Operators Care About Regulatory Agencies in Developing Countries?* Policy Research Working Paper 8045, World Bank, Washington, DC.

Boland, J. J., & Whittington, D. (1998). *The Political Economy of Increasing Block Tariffs for Water in Developing Countries*. EEPSEA Special and Technical Paper Sp199801t3, Economy and Environment Program for Southeast Asia, Laguna, Philippines.

CAF (Corporación Andina de Fomento). (2014). *Encuesta de hogares 2013: Principales resultados*. Retrieved from http://scioteca.caf.com/handle/123456789/409.

Clarke, G. R., Kosec, K., & Wallsten, S. (2009). Has Private Participation in Water and Sewerage Improved Coverage? Empirical Evidence from Latin America. *Journal of International Development, 21*(3), 327–361.

Ducci, J., & García Merino, L. J. (2013). *Principales indicadores financieros de entidades prestadoras de servicio de agua potable y saneamiento en América Latina y el Caribe*. Washington, DC: Inter-American Development Bank.

Estache, A., Gómez-Lobo, A., & Leipziger, D. (2001). Utilities Privatization and the Poor: Lessons and Evidence from Latin America. *World Development, 29*(7), 1179–1198.

FAO. (2016). AQUASTAT Website. Food and Ariculture Organization of the United Nations (FAO). Website. Retrieved July 31, 2017.

Fay, M., Andrés, L. A., Fox, C. J. E., Narloch, U. G., Straub, S., & Slawson, M. A. (2017). *Rethinking Infrastructure in Latin America and the Caribbean: Spending Better to Achieve More*. Washington, DC: World Bank.

Ferro, G., & Lentini, E. (2013). *Políticas tarifarias para el logro de los objetivos de desarrollo del milenio (ODM): Situación actual y tendencias regionales recientes*. Washington, DC: Inter-American Development Bank.

Ferro, G., Romero, C. A., & Covelli, M. P. (2011). Regulation and Performance: A Production Frontier Estimate for the Latin American Water and Sanitation Sector. *Utilities Policy, 19*, 211–217.

Foster, V. (2005). *Ten Years of Water Service Reform in Latin America: Toward an Anglo-French Model*. Water Supply and Sanitation Sector Board Discussion Paper 3 (32027), World Bank, Washington, DC. Retrieved from http://documents.worldbank.org/curated/en/962281468269682959/pdf/320270WSSServiceReform.pdf.

Galliani, S., Gertler, P., & Schargrodsky, E. (2005). Water for Life: The Impact of Privatization of Water Services on Child Mortality. *Journal of Political Economy, 113*(1), 83–120.

Gamper-Rabindran, S., Khan, S., & Timmins, C. (2010). The Impact of Piped Water Provision on Infant Mortality in Brazil: A Quantile Panel Data Approach. *Journal of Development Economics, 92*(2), 188–200.

Global Water Intelligence. (2016). Global Water Intelligence 2016 Tariff Survey, October.

Grafton, R. Q., Ward, M. B., To, H., & Kompas, T. (2011). Determinants of Residential Water Consumption: Evidence and Analysis from a 10-Year Country Household Survey. *Water Resources Research, 47*(8).

Herrera, V., & Post, A. E. (2014). Can Developing Countries Both Decentralize and Depoliticize Urban Water Services? Evaluating the Legacy of the 1990s Reform Wave. *World Development, 64*, 621–641.

IBNET (International Benchmarking Network). Retrieved from https://www.ib-net.org.

Jiménez, R., Serebrisky, T., & Mercado, J. (2016). What Does 'Better' Mean? Perceptions of Electricity and Water Services in Santo Domingo. *Utilities Policy, 41*, 15–21.

Joint Monitoring Programme of the World Health Organization and UNICEF. (2016). *WHO/UNICEF Joint Monitoring Programme (JMP) for Water Supply and Sanitation*. Retrieved from https://washdata.org.

Lanau, S. (2017). *The Growth Return of Infrastructure in Latin America*. Working Paper 17/35, International Monetary Fund, Washington, DC.

Lentini, E. (2015). *El futuro de los servicios de agua y saneamiento en américa latina: Desafíos de los operadores de áreas urbanas de más de 300.000 habitantes*. Inter-American Development Bank and Corporación Andina de Fomento (CAF).

Molinos-Senante, M., & Donoso, G. (2016). Water Scarcity and Affordability in Urban Water Pricing: A Case Study of Chile. *Utilities Policy, 43*, 107–116.

Nauges, C., & Whittington, D. (2017). Evaluating the Performance of Alternative Municipal Water Tariff Designs: Quantifying the Tradeoffs Between Equity, Economic Efficiency, and Cost Recovery. *World Development, 91*, 125–143.

Schady, N. (2015). *Does Access to Better Water and Sanitation Infrastructure Improve Child Outcomes? Evidence from Latin America and the Caribbean*. Working Paper IDB-WP-603, Inter-American Development Bank, Washington, DC.

Sirtaine, S., Pinglo, M. E., Guasch, J. L., & Foster, V. (2005). How Profitable Are Private Infrastructure Concessions in Latin America? Empirical Evidence and Regulatory Implications. *Quarterly Review of Economics and Finance, 45*(2), 380–402.

Whittington, D., Nauges, C., Fuente, D., & Wu, X. (2015). A Diagnostic Tool for Estimating the Incidence of Subsidies Delivered by Water Utilities in Low-and Medium-Income Countries, with Illustrative Simulations. *Utilities Policy, 34*, 70–81.

WHO (World Health Organization). (2016). UNICEF Water and Sanitation Coverage. Retrieved from http://www.who.int/topics/drinking_water/en/.

Zetland, D., & Gasson, C. (2013). A Global Survey of Urban Water Tariffs: Are They Sustainable, Efficient and Fair? *International Journal of Water Resources Development, 29*(3), 327–342.

9

Africa

Successes and Failures of Water and Sanitation Governance Choices in Sub-Saharan Africa (1990–2017)

Antonio Estache

1 Introduction

The main purposes of this chapter are (1) to provide a snapshot as of 2017 of the performance of the water and sanitation sector (WSS) across Sub-Saharan Africa (SSA), (2) to review the wide range of institutional and governance reform experiences in the region since the early 1990s, and (3) to show that, despite the significant progress achieved in the last 15 years or so as part of the efforts to deliver the Millennium Development Goals (MDGs), the reforms are still far from delivering as much as needed.

In SSA, as of 2017, the policy debate on WSS is still largely centered on the need to provide an affordable quality access to a significant share

A. Estache (✉)
European Center for Advanced Research in Economics and Statistics (ECARES), Université libre de Bruxelles, Bruxelles, Belgium

S. Porcher, S. Saussier (eds.), *Facing the Challenges of Water Governance*, Palgrave Studies in Water Governance: Policy and Practice,
https://doi.org/10.1007/978-3-319-98515-2_9

of the population unable to get it or to afford it in many of the countries.[1] It is thus mostly about investment and its financing, even if technological choices, unaccounted-for water (UFW), implementation mismanagement, and other efficiency-related issues are also quite present in the policy debates. The various waves of reforms implemented for almost 30 years now have not yet delivered WSS performances comparable to those achieved in other regions with similar or equivalent reforms. The sector is still underinvesting and still faces constraints in its efforts to fund investment requirements.[2] The international community has been quite involved in supporting SSA efforts through multiple initiatives and efforts to mobilize resources with some success but also with many limitations in terms of their impact on the ground.[3]

To summarize the evidence on the success and failures of governance and other reforms, the chapter is organized as follows. Section 2 reviews the major stylized facts. This is not a straightforward exercise. Although some of the basic data on access rates is common to all sources, most of the data on investment needs and financing sources are based on assumptions which vary across institutions. Moreover, there is often little transparency on the details of the assumptions. On other dimensions such as UFW, the evidence is mostly on country-specific studies with very little strict comparability across countries available from public sources of information.[4] This is why, for many of the indicators reported across performance dimensions, ranges rather than point estimates are provided. Section 3 summarizes the various experiences in governance, including regulatory reforms, conducted in the region. Section 4 is a brief survey of the relatively narrow technical literature assessing the impact of these reforms. This literature is quite modest in comparison

[1] Although policy discussion also includes a rebirth of old and recurring debates on the need to do better on irrigation to improve land productivity in the region, this chapter will not address the topic. See, for instance, Woodhouse et al. (2017).

[2] Hutton and Varughese (2016).

[3] Sambu (2016) and Ndikumana and Pickbourn (2017).

[4] The latest comparable information dates from the Africa Infrastructure Country Diagnostic (AICD)—summarized in Foster and Briceno-Garmendia (2009). It is based on 2005–2007 data for most dimensions and this is by now largely out of date.

with the many case studies provided by international organizations but is important as it provides statistically robust tests of the insights from the less technical studies. Section 5 concludes.

2 The Stylized Facts

On an annual basis, SSA's investment needs in WSS for the foreseeable future are estimated between 1% and 2% of GDP on average (depending on the assumption made for technology and growth). Matching maintenance cost of the sectors increases these figures by 25% to 35%. This roughly adds up to total financing requirements of around 2.5% of GDP—this is the bulk part figure anchoring many of the policy discussions among experts. There is some dispersion across countries as national coverage rates vary significantly as illustrated in Figs. 9.1 and 9.2 for water and sanitation access rates, respectively. Overall, this adds up to estimations in 2017 prices as low as US$25 billion and as high as US$50 billion annually for the region.

Current expenditure levels are well below the needs, even accounting for their current levels of inefficiency. IMF data reported by Gutman et al. (2015) on a sample of 25 SSA countries implies that, on average, public expenditure in the sector is currently at around 0.9% of GDP (i.e. 20% of the total spent on infrastructure), even if 40% of the countries of the sample spend more, up to 2% of GDP. According to WHO (2014), over 80% of these public expenditures are spent in urban areas, yet about

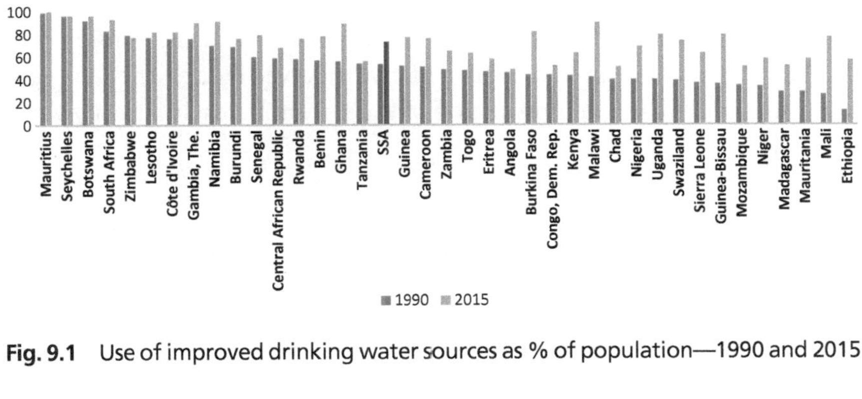

Fig. 9.1 Use of improved drinking water sources as % of population—1990 and 2015

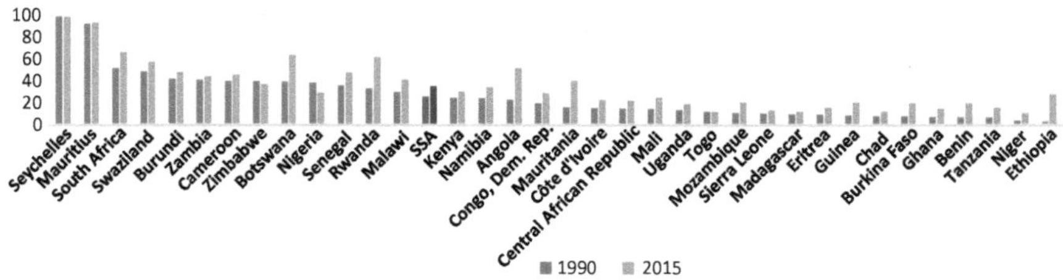

Fig. 9.2 Use of improved sanitation services as % of population—1990 and 2015. Source: Joint Monitoring Programme (JMP)

half of the WSS needs are still in rural areas.[5] For a sample of nine countries covered by WHO (2014), household contributions to the recovery of these costs ranged from 6% to 66%, and for only four out of these countries did external financing reach 20% or more.[6]

Raising the public resources needed continues to be a problem for over 80% of the countries of the region since average income levels are still quite low and the tax basis quite narrow. For most countries, getting access to private funding in this sector has not been a serious alternative so far. According to the World Bank PPI database, between 1990 and 2016, private funding has generated investments adding up to US$425 million, and most of this has been captured by only five countries (South Africa, Nigeria, Kenya, Ghana, and Uganda).[7] That's about 16 million per year for a bit more than a project per year. It's also less than 0.1% of the US$4 billion in funds committed to the sector in 2013 by the Infrastructure Consortium for Africa (ICA 2014).

Growing sources of additional financing include institutional investors such as pension funds, hedge funds, and wealth funds. And they also include Chinese investments. These have become one of the main players in the construction and financing of hydropower dams. They are also increasingly active in the construction of water treatment stations, city water

[5] World Health Organization (2015), INVESTING IN WATER AND SANITATION: INCREASING ACCESS, REDUCING INEQUALITIES GLAAS 2014 findings—Special report for Africa.

[6] Burkina Faso, Ghana, Lesotho, and Madagascar.

[7] http://ppi.worldbank.org/snapshots/sector/water-and-sewerage.

supply systems, and rural water well drilling.[8] In 2010, they were present in 30 African countries.[9] Investment funds interested specifically in water are promising new financing options but are not yet a reasonably predictable source of funding for the region. The first public-private water fund was launched in Kenya in 2015 (the Upper Tana-Nairobi Water Fund) to finance upstream water conservation through activities such as watershed protection and reforestation.[10] The US$15 m the fund was aiming at illustrates the modest size of this expected source of financing.

The difficulties in finding enough financing to meet the investment needs and match operation and maintenance (O&M) expenditures explain why as of 2015, according to the World Development Indicators (WDI), about 319 million people in the region were without access to improved reliable drinking water sources. In other words, only 68% of SSA population had access to clean water sources. This is the coverage rate achieved by lower middle-income countries in 1990 (i.e. a quarter of a century ago!). These lower middle-income countries have now reached 89% of coverage. Figure 9.1 shows for water that the best performers have reached almost full coverage, while the worse ones (Chad, Ethiopia, Niger, or Mozambique, for instance) lag significantly, despite significant progress since 1990. Figure 9.2 provides a very explicit sense of the extent to which sanitation continues to be a major problem across the region with only a few exceptions. Both figures also point to the dispersion of progress across countries.

The central fact emerging from these statistics is that, almost 20 years within the twenty-first century, 12 countries in SSA still suffer from water access rates below 60%. They also imply that only one person in three in the region has access to improved sanitation facilities as compared to one person in two in lower middle-income countries, for instance. Figures 9.3 and 9.4 show how different progress has been between urban and rural areas, in favor of urban population. But the effort has not really been very effective enough to adjust to the fast pace of urbanization in the region. Partially, this is due to the endogeneity of the urbanization process or in

[8] Tukic and Burgess (2016).

[9] Godfrey and Ross (2016).

[10] https://www.nature.org/ourinitiatives/regions/africa/explore/nairobi-water-fund.xml.

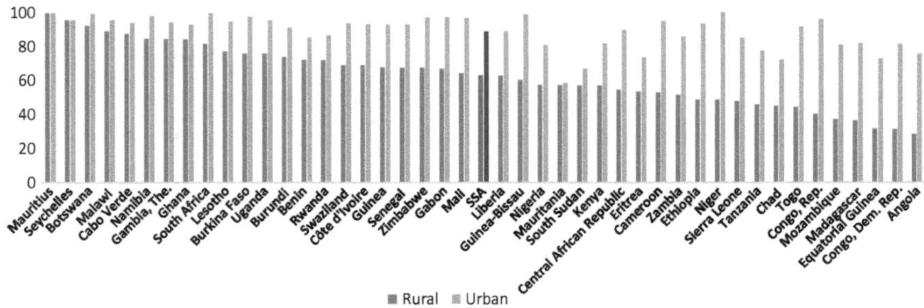

Fig. 9.3 Urban and Rural access to water as % of population—2015. Source: Joint Monitoring Programme (JMP)

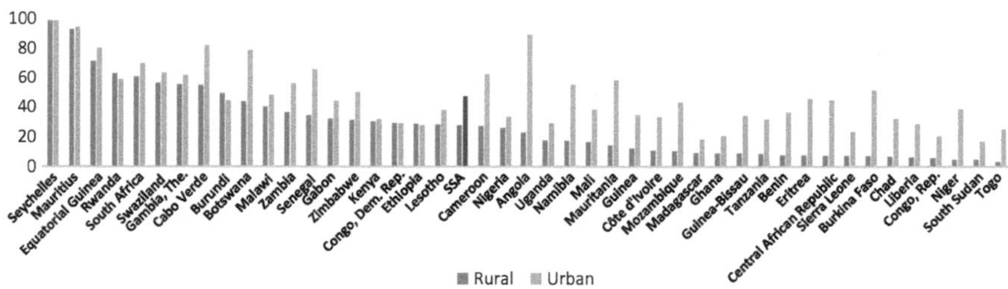

Fig. 9.4 Urban and rural access to sanitation as % of population—2015. Source: Joint Monitoring Programme (JMP)

particular its speed. The stronger the rural-urban access to public services gap, the stronger is the incentive to migrate. All the evidence points to a tripling of SSA's urban population by 2050. However, this will not guarantee them with access to improved water sources. The share of SSA's urban population with access to water piped to their home has actually declined, from 43% in 1990 to 33% in 2015, according to WHO (2015).

Much of the high-level policy discussions of these statistics have, so far, relied on the very broad definition of access to clean water adopted here as well. But it is important to highlight that, with this broad definition, many of those credited with access to improved water sources do not have home connections and need to rely on shared connections. Quality of service continues to be a major issue in the region, and in most countries, regulators and other authorities lack the means to enforce quality requirements.[11] This aggregate data also hides that the poor are the main victims of the slow progress. In a sample of 35 SSA countries monitored by UNICEF, the poorest rural quintile has no access to piped water, and open defecation is the only option for over 60% of households.[12] In contrast, over 90% of the richest urban quintile can rely on improved sanitation and drinking water.

To complete the snapshot, it is important to point out that the slow progress of the region can also be partially blamed on technological choices made by the authorities and the suppliers. The financing requirements could be cut by improving the technical performance of many of the operators in the region as well as differentiating the technological options to improve the matching between preference and ability to pay. Rural areas have been more effective on this front simply because they often have no choice when they can no longer afford to wait for long undelivered promises by central authorities to deliver access through connections to networked facilities.[13]

To illustrate the scope for improvements, consider UFW due to both physical/technical (i.e. failures in trunk connections or aging pipes) and non-physical (i.e. underbilling or illegal connections) reasons. It continues to be a challenge in the region. Sharma (2008) suggests that, by the mid-2000s, it averaged at around 39% in SSA but could reach as much as 60% in many capital cities of the region. This compared to about 15% in the USA, for instance. A recent diagnostic of UFW in South Africa shows that the situation has not changed too much since the mid-2000s. Based on a sample covering about half of the country's municipalities, non-revenue water was estimated for the country as a whole at 36.8%, of which 25.4% were linked to physical leakage.[14] Similar assessments have been conducted for other countries and report similar results in the region, although they are often more addressed in technical consulting reports than in policy-oriented academic publications.

---

[11] Rimi Abubakar (2016) provides a powerful case study in Nigeria.

[12] UNICEF (2015).

[13] Many of these issues are often linked to poor implementation and poor coordination of otherwise-reasonable policy decisions. See World Bank (2017) for a series of detailed case studies, including some on SSA.

[14] McKenzie et al. (2012).

To address these concerns, there are institutional and technological options. At the institutional level, there is enough informal evidence to be concerned with a bias built in international financing sources. It is indeed still easier to get money for new plants that will use a huge amount of the potential financial resources available than work harder on UFW and water losses. The cost-effectiveness of solutions is not always accounted for in the evaluation processes for a given level of benefit.

At the technological level, innovations are increasingly helping reduce the non-physical losses. For instance, Durban City's eThekwini is often cited as the most successful experiment so far to reduce the risks of failures to collect bills. The utility adopted flow meters that get switched into low-pressure mode when customers fail to pay their bills. The meter can limit the quantity that flows through the valves to very small quantities that would add up to the volume the utility is obligated to, but no more.

Other technical issues are related to the cost of producing and delivering water in a region that is the world's second-driest continent (after Australia) with only 9% of global renewable water resources to support 15% of the global population.[15] They also include the high cost of sanitation as a result of the lack of treatment tradition for many of the industries.[16] For example, in Nigeria, Taiwo et al. (2012) report that less than 10% of industries treat their effluents before discharging them into surface water. Taiwo (2011), again, shows that in many SSA countries, rain washes municipal solid wastes and other pollutants into rudimentary drainage systems and, subsequently, rivers. The upshot is, as suggested by Li et al. (2011), that pollutant concentrations observed in effluents can be as much as five times higher than observed in Europe.

But not all challenges are linked to technical issues. Some can also be blamed on policy decisions and in particular on the pricing strategy adopted in most countries. In 2007, underpricing resulted in a loss of over 60% of utilities revenues.[17] This is anchored, at least partially, in the choices made for tariff design. Based on the information available for 25 SSA countries on 2016 tariff structures, most countries of the region

[15] UNEP (2010).

[16] For a detailed discussion, see Wang et al. (2014).

[17] Foster and Briceno-Garmendia (2009).

(22 countries for this sample) have adopted increasing block tariffs (IBTs) (without a fixed component).[18] None seems to have adopted explicitly two-part tariffs in the region as the three other countries of the sample are split between a linear, flat, and decreasing block tariff. The intention is likely to have been to protect the affordability of a minimal use of water by all users. Moreover, few countries account for fixed costs in their billing in the region. Cost recovery is usually limited at recovering O&M expenditures, implying that the capital expenditures are often largely subsidized.[19] The intentions to care for equity are laudable, but the implementation of the increasing block approach may not be sending the right signal to consumers with a capacity to consume and may be regressive rather than progressive as expected since connection subsidies go to those with connection and this excludes the poorest.[20] According to van den Berg and Danilenko (2011), about 90% of the population with piped water access belongs to the richest 60% of the population. This leaves only 10% of those with access with a justified subsidy. Ideally, it seems that a more efficient and equitable solution would combine the two-part tariff approach with the IBTs (i.e. fixed charge + various consumption blocks with different tariffs and an exception for the poorest for whom the fixed part would be lifted).

Increasingly, the region has been trying to address both the risks of revenue losses from non-payment for consumption and the capacity to pay constraint of the poor by adopting pre-payment solutions. Pre-paid meters reduce the risks of arrears or debt, as customers pay in advance for a specified amount of water. This, in turn, improves revenue to fund wider coverage and limits the need to inflate tariffs to compensate for revenue losses. But there is also the risk that the approach undoes some of the social purpose of the increasing block approach so common in the

---

[18] Based on data collected from the IBNET and GWI website, the two main sources of comparable information on tariffs for SSA.

[19] In rural and/or non-networked water supply, users are often required to pay around 5–10% of capital costs. Often this can be paid in labor or local materials. For sanitation, users are mostly required to pay 100% of capital costs and O&M, and they usually cannot afford it.

[20] See for multiple quantitative examples Angel-Urdinola and Wodon (2012), Banerjee et al. (2008), Bardasi and Wodon (2008), and Estache and Wodon (2014), and for powerful documentation of 17 cases studies, see Heymans et al. (2016).

region.[21] This is because the flat rate built in the pre-payment is often higher than the price of the service built in the first blocks of the increasing block approach. South Africa's decision to rely on pre-paid but with IBT pricing seems to be the most desirable combination of efficiency and equity. This can be done relatively easily on the basis of the consumer ID and reloading dates information.

Note that the increased use of pre-paid meters also implies high setup costs, including high capital expenditure on metering devices and high recurrent costs linked to sales, repairs, and monitoring. In many instances, their sustainability depends on the margin for cross-subsidies. This is not very different from the more traditional approach to cross-subsidies between urban and rural consumers or industrial and residential consumers. How much margin the service providers should be given to implement these cross-subsidies is a recurrent regulatory concern. In all cases, the proper decision has to take into account the fiscal limits on direct subsidies and the distortions that cross-subsidies can lead to, for instance, in terms of investment decisions.[22]

The human implication of these very dry statistics is that mortality and morbidity rates linked to underperformance in the water sector continue to be much higher than in any other region of the world. For instance, mortality rates attributed to unsafe water, sanitation, and hygiene are estimated by the WHO (2014) at 39.52 per 100,000 people, 3.3 times more than in the Middle East or 12 times more than in Latin and Central America. And unless the awareness of the risks improves, these statistics are likely to get worse. The estimated number of people practicing open defecation has increased in the region from 181 million in 1990 to 229 million in 2015, adding to the complexity of the sanitation challenge in the region. The poor sector performance also has a dramatic impact on the productivity of the region. Research conducted by international organizations suggests that for every $1 invested in water and sanitation, an average of at least $4 is returned in increased productivity.[23]

---

[21] Heymans et al. (2014).

[22] Estache et al. (2006) and Auriol and Blanc (2009).

[23] UNICEF (2015).

Despite the reasonable frustration with the size of the challenges ahead, it is also hard to ignore that solid progress has been achieved. Indeed, SSA had a coverage rate of 47% in 1990. This implies an increase of 1.8% per year on average since. Figure 9.1 shows that 11 countries have already reached access rates over 90%. During the MDG period, 427 million people are better off. Between 2000 and 2010 alone, 84 million urban Africans gained access to improved water supply and 42 million to improved sanitation, an impressive 3.9% average increase in access over the decade. It is impressive indeed. But once this good news has been acknowledged, the hard reality has to be used to motivate the additional efforts needed. The number of people without access to these basic services continues to be excessive.[24] Funding continues to be a challenge for too many countries. Finally, income levels in the region are such that it is still unrealistic to expect the lower-income classes to be able to get to pay cost recovery tariff. And this implies that the economic regulation of the sector has to address jointly efficiency concerns (i.e. underinvestment, excess costs, mismatch between the supply and demand for quality, and excess physical and non-physical losses), equity concerns (i.e. the penalization of poor and rural areas), and fiscal concerns (i.e. the need to be able to minimize the largely unavoidable subsidies to the sector).

3 The Sector Organization and Its Regulation

The characterization of the organization and institutions of the WSS is not an easy exercise in any context.[25] It may be particularly complex in the African context as a result of the multiplicity of experiences across, and often within, countries spread over almost 30 years. This is partially linked to the diversity of the dimensions that drive the design of institutions and to the efforts, often too modest, to match these to local

[24] And it may be useful to remind the reader that water access and consumption are quite different matters. The average water consumption in Rwanda is 15 liters a day. In the USA, it is almost 40 times as much (575 liters per day) (see https://20liters.org/clean-water/?gclid=EAIaIQobChMImquFt7u-1QIVhj4bCh2gvgr1EAAYBCAAEgL1fvD_BwE).

[25] Maziotis et al. (2013).

constraints and preferences. This diversity reflects differences in legal traditions (civil vs. common law), accounting tradition, development levels (low-income vs. low middle-income levels), country size, and ethnic homogeneity, for instance.[26] It also reflects differences in market structures. For instance, in the mid-2010s, there were 130 local utilities in Tanzania, 103 in Kenya, 18 in Zambia, 15 in Mozambique, but only 1 each in Rwanda, Burundi, and Lesotho.[27] There are however a number of common characteristics to be highlighted to provide a good sense of the way the sector is organized in SSA, to define, implement, and enforce policies and the supporting regulations.

As in many other developing regions, national water and/or environment ministries lead the formulation of water-related policies. They often need to coordinate with public works or finance ministries, in particular in the context of efforts to rely on large-scale private options for the delivery of water services. They also need to coordinate with agriculture ministries in the management of irrigation or health ministries to coordinate the monitoring of water quality and define effluent standards, for instance. Many of these ministries create separate agencies which are often specialized in a specific activity, such as monitoring, enforcement, or implementation.

One of the main innovations launched in the late 1990s has been the decision to create separate regulatory agencies for WSS utilities, largely stimulated by international organizations. As of 2017, such separate utilities regulators had been adopted by 23 of the countries (48%) of the region, and a few more have it on the agenda as a signal of their willingness to minimize the risks of political interference in regulatory decisions.[28] But, as discussed in the next section, this has never been a required nor a sufficient condition to attract private financing into the sector.

Note that institutional changes have not only been designed to attract private investment. Since the mid-1990s, the region has internalized the importance of an integrated approach to water resource management. As

[26] Guerriero (2011).

[27] ESAWAS (2017).

[28] Bertoméu-Sánchez et al. (2017).

a result, the relevance of the scarcity of water as a resource has been formalized in water laws which are quite comparable across the region.[29] The change is likely to have an impact on the regulation of water providers in the near future as competition for access to resources is likely to change as coverage improves and scarcity for other uses increases for natural reasons (including climate change). For now, the growing adoption of integrated water management approaches has not yet had a significant impact as residential water consumptions continue to have priority over other users in the broad allocation of resources. Moreover, the implementation and enforcement of the law are still lagging in many countries. Budgetary constraints, poor planning and coordination, information gaps have all been blamed for the slow progress. But there is little doubt that water providers' regulation will have to adjust and prices as well as the opportunity cost of water starts being part of the decision-making process.

The third increasingly standard element of the institutional framework of many countries in SSA which is of relevance to the operation of WSS and to its regulation is the inclusion of the requirement of consultation processes with stakeholders and public participation in water planning, management, and regulatory decisions.[30] These processes can be broad or narrow. In 2017, for instance, South Africans launched a global consultative process to agree on the values of water and ensure a more equitable use of the available resources. But narrow consultations prior to tariff increases are indeed now quite common in the region, and it is having an impact. In Maputo, Mozambique, focus group discussions with residents and interviews with key actors were used to identify the demand for flexible payment options to the urban poor in order to increase the affordability of connection charges.[31] It allowed the provider to expand its customer basis as well. While these examples are success stories, the evidence on the effectiveness of participatory approaches is however still mixed.[32] For instance, it is not uncommon to observe that the connected households are usually the politically influential urban middle class,

[29] Global Water Partnership/GWP (2015).

[30] Oxfam (2016).

[31] Jimenez-Redal et al. (2014).

[32] Mansuri and Rao (2013).

which is not only more potentially keen to block tariff increases when consulted but also better organized to do so. It is, however, likely to become a standard regulatory tool in most of the region because it also serves as a communication tool, and based on a recent meta-analysis, bottom-up approaches to consultation processes are likely to deliver desirable outcomes in terms of most performance indicators.[33]

A fourth change in the sector is the increased role of local authorities in the design and implementation of policies.[34] Over 80% of the countries of the region have implemented at least some form of shift of responsibilities in the sector to subnational authorities, although often more in the form of devolution rather full decentralization. But the degree of devolution has varied a lot across countries so that broad lessons are difficult to draw. The main one is that success has been mixed. The failures of a full transfer of decision-making authority can often be blamed on the reluctance of the central authorities to give up responsibilities they were traditionally controlling or to provide sufficient technical assistance to allow the local authorities to learn enough about their new mandates (e.g. Malawi or Mozambique). This is partly linked to a reluctance to transfer the resources and assets needed to match a devolution of responsibilities to subnational authorities. But failures have also resulted from the undermining of the ability of subnational governments to deliver by competition from other actors with interests not necessarily aligned enough to lead to coherent decisions, such as user committees, local water forums, or small private operators, making the design of regulation of water services in small towns more challenging than needed.[35]

Besides these dimensions defining how policies are decided and monitored, a characteristic of the sector is the extent to which it can rely on private actors to finance, develop, and/or manage its activities. As of 2017, the World Bank PPI dataset implies that 80% of the countries have at least some type of contract with private operators whether at the national or subnational level.[36] Although most casual observers could

[33] Narayanan et al. (2017).

[34] Herrera and Post (2014).

[35] Jaglin et al. (2011).

[36] Many of these contracts are actually quite small in terms of investment commitment, but they do represent a private sector presence at least in management of activities.

believe that these public-private partnerships (PPPs) started in the 1990s, the first large ones in the region took place in Ivory Coast, in 1959, with the implementation of an urban water "affermage", followed by a similar contract in 1960 in Senegal. Table 9.1 provides an historical perspective of the large-scale deals reported in that database since 1993 when the most recent wave of PPPs in SSA started.

Table 9.1 Large-scale public-private partnerships in SSA since 1993

Year	Country	Contract type	Contract period	Government level	Separate regulatory agency
1993	South Africa	Lease	10	Local	No
1996	Senegal	Lease	10	National	No
1998	Uganda	Management	3	National	No
1999	Kenya	Management	5	Local	Yes (2003)
	Mozambique	Lease	15	National	Yes (1998)
	Mozambique	Management	15	National	
	South Africa	BOT	20	Local	No
	South Africa	ROT	30	Local	
2000	Mali	Concession (ceased after 5 years)	20	National	Yes (2000)
	South Africa	BOT	30	Local	No
2001	Namibia	Lease	20	Local	Yes (2017)
	Niger	Lease	10	National	Yes (1999)
	South Africa	Management	5	Local	No
	Zambia	Management	4	National	Yes (1997)
2002	Congo Rep.	Management	2	National	No
	Uganda	Management	2	National	No
2003	Tanzania	Canceled lease	10	National	Yes (2001)
2005	Ghana	Management	5	National	Yes (1997)
2006	Senegal	Lease	6	National	No
	South Africa	Management	6	Local	No
2007	Cameroon	Lease	10	National	No
	Sudan	BOT	13	National	No
2008	Ivory Coast	Lease	15	National	Yes (1973)
	Mauritius	Management	7	National	No
2012	Ghana	BOT	25	National	Yes (1997)
2013	Congo Rep.	Management	4	National	No
2015	Senegal	Lease	10	National	No

Source: Author compilation from various sources, including the World Bank PPI database

Table 9.1 points to at least eight insights.[37] First, the number of countries with deals large enough to be significant represents about a third of the countries of the region. In other words, two-thirds of the countries have not been interested or have not had the opportunity to try the option. Second, one country (South Africa) signed almost a quarter of the deals over the 22 years' period covered by the database. Third, over a quarter of the deals (including all the South African ones) were signed with local governments. Fourth, 39% of the contracts were management contracts, 39% lease/*affermage* contracts, and 21% Build-Operate-and Transfer (BOT)/concessions, implying that massive private investment has not been the main result of the contracts signed in about 80% of the cases. Fifth, the average duration of contracts was 11 years, but it spreads from 2 to 30 years. Sixth, the craving for BOT has been decreasing during the period. Seventh, the existence of a separate regulatory agency was not needed for many of the contracts to attract private actors. And eighth, the same type of contract can be monitored by either separate regulatory agencies or ministries. In both cases, the contract serves as a key, if not the key, regulatory instrument.

What the table does not show is the complexity of the matching institutional designs needed to allow the contracts to function. The contracts in Senegal and Niger, which follow a similar model, provide a useful illustration. They follow an *affermage* model (based on the old French contract type) which also includes the creation of a state-owned asset holding company. This company takes over asset ownership, infrastructure development, service of debt, monitoring of the service quality, and development of public awareness about the sector reform on behalf of the government. Government (sometimes shared with regulator) retains policy, tariff setting, and water resource management responsibilities. The holding company enters the *affermage* for a specific duration with a professional operator, enjoying exclusivity over water services provision in its service area. A matching performance contract can also be introduced, covering technical and commercial performance obligations, financial incentives or penalties, and responsibilities regarding the rehabilitation of water systems.

[37] Note that the table does not report a number of deals which involved both water and electricity utilities. These include Cape Verde, Chad, Comoros, Gabon, Gambia, Guinea-Bissau, Madagascar, Rwanda, and São Tomé. Most were management contracts.

The table is also somewhat misleading as it does not recognize the significant role of private actors in other ways, notably in rural areas and smaller cities. A few African countries (including Cape Verde, Cote d'Ivoire, Gabon, Mozambique, Niger, and Senegal, for instance) work with private actors through contractual arrangements other than service and management contracts (Mwanza 2010). In most countries, also, small and often informal enterprises end up meeting the demand for water and sanitation services from households that the larger utilities are unable or unwilling to serve. Many of their clients are the poorer customers in smaller cities and peri-urban and remote areas. And their technological solutions are as diverse as wells, public stand posts, water kiosks, informal distribution networks, tankers, and small-scale vendor. As explained by Auriol and Blanc (2009), these facts imply that the poor end up paying more than they should and the rich much less, since the utilities serving the rich tend to be subsidized in the region, while alternative suppliers are not.

The final main institutional characteristic is the form of regulation adopted in the region.[38] Wherever alternative providers control the market (and these are particularly important providers for rural areas), self-regulation tends to be the norm. Self-regulation is also still quite common for larger utilities, as illustrated by the fact that about half of the countries have decided to maintain the regulatory decisions within a ministry or similar government structures with enough margin for political control. It is most often used by municipalities, ministries, or state-owned companies. Since the mid-1990s, however, formal regulatory mechanisms have increasingly been adopted to supervise both public and private operators.

In many of the deals, prices and their evolution are defined by contract, explicitly or implicitly, as often the competition for the market was decided based, partially, on the lowest bid. This implies that price caps are the most common form of price regulation—even if these are often turned into hybrid regimes through renegotiations that include an

[38] Banerjee et al. (2008) provide the most systematic review of the regulatory context in the region. It is however a bit dated now, and unfortunately there is no recent dataset reporting comparable information.

increasing number of cost components protected by automatic cost pass-through rules.[39] The time between reviews ranges from one to five years, but unexpected adjustments have turned price regulation closer to a hybrid form, notably when some of the imported input prices (e.g. chemicals) were increasing, occasionally as a result of exchange rate depreciations (e.g. Nigeria). The most recent change in the sector is the fact that price increase is increasingly anchored in consultation processes as recently happened in Kenya (2014) or Niger (2017).

Contracts have, de facto, become one of the main regulatory tools to define all rights and obligations. But there is a diversity of models. Regulation based on performance review or contracts is quite common in the region—a lasting inference of the popular contract plans of the 1970s. Successful examples include Uganda and Burkina Faso. Performance contracts are also built in the Niger and Senegal models of PPP and have been, until recently, monitored and enforced by a holding rather than an independent regulator. Explicit performance commitments are also part of most management contracts. More modern approaches have been adopted in most cases involving leases, BOT/BOO/ROT. Contracts are more detailed and are complemented by some discretionary power granted to regulatory authorities.

4 A Survey of Recent Evidence on Governance Reforms and Performance

One the most remarkable changes in terms of performance assessment in the region since the mid- to late 2000s is the effort led by national authorities or groups of national authorities to conduct performance audits.[40] Many of these are anchored in partial indicators and provide useful insights, but they do not offer analytically robust test of the impact of changes in institutions, regulations, and contract as their main purpose is simply to audit the evolution of performance.

[39] Banerjee et al. (2008) argue that most of the contracts were indeed price caps.

[40] For example, Water Services Regulatory Board (2016).

The region has also attracted the interest of academics, NGOs, policy advisors, and other observers of SSA's WSS challenges more broadly. All of the actors have expressed a wide range (and a large volume) of opinions on how the sector has performed as a result of the institutional changes summarized in Sect. 3.[41] Many of the assessments have some ideological tone, in particular when discussing the relative role of the private and public sectors. As often when trying to explain failures, each side to the debates is quite effective at selecting cases and indicators providing at least partial evidence to argue their position. The bigger picture may be that neither side has the answer to speed up progress. Almost 30 years of trial and errors show that there is no simple recipe to improve performance fast. But this experience also shows that there are a few pragmatic general insights on which there is convergence.

A recurring cross-cutting message from the more recent diagnostics of the failures of the sector on which most seem to agree is the continued difficulty central governments face when trying to build up alone the sustainable institutions needed to spend the resources and operate the WSS assets properly. This inability has several drivers. First is a lack of adequate operational structures with dedicated budgets to finance such basic operational inputs as salaries, transportation, supplies, or training. Second is the atomization of mandates in a sector which increases coordination costs in an environment in which the basic administrative skills continue to be limited. Third is an underuse of the comparative advantages in the allocation of mandates. Fourth, there is an intra-firm lack of incentives to perform as demonstrated for Uganda by Mugisha (2007). And last but not least, the failure to account for the relevance of the political economy of reforms which involve all levels of governments in a sector particularly politically sensitive.[42] For most of these detailed dimensions, data is limited, and case studies are quite useful to identify concerns and possible solutions.

But there are also many other dimensions of the sector on which there is disagreement. This section surveys the most technical assessments of the main broad reform options on which there are disagreements (i.e. PPPs,

[41] See Woodhouse and Muller (2017) for a recent overview.

[42] See Almer et al. (2017) for recent evidence on water riots in SSA.

regulatory choices, contract types, and decentralization) anchored in econometrics, data-envelopment analysis (DEA), general equilibrium models, and other quantitative techniques. This leaves out the large literature documenting case studies and other qualitative studies. These are useful in pointing out the relevance of control/contextual variables in their assessments of changes in specific indicators but make it hard to distinguish the relative importance of these factors from those specific to the WSS choices as drivers of outcomes. A recent meta-analysis by Thillairajan et al. (2016) of infrastructure reform experiences shows that this confusion tends to lead to an optimism bias in the interpretation of the results. They demonstrate that in assessments of the impact of various forms of PPP, qualitative studies show a high proportion of positive evidence as compared to quantitative studies. In other words, quantitative studies are more likely to indicate no significant impact of a reform on access rates and service quality than qualitative studies.

Note that the technical assessments face their own share of limitations. Most have been quite constrained by data quantity and quality, and many of the results are closer to establishing correlations rather than causality between institutional and regulatory changes and outcomes. But most provide reasonably robust correlations on the performance of WSS companies and their drivers for the samples they are tested on. This produces partial evidence on the relevance of ownership, regulation, contract type, separate regulatory agencies, and decentralization. We also report the results of studies of the impact of foreign aid in WSS, since this is a strong source of financing.

The following focuses on quantitative and theoretical technical research covering explicitly SSA since the mid-2000. The impact indicators are the access rates, the efficiency, the equity/affordability, and the financial viability of the sector. This evidence is summarized in Table 9.2. Overall, the table reveals that there are not many technical assessments. Most of the research on SSA is indeed conducted through case studies, and most of the many quantitative broad assessments of the reform options based on panel data do usually not test for an SSA-specific effect.[43]

[43] It is notable that there is no technical assessment of the impact of reforms on the accountability of the service providers in a region in which one of the major issues has been the political instrumentalization of public enterprises and the capture of politicians by private agents.

Table 9.2 Synthesis of the empirical literature on the impact of WSS policies and institutions in SSA (2005–2015)

	Access	Efficiency	Equity/affordability	Financial viability
Ownership	Mbuvi and Tarsim (2011: public does better at delivering access (Uganda)	• Kirkpatrick et al. (2006): private sector has lower water losses, but other measures suggest smaller or no differences (cross section) • Mbuvi and Tarsim (2011: public do better PPP deals (Uganda) • Mande Bafua (2015): the impact of ownership is sensitive to the type of regulation in place (panel)	• Estache and Grifell-Tatjé (2013): rural poor gained much less and taxpayers lost, while foreign workers and investors benefited much more than locals (Mali)	• Mbuvi and Tarsim (2011): public firms have done better than PPPs in Uganda in raising revenue (Uganda)
Regulation and contract type		• Mbuvi et al. (2012): no difference between performance contracts and other forms of regulation based on a one-year snapshot • Mande Bafua (2015): performance contracts actually help relatively more than other contracts when more details and more countries are picked up over a five-year period	• Boccanfuso et al. (2009): price reforms linked to restructuring were regressive (Senegal) • Estache and Wodon (2014): poor targeting of pricing and subsidies continues to be an important driver of inequity in service affordability when access is in theory available • van den Berg and Danilenko (2017): smaller utilities typically charge higher water rates than larger ones (panel)	• Marson and Savin (2015): access improvements increase for relatively low levels of capital cost recovery but drop beyond a certain threshold (panel)

(*continued*)

Table 9.2 (continued)

	Access	Efficiency	Equity/affordability	Financial viability
Separate regulatory agency		• Mbuvi et al. (2012): slight positive effect but only when service connectivity to pipe is considered and not robust to bootstrapping tests (cross section) • Mande Bafua (2015): no clear impact (panel) • van den Berg and Danilenko (2017): in low-income countries, a regulator results in better customer performance but does not extend to operational performance (panel)		• Bertoméu-Sánchez et al. (2017): separate agency is more able to attract private interest but not by much and much less than in other regions (cross section) • van den Berg and Danilenko (2017): utilities with regulators do less well financially
Decentralizing	• van den Berg and Danilenko (2017): coverage lags with decentralization	• van den Berg and Danilenko (2017): district- or municipal-based service delivery shows better results on customer orientation but has a negative impact on labor productivity		

(continued)

Table 9.2 (continued)

Foreign aid	• Ndikumana et al. (2017): positive effect (panel) • Hopewell et al. (2014): no association for large systems (panel)	• Ndikumana et al. (2017): aid effectiveness is linked to the capacity of countries to absorb the financial support

On access. It is quite remarkable that, in the last ten years, technical studies on SSA hardly cover the impact of institutional changes on access rates. This is in sharp contrast to the clear focus on access of many case studies looking at correlations between improvements in regulation, contracts, or separate agencies and reductions in SSA's coverage gaps. The broad picture is that ownership matters, decentralization is not a sure bet, and foreign aid is likely to work. However, context and measurement methods matter as sample choice and treatment techniques tend to have an impact on outcome identified.

More specifically, on access, Mbuvi and Tarsim (2011), with a DEA treatment of a Uganda-specific dataset, find that public firms were better at meeting access needs between 2005 and 2007 than PPP. Considering a much larger dataset of 14 countries, and also relying on a DEA, van den Berg and Danilenko (2017) find that decentralization did not matter to coverage for the low-income group in the 2010–2013. The two papers focusing on the impact of foreign aid on access rates in the sector across the region disagree. Hopewell and Graham (2014) rely on a simple Ordinary Least Squares (OLS) regression on data from 31 cities in SSA between 2000 and 2010 and find no correlation between official development assistance allocated to WSS (large systems) and access. They do find a negative and significant association between aid and the prevalence of open defecation. Ndikumana and Pickbourn (2017) address the possible endogeneity issues raised by the Hopewell and Graham results. They rely on a panel of 29 SSA countries over the 1990–2010 period. They find that increased aid targeted to the supply of water and sanitation is associated with increased access. This more robust result confirms findings for other regions and suggests that aid continues to be an essential part of the efforts to close the financing gap of the sector.

On efficiency. The papers on efficiency test the impact of ownership, of contract type, of regulation type, of the creation of a separate regulator, and of foreign aid. On most reform options, the results are uncertain. To some extent, this can be blamed on the fact that many of the institutional details picked up by case studies are not easy to model because they can seldom be turned into data to be used in technical papers.

On ownership, regulation, and contracts, the results are quite sensitive to samples and methods. Kirkpatrick et al. (2006) conducted a cross-sectional analysis (for the year 2000) based on 110 water utilities in SSA, among which 14 utilities reported private sector involvement. They are the first to show how much the impact of private operators depends on the methods used to treat the data. While the DEA results favor private operators, the stochastic cost frontier analysis favors state-owned utilities, although statistically insignificantly. Mbuvi and Tarsim (2011) focus on two heterogeneous urban water utility groups in Uganda (10 public-public owned, 17 public-private owned) between 2006 and 2007 and treat the data with a DEA. They find public firms to be doing better. Mande Bafua (2015) fuels the doubt with an econometric analysis covering the 2000 and 2005 period for a sample of 27 utilities in 17 SSA countries. His results imply that during that period, combining private sector participation in management and operations with regulation by either an independent regulatory agency or a performance contract and combining public management with regulation by either means are both associated with a positive but non-significant effect on technical efficiency. However, water utilities under public management regulated by either means seem more efficient than water utilities with private sector participation in operation and management when regulated by either means.

On the role regulators, Mbuvi et al. (2012) cover 51 water utilities in 2006 using the DEA techniques as well. The effect of separate regulatory agencies on efficiency had not been significant for that year when robustness checks are conducted, neither is regulation by a performance contract. Mande Bafua (2015) confirms with another sample, finding that performance had not been influenced by the existence or absence of a separate regulator during that period. Van den Berg and Danilenko (2017) find that getting into more details may help. They show that, for low-income SSA countries, having a regulator improves customer performance, but it has no impact on operational performance for their sample. Note that Ndikumana and Pickbourn (2017) contribute to the assessment of the relevance of regulatory institutions when they show that the effectiveness of aid for WSS is sensitive to the efficiency of aid utilization. They highlight that there may be a case to try and improve the absorptive capacity after a certain level of aid.

Specifically, on contract types and regulation, Mbuvi et al. (2012) find no different impact between performance contracts and other forms of regulation for the single year they considered for their one-year sample.

In contrast, Mande Bafua (2015) finds a difference for the five-year period he considers. Regulating water utility operations via performance contracts led to higher technical efficiency compared to control by an independent regulatory agency. He found no evidence that the form of regulation leads to a statistically significant difference between the technical efficiency of publicly and privately owned utilities.

Finally, on decentralization, van den Berg and Danilenko (2017), relying on an econometric treatment of a new dataset, find that district- or municipal-based service delivery shows better results on customer orientation. However, decentralization also tends to have a negative impact on labor productivity.

On equity and affordability. Despite the recurring policy and political concerns for the need to protect the poorest and to treat them fairly as part of policy reforms, only four technically robust papers have covered equity/affordability concerns in the region. They complement the very useful research focusing on case studies (World Bank 2017) and/or basic descriptive statistics and/or correlations (e.g. Banerjee and Morella 2011). The first of these technical papers is van den Berg and Danilenko (2017). They show that the unbundling of the sector leads to smaller firms and these in turn rely on higher tariffs. This is related to the existence of economies of scale in the production of water in the region (also documented by some of the efficiency studies). This observation is consistent with Auriol and Blanc (2008), who show conceptually why it makes sense to expect that PPPs generally fill the gap of deficient public providers, at a high cost for end users in the SSA context. Indeed, they provide water to the poorest part of the population, while public firms serve the richest part. Moreover, they are often unregulated and apply large markups, while public water is subsidized. This is why the average price per cubic meter of water from PPPs can be close to ten times higher than that of public utilities. Boccanfuso et al. (2009) find that the specific form of price regulation and price rebalancing as a result of reform has broader impacts. In the context of Senegal, relying on a general equilibrium model to track all the effects of any change, they find that most of the gains from their water reforms accrued to the highest income classes, while the poor have seen little or only slow changes. Estache and Wodon (2014), working through a wide range of incidence analysis, show the extent to which pricing and subsidies mistargeting, two common regulatory failures, explain why, even when access is in theory available, the service is too often unaffordable for the poorest. Estache and Grifell-Tatjé (2013), focusing on a decomposition of the sources of efficiency gains achieved by reforms, looked at Mali's brief privatization experience and found a comparable result. While most users, intermediate suppliers, investors, and workers did benefit, the poor rural users gained much less, and taxpayers lost. They also show that skilled foreign workers and investors benefited much more than locals and that the firm's owners captured a large share of the rent they helped create (probably through transfer pricing as they controlled cost data for key intermediate inputs).

On financial viability. The financial viability criteria is seldom tackled explicitly in the research of the last ten years, even if it has continued to be a highly publicized concern justifying in many ways the efforts to commercialize the large public utilities. The paper by Mbuvi and Tarsim (2011) finds that this commercialization paid off in Uganda. Public firms have done better than PPPs in raising revenue in their sample. Having a separate regulator has a less clear effect. While van den Berg and Danilenko (2017) find that utilities with regulators do less well financially, Bertoméu-Sánchez et al. (2017), focusing on a logit treatment of a cross section of countries cutting across regions in 2016, find that the odds of private participation in WSS in SSA are 0.04 and 0.08 times higher with a regulator. Since PPPs are unlikely to be signed on by private investors unless the bottom line is positive, this insight implies that having a regulator does indeed improve the chance of having a positive bottom line, but not very significantly so.[44] With respect to decentralization, van den Berg and Danilenko (2017), once again, provide the only result. They find, for their sample, that it made no difference to the financial and operational performance of utilities. However, for customer performance, there are statistically significant differences: district- or municipal-based service delivery shows better results.

[44] Note that they also show that this is quite different to what is observed in other regions. When everything else is held constant, the odds are about 17 times higher for a European or Central Asian country with a regulator and up to 43 times higher for a country in Latin America and the Caribbean.

The result most supportive, and yet most conflictive, of the importance of regulation is however provided by Marson and Savin (2015). Indeed, they show that it can be good for financial viability, but they also document a very real trade-off between social and financial concerns. This trade-off has already been identified early in the reform processes by various authors (e.g. Bayliss and Fine 2007). Marson and Savin (2015) increase the robustness of the conclusions reached by Bayliss and Fine (2007) but also Gerlach and Francey (2010) and Bayliss (2014) who had already documented the interactions between water coverage and the financial performances of utilities and in particular the degree of cost recovery. Marson and Savin (2015) present a solid econometric treatment of data on 25 SSA countries from 1996 to 2012. They find important access increases for relatively low levels of capital cost recovery but simultaneously a deterioration of access performances beyond a certain threshold. Van den Berg and Danilenko reach similar conclusions with their own dataset. For their sample, they find that where supply is delivered by utilities with good financial performance, on average, users spent 3.2% of GNI per capita on water supply services, compared to 2.3% for those served by utilities with weak financial performance. These results are important not only analytically, as they inform about the existence of trade-offs, but also at the policy level, because they point to the need to take a decision on the trade-off between social and financial objectives. And this is, implicitly, a debate on the case for targeted subsidies in the sector already identified in Estache and Wodon (2014). Their results suggest that the privatization wave and now the commercialization and private small-scale providers give priority to financial performance.[45] The real agenda should now be to reconcile the two by refocusing the debate on the scope for subsidies and technology choices. And if fiscal constraints are binding, the debate should be about technology choice, tariff structures, and the margin for cross-subsidies between the haves and the have-nots, if only for ethical reasons.

[45] It is remarkable that the narrative has made it to the NGOs' message as well. See, for instance, Rusca and Schwartz (2012).

5 Concluding Comments

This chapter shows that SSA still has a water and sanitation problem, despite the significant improvement conducted under the MDG umbrella. Likewise, reaching the universal water access and adequate sanitation targets promised by the Sustainable Development Goals (SDGs), agreed upon in 2016 by the international community, seems to be unrealistic, based on the current record. More, faster, and better is needed if SSA is to reach those targets.

Financing constraints are central to the challenge as the financing gap continues to be significant. The slow internalization of strategic mistakes in reform choices, contract choices/designs, and implementation is just as important an explanation for the slow progress. The uncertain handling of the fast and accelerating urbanization process in the region is an increasingly important factor as well. Currently, approximately 409 million Africans or 40% of the continent's population lives in urban areas, more than twice the numbers in 1990. By 2030, that percentage of the urban population will rise to half the continent's population, or some 654 million people (Jacobsen et al. 2012). Over the next quarter century, demand for water in Africa is projected to nearly quadruple—the world's fastest rate (Jacobsen et al. 2012). Rising industrial and commercial demand linked to industrialization and rising incomes for some has nourished expectations of higher quantities and better quality of water, increasing pressure for improved water management policies and implementation.

How does SSA get there and how are its poor protected in the process? For the 12 countries with access rates below 60%, is it realistic to expect to be able to achieve an improvement at a rate of 2.6% in a region in which the average growth rate of population is at about 2.5%? In other words, will the international community be willing to contribute to the financing of an increase of about 5% on access on an annual basis until 2015? Will the local governments be keen to rely more systematically on cross-subsidies from those already connected (and who benefited from subsidies in the past in view of the low cost recovery rates) to those without access? Will countries (and their advisors) work harder at matching regulatory designs and contractual forms to the local institutional capacity, commitment, and fiscal constraints?

These are not simple questions. But the answer the national governments, the users, and the donors (officials, NGOs, and philanthropists) will give to these questions will determine the odds of meeting the SDG targets because they will determine the ability of countries to design implementation strategies that meet the implementation capacity of the countries at the national and at the subnational level. Or to put it more concretely, they drive the odds of making sure that the 319 million people in the region without access to improved reliable drinking water sources in 2015 and the 229 million people with no other choice but to practice open defecation stand to experiment what has seemed to be so normal to the reader of this chapter for so long: drink tap water and flush a toilet.

References

Almer, C., Laurent-Lucchetti, J., & Oechslin, M. (2017). Water Scarcity and Rioting: Disaggregated Evidence from Sub-Saharan Africa. *Journal of Environmental Economics and Management, 86*(C), 193–209.

Angel-Urdinola, D., & Wodon, Q. (2012). Does Increasing Access to Infrastructure Services Improve the Targeting Performance of Water Subsidies? *Journal of International Development, 24*(1), 88–101.

Auriol, E., & Blanc, A. (2008). Capture and Corruption in Public Utilities: The Cases of Water and Electricity in Sub-Saharan Africa. *Utilities Policy, 17*(2), 203–216.

Auriol, E., & Blanc, A. (2009). Capture and Corruption in Public Utilities: The Cases of Water and Electricity in Sub-Saharan Africa. *Utilities Policy, 17*(2), 203–216.

Banerjee, S., & Morella, E. (2011). *Africa's Water and Sanitation Infrastructure: Access, Affordability, and Alternatives*. Washington, DC: World Bank Directions in Development Series.

Banerjee, S., Skilling, H., Foster, V., Briceño-Garmendia, C., Morella, E., & Chfadi, T. (2008). *Ebbing Water, Surging Deficits: Urban Water Supply in Sub-Saharan Africa*. AICD Background Paper 12. Retrieved from http://documents.worldbank.org/curated/en/202131467990376308/pdf/482150ESW0P1111port0on0Water0Review.pdf.

Bardasi, E., & Wodon, Q. (2008). Who Pays the Most for Water? Alternative Providers and Service Costs in Niger. *Economics Bulletin, 9*(20), 1–10.

Bayliss, K. (2014). The Financialization of Water. *Review of Radical Political Economics, 46*(3), 292–307.

Bayliss, K., & Fine, B. (2007). *Privatization and Alternative Public Sector Reforms in Sub-Saharan Africa: Delivering on Electricity and Water*. London: Palgrave Macmillan.

Bertoméu-Sánchez, S., Camos, D., & Estache, A. (2017). *Do Private Water Utility Operators Care About Regulatory Agencies in Developing Countries?* Policy Research Working Paper Series 8045, The World Bank.

Boccanfuso, D., Estache, A., & Savard, L. (2009). Water Price Reforms in Senegal: Distributional Impact Analysis. *Journal of African Development, African Finance and Economic Association, 11*(1), 11–39.

Camos, D., & Estache, A. (2017). Regulating Water and Sanitation Network Services—Accounting for Institutional and Informational Constraints. In R. Damania (Ed.). *Uncharted Waters: The Economics of Water Scarcity and Variability*. OUP.

ESAWAS. (2017). Regional Benchmarking of Large WSS Utilities 2015/16, Lusaka, Zambia.

Estache, A., & Grifell-Tatjé, E. (2013). How (Un)even Was the Distribution of the Impacts of Mali's Water Privatization Across Stakeholders? *Journal of Development Studies, 49*(4), 483–499.

Estache, A., & Wodon, Q. (2014). *Infrastructure and Poverty in Sub-Saharan Africa*. New York: Palgrave Macmillan.

Estache, A., Laffont, J.-J., & Zhang, X. (2006). Universal Service Obligations in LDCs: The Effect of Uniform Pricing on Infrastructure Access. *Journal of Public Economics, 90*(6–7), 1155–1179.

Foster, V., & Briceno-Garmendia, C. (Eds.). (2009). *Africa's Infrastructure: A Time for Transformation*. Washington, DC: World Bank. Retrieved from http://documents.worldbank.org/curated/en/2009/01/11487313/africas-infrastructure-time-transformation.

Gerlach, E., & Francey, R. (2010). Regulating Water Services for All in Developing Economies. *World Development, 38*(9), 1229–1240.

Global Water Partnership. (2015). *Integrated Water Resource Management in Eastern Africa: Coping with Complex Hydrology*. Technical Paper.

Godfrey, S., & Ross, A. (2016). China and Africa: The New Water World. *Waterline, 35*(1), 12–17.

Guerriero, C. (2011). Legal Traditions and Economic Performances: Theory and Evidence. In B. Bouckaert & G. De Geest (Eds.), *Encyclopedia of Law and Economics* (Vol. VII, 2nd ed., pp. 144–163). Ch. 9. Cheltenham: Edward Elgar.

Gutman, J., Sy, A., & Chattopadhyay, S. (2015). Financing African Infrastructure: Can the World Deliver? Brookings, March. Retrieved from https://www.brookings.edu/wp-content/uploads/2016/07/AGIFinancingAfricanInfrastructure_FinalWebv2.pdf.

Herrera, V., & Post, A. E. (2014). Can Developing Countries Both Decentralize and Depoliticize Urban Water Services? Evaluating the Legacy of the 1990s Reform Wave. *World Development, 64*, 621–641.

Heymans, C., Eales, K., & Franceys, R. (2014). *The Limits and Possibilities of Prepaid Water in Urban Africa: Lessons from the Field*. The World Bank.

Heymans, C., Eberhard, R., Ehrhardt, D., & Riley, S. (2016). *Providing Water to Poor People in African Cities Effectively: Lessons from Utility Reforms*. Water and Sanitation Program, The World Bank.

Hopewell, M. R., & Graham, J. P. (2014). Trends in Access to Water Supply and Sanitation in 31 Major Sub-Saharan African Cities: An Analysis of DHS Data from 2000 to 2012. *BMC Public Health*. Retrieved from https://bmc-publichealth.biomedcentral.com/track/pdf/10.1186/1471-2458-14-208.

Hutton, G., & Varughese, M. (2016). *The Costs of Meeting the 2030 Sustainable Development Goal Targets on Drinking Water, Sanitation, and Hygiene*. Technical Paper, Water and Sanitation Program (WSP), World Bank, Washington, DC.

Infrastructure Consortium for Africa. (2014). *Infrastructure Financing Trends in Africa: ICA Report 2013*. Tunis, Tunisia: ICA Secretariat, African Development Bank. Retrieved from http://www.icafrica.org/fileadmin/documents/Annual_Reports/ICA-2013-INFRA-FIN-TRENDS-AFRICA-2013-FINAL-WEB.pdf.

Jacobsen, M., Webster, M., & Vairavamoorthy, K. (Eds.). (2012). *The Future of Water in African Cities: Why Waste Water? Directions in Development*. Washington, DC: World Bank.

Jaglin, S., Repussard, C., & Belbéoch, A. (2011). Decentralisation and Governance of Drinking Water Services in Small West African Towns and Villages (Benin, Mali, Senegal): The Arduous Process of Building Local Governments. *Canadian Journal of Development Studies/Revue canadienne d'études du développement, 32*(2), 119–138.

Jimenez-Redal, R., Parker, A., & Jeffrey, P. (2014). Factors Influencing the Uptake of Household Water Connections in Peri-Urban Maputo, Mozambique. *Utilities Policy, 28*, 22–27.

Kirkpatrick, C., Parker, D., & Zhang, Y.-F. (2006). An Empirical Analysis of State and Private-Sector Provision of Water Services in Africa. *The World Bank Economic Review, 20*(1), 143–163.

Li, F. T., Wang, H. T., & Mafuta, C. (2011). Current Status and Technology Demands for Water Resources and Water Environment in Africa. In L. F. Ting (Ed.), *Research on Water Resources of African Typical Areas*. Beijing: Science Press.

Mande Bafua, P. (2015). Efficiency of Urban Water Supply in Sub-Saharan Africa: Do Organization and Regulation Matter? *Utilities Policy, 37*, 13–22.

Mansuri, G., & Rao, V. (2013). *Localizing Development: Does Participation Work?* Washington, DC: World Bank.

Marson, M., & Savin, I. (2015). Ensuring Sustainable Access to Drinking Water in Sub Saharan Africa: Conflict Between Financial and Social Objectives. *World Development, 76*, 26–39.

Maziotis, A., Calliari, E., & Mysiak, J. (2013). *Robust Institutions for Sustainable Water Markets: A Survey of the Literature and the Way Forward*. Venice: Fondazione Eni Enrico Mattei, 58–2013.

Mbuvi, D., & Tarsim, A. (2011). *Managerial Ownership and Urban Water Utilities Efficiency in Uganda*. UNU-MERIT Working Paper Series, #2011-036.

Mbuvi, D., De Witte, K., & Perelman, S. (2012). Urban Water Sector Performance in Africa: A Step-Wise Bias-Corrected Efficiency and Effectiveness Analysis. *Utilities Policy, 22*, 31–40.

McKenzie, R., Siqalaba, Z. N., & Wegelin, W. A. (2012). *The State of Non-Revenue Water in South Africa*. Report to the Water Research Commission by WRP Consulting Engineers (Pty) Ltd WRC, Report No. TT 522/12, August.

Mugisha, S. (2007). Effects of Incentive Applications on Technical Efficiencies: Empirical Evidence from Ugandan Water Utilities. *Utilities Policy, 15*, 225–233.

Mwanza, D. (2010). *Roles and Institutional Arrangements for Economic Regulation of Urban Water Services in Sub-Saharan Africa*. Doctoral thesis. Loughborough University, UK.

Narayanan, S., Thillai Rajan, A., Jebaraj, P., & Elayaraja, M. S. (2017). Delivering Basic Infrastructure Services to the Urban Poor: A Meta-Analysis of the Effectiveness of Bottom-Up Approaches. *Utilities Policy, 44*, 50–62.

Ndikumana, L., & Pickbourn, L. (2017). The Impact of Foreign Aid Allocation on Access to Social Services in sub-Saharan Africa: The Case of Water and Sanitation. *World Development, 90*, 104–114.

Oxfam. (2016). *Facilitation Active Civic Engagement: From Consultation to Participation: Learning from Local Communities Through Work with Water, Sanitation and Hygiene (WASH) in South Africa and Zambia*. Retrieved from http://water.oxfam.org.uk/en/fair-access-en/facilitating-active-civic-engagement/.

Rimi Abubakar, I. (2016). Quality Dimensions of Public Water Services in Abuja, Nigeria. *Utilities Policy, 38*, 43–51.

Rusca, M., & Schwartz, K. (2012). Divergent Sources of Legitimacy: A Case Study of International NGOs in the Water Services Sector in Lilongwe and Maputo. *Journal of Southern African Studies, 38*(3), 681–697.

Sambu, D. (2016). Impact of Global Initiatives on Drinking Water Access in Africa. *African Geographical Review, 35*(2), 151–167.

Sharma, S. (2008). *Performance Indicators of Water Losses in Distribution System*. mimeo, UNESCO-IHE, Institute for Water Education.

Taiwo, A. M. (2011). Composting as a Sustainable Waste Management Technique in Developing Countries. *Journal of Environmental Science and Technology, 4*, 93–102.

Taiwo, A. M., Olujimi, O. O., Bamgbose, O., & Arowolo, T. A. (2012). Surface Water Quality Monitoring in Nigeria: Situational Analysis and Future Management Strategy. In K. Voudoris (Eds.), *Water Quality Monitoring and Assessment* (pp. 301–322). Retrieved from www.intechopen.com/books/water-quality-monitoring-andassessment/surface-water-quality-monitoring-in-nigeria-situational-analysis-and-future-managementstrategy.

Thillairajan, A., Mahalingam, A., & Deep, A. (2016). *Impact of Private-Sector Involvement on Access and Quality of Service in Electricity, Telecom, and Water Supply Sectors: A Systematic Review of the Evidence in Developing Countries*. London: EPPI-Centre, Social Science Research Unit, Institute of Education, University of London.

Tukic, N., & Burgess, M. (2016). China's Role in Africa's Water Sector: Mapping the Terrain. *Waterline, 35*(1), 18–36.

UNEP. (2010). *Africa Water Atlas*. Nairobi.

UNICEF. (2015). *Water, Sanitation and Hygiene: The Case for Support*, December. Retrieved from https://www.unicef.org/publicpartnerships/files/WASHTheCaseForSupport.pdf.

van den Berg, C., & Danilenko, A. (2011). *The IBNET Water Supply and Sanitation Performance Blue Book*. Washington DC: World Bank, IBRD.

van den Berg, C., & Danilenko, A. (2017). *Performance of Water Utilities in Africa*. Washington, DC: The World Bank.

Wang, H., Wang, T., Zhang, B., Li, F., Toure, B., Omosa, I. B., Chiramba, T., Abdel-Monem, M., & Pradhan, M. (2014). Water and Wastewater Treatment in Africa—Current Practices and Challenges. *Clean Soil Air Water, 42*, 1029–1035.

Water Services Regulatory Board. (2016). *Impact: A Performance Review of Kenya's Water Services Sector 2014–2015, Issue No. 9*. Nairobi, Kenya. Retrieved from https://wasreb.go.ke/downloads/WASREB_Impact_Report9.pdf.

WHO. (2014). Progress on Drinking Water and Sanitation Joint Monitoring Programme Update 2014. Retrieved from http://www.who.int/water_sanitation_health/publications/2014/jmp-report/en/.

Woodhouse, P., & Muller, M. (2017). Water Governance—An Historical Perspective on Current Debates. *World Development, 92*, 225–241.

Woodhouse, P., Veldwisch, G. J., Venot, J.-P., Brockington, D., Komakech, H., & Manjichi, A. (2017). African Farmer-Led Irrigation Development: Re-framing Agricultural Policy and Investment? *The Journal of Peasant Studies, 44*(1), 213–233.

World Bank Group. (2017). *Reducing Inequalities in Water Supply, Sanitation, and Hygiene in the Era of the Sustainable Development Goals: Synthesis Report of the Water Supply, Sanitation, and Hygiene (WASH) Poverty Diagnostic Initiative*. Washington, DC: World Bank. Retrieved from https://openknowledge.worldbank.org/handle/10986/27831.

World Health Organization. (2015). *Investing in Water and Sanitation: Increasing Access, Reducing Inequalities—GLAAS 2014 Findings—Special Report for Africa*. Geneva. Retrieved from http://www.who.int/water_sanitation_health/glaas/glaas2014-africa-region.pdf.

10

CEE
Water Services in Selected Central and Eastern European Countries

Michael Klien and Maria Salvetti

1 Introduction

A crucial component to understand the current challenges in the water and sanitation sector (WSS) in Central and Eastern Europe (CEE) is the breakdown of the Soviet Union and Communism in the late 1980s, which represents the starting point for a number of important transformation processes.[1] Of particular relevance here is that these events triggered a wave of decen-

[1] In this chapter we use CEE to refer to the following countries: Albania, Bosnia and Herzegovina, Bulgaria, Croatia, Czech Republic, Hungary, Kosovo, FYR Macedonia, Moldova, Montenegro, Romania, Serbia, Slovakia, Slovenia, Ukraine. While not all countries in CEE were part of the Soviet Union or the Eastern Bloc, for example, socialist Yugoslavia and its successor states, the close political and economic integration had strong effects on all countries in the region.

M. Klien
Austrian Institute of Economic Research (WIFO), Wien, Austria
e-mail: michael.klien@wifo.ac.at

M. Salvetti (✉)
Sorbonne Business School, Paris, France
e-mail: Maria.Salvetti@univ-paris1.fr

S. Porcher, S. Saussier (eds.), *Facing the Challenges of Water Governance*, Palgrave Studies in Water Governance: Policy and Practice,
https://doi.org/10.1007/978-3-319-98515-2_10

tralization reforms, affecting virtually all countries in the region. Due to the decentralization reforms during the 1990s, the number of municipalities increased sharply in most CEE countries. These decentralization reforms also affected water provision. Apart from infrastructure ownership, which in many cases was transferred to the local level, the service provision was also shifted from national and regional levels to the local government.

At the same time, and following the collapse of the socialist system, the overall economy in many CEE countries was reorganized from a state-centered approach toward more private sector involvement. Albeit in most cases to a lower extent as well as with large differences across countries, also the water sector of the region experienced an increase in private sector participation (PSP) after 1990.

It is important to bear in mind that these transformation processes were not targeted specifically toward the WSS and impacted the regional water sector along with many other areas of public service provision as well as the overall economy. Consequently, a number of reforms specific to the water sector ensued, in some cases (partially) reversing the initial reactions after the collapse of the Soviet Union. The current trend of regional aggregation reforms can be interpreted as a (partial) reversal of the initial decentralization reforms. On top of that, since the initial reforms were not designed to address the specific problems of the water sector, despite some progress, many deficiencies which were already present in the 1990s—such as access in rural areas—are still present today.

The situation in most CEE countries in 1990 can be characterized as a so-called low-level equilibrium[2]. Water provision was characterized by low quality in the sense of high intermittency rates, reduced service continuity, and low treatment standards coupled with low tariffs and costs. In many cases there was no or limited metering, or the associated tariffs were much too low for any significant cost recovery (see Maslyukivska et al. 2003; World Bank 2015). Due to the lack of financial resources, the infrastructure was poorly maintained, further adding to the low quality of the service. At the same time, customers were unwilling to accept increased tariffs for a low-quality service. This vicious circle constitutes the low-level equilibrium of the WSS in CEE countries in the early 1990s.

[2] See Savedoff and Spiller (1999), for an in-depth presentation of the concept of low-level equilibria in the water sector.

The EU accession and particularly the adoption of the acquis communautaire in the form of the Water Framework Directive (WFD[3]), the Drinking Water Directive (DWD[4]), as well as the Urban Waste Water Treatment Directive (UWWTD[5]) can be considered a push out of this low-level equilibrium because compliance with EU regulations requires higher quality through specified minimum quality standards. At the same time, it was clear that this higher quality would require investments and therefore at least to some extent higher tariffs. While some of these adaptation costs, particularly on the investment side, would be compensated through targeted EU funding, improvement of the revenue base is crucial. This explains why EU accession has to be understood as a main influencing factor in the water sector in Central and Eastern Europe.

To face the challenge of rising tariff pressure to improve service quality, access, and infrastructure maintenance, two overarching trends have emerged in the CEE countries over the last two decades. On the one hand, the use of dedicated regulatory bodies to monitor and incentivize water utilities has become a frequent phenomenon in the region. On the other hand, and with a focus to overcome possible disadvantages from territorial fragmentation on efficiency and performance of water utilities, a number of countries have embarked on aggregation reforms. Although the primary goal of such reforms is not necessarily economic (see World Bank 2017), the typical expectation was to reap cost savings through economies of scale.

The structure of this chapter is the following. The next section will sketch the main challenges for the water sector in CEE countries. In particular the questions of service quality, access, but also affordability play a key role in this analysis. In Sect. 3, the role of EU accession for governance and regulatory policies is discussed, followed by a presentation of the two main reform trends in the region's WSS: aggregations and regulatory bodies. Section 4 outlines the role of private sector participation, before Sect. 5 briefly discusses the WSS performance across CEE countries. Section 6 concludes.

---

[3] Directive 2000/60/EC of the European Parliament and of the Council establishing a framework for the Community action in the field of water policy.

[4] Council Directive 98/83/EC of 3 November 1998 on the quality of water intended for human consumption.

[5] Council Directive 91/271/EEC concerning urban waste water treatment.

2 Challenges for the Water Sector in CEE Countries

In 2015, the Danube Water Program from the World Bank produced a report analyzing the state of the water sector in 16 countries of the Danube River Basin (World Bank 2015). The information and data used to describe the challenges of the water sector in CEE countries are taken from this report.

Water Resources and Provision

The Danube River Basin spans over 801,463 km^2 and encompasses 19 countries including all 16 CEE countries. The area is rich with renewable water resources, but there are important differences between countries with Croatia enjoying more than 24,000 m^3 per capita per year, while Czech Republic experiences water stress with 1250 m^3 per capita per year (water stress accounting for less than 1700 m^3 per capita per year, according to Falkenmark indicator) (Fig. 10.1).

The effects of climate change in the region induce more frequent droughts and floods, as well as more frequent extreme rainfalls. These phenomena are expected to impact WSS services through water quantitative shortages and quality issues or damages to residential and industrial WSS infrastructure.

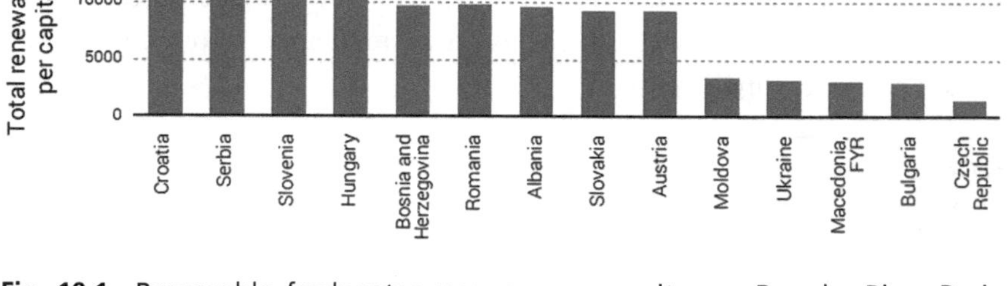

Fig. 10.1 Renewable freshwater resources per capita per Danube River Basin country. Source: State of the Sector report (World Bank 2015)

Groundwater is the main source for water supply in CEE countries as 70% of drinking water is produced from groundwater (Fig. 10.2). However the share of surface water in drinking water production varies a lot across countries and can represent up to 70% in Bulgaria, for instance. Industrial and domestic uses are predominant with agriculture representing from 0% to 40% of total water consumption (Fig. 10.3). However, due to the increase in metering level and in tariffs, and the decrease in industrial activities throughout the region, the water consumption per capita dropped over the past 15 years, reaching an average of 122 liters per capita per day in 2012 and 113 liters per capita per day for CEE EU Member States. The quality of drinking water is rather high in the region with an average of 93% of samples tested in full compliance with quality standards and 96% for CEE EU Member States.

Access

The level of access to WSS services in CEE countries is high in comparison with the rest of the world, with an average of 83% for piped water supply and 79% for flush toilet. However the situation appears more

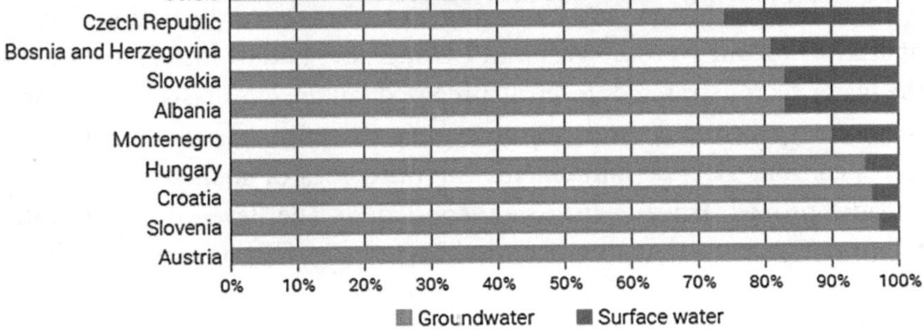

Fig. 10.2 Ratio between groundwater and surface water as drinking water source. Source: State of the Sector report (World Bank 2015)

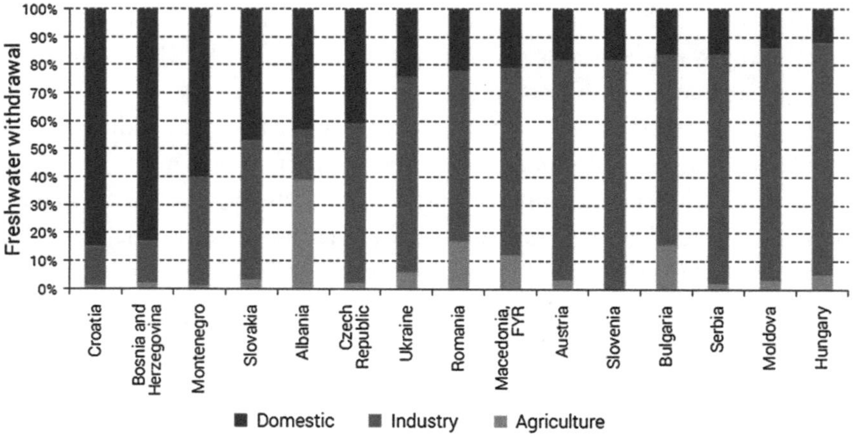

Fig. 10.3 Freshwater withdrawal distribution per usage

nuanced when comparing urban and rural situations. The access to piped water ranges from 86% to 100% in urban areas and from 24% to 100% in rural areas (Fig. 10.4).

The situation is even more contrasted for access to sanitation services which ranges from 75% to 100% in urban context versus 6% to 100% in rural context. Moreover the access rate to piped public water services amounts to 74% in CEE countries and 83% in CEE EU Member States, while access to public sewers reaches 66% in CEE countries and 67% in CEE EU Member States.

Most of CEE countries have prepared water strategies which mainly focus on increasing WSS coverage, improving protection of waters from point source pollution, and achieving cost recovery and sustainability of operation within 10 to 25 years. In EU member and candidate countries, the water sector strategies recently prepared lay the emphasis on EU compliance requirements, and objectives aligned with EU water directives transposition. To a certain extent, the outcomes of such strategies have already proved fruitful in some countries. The level of wastewater treatment coverage has significantly improved in CEE countries over the past 15 years (Fig. 10.5). This trend has clearly been driven by EU UWWTD requirements as the most dramatic improvements have been achieved in CEE EU member countries.

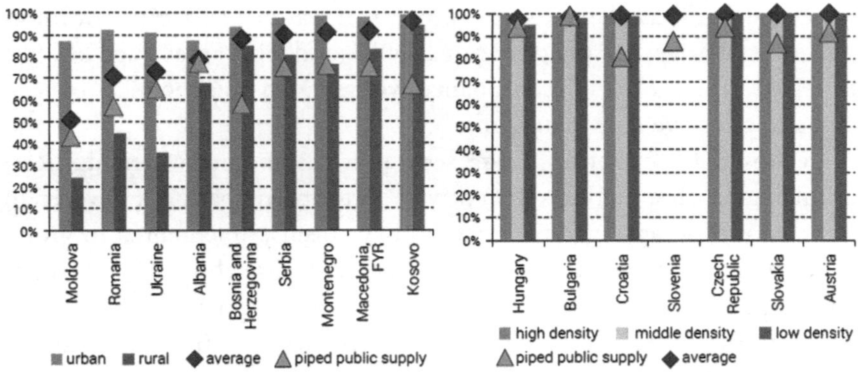

Fig. 10.4 Percent of population with piped water by location and type of provision. Source: State of the Sector report (World Bank 2015)

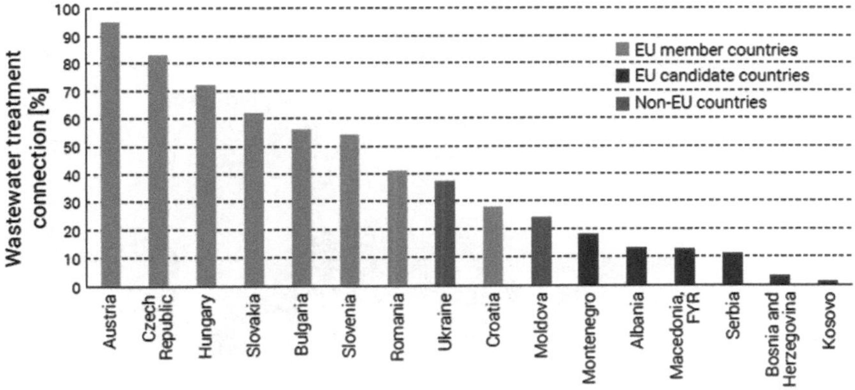

Fig. 10.5 Wastewater treatment coverage in the Danube region, 2012. Source: State of the Sector report (World Bank 2015)

In 2000, about 35% of the total population of the region was connected to a wastewater treatment plant. This situation evolved gradually so that, in 2012, 45% of the population of CEE countries was connected to a wastewater treatment plant, against only 9% in non-EU member countries. Huge differences persist between countries though, with values as high as 95% for Austria and as low as 2% for Kosovo. In CEE EU Member States, this proportion reaches 62% of the population, thus clearly showing the impact of EU environmental legislation and of EU structural funds available for investments in wastewater treatment infrastructure. Wastewater treatment quality is good with an average of 79% of tested samples in full compliance with BOD5 (biochemical oxygen demand after five days) standards.

In most CEE countries, water service provision and ownership lie at municipal level, sometimes with private sector participation. However, in some countries, aggregation processes have been implemented to group WSS services at regional level. As a whole, there are about 10,000 water service providers supplying 74% of the region's population (Fig. 10.6). Among these providers, 30.4% are municipal services, 6.3% are regional ones, and 0.8% are private operators.

As a result of this scale diversity among providers, the average population served by water utilities varies a lot across countries, ranging from 1400 customers in Austria to 200,000 in Hungary where an aggregation

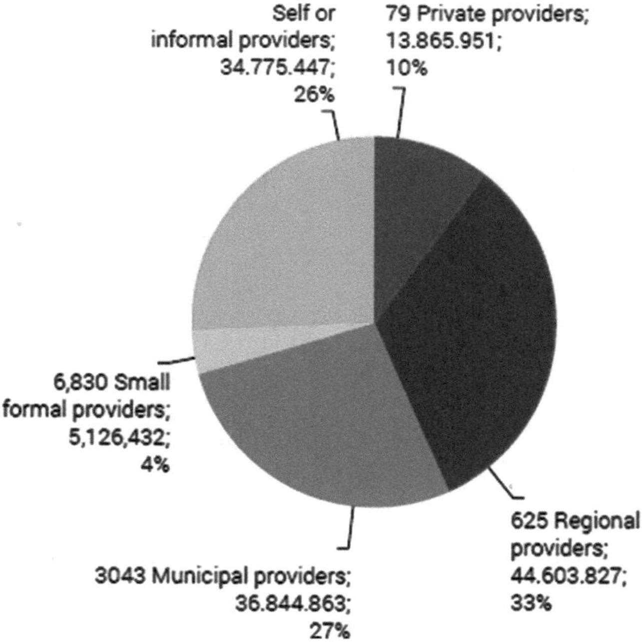

Fig. 10.6 Water service providers and population served in the region. Source: State of the Sector report (World Bank 2015)

reform took place in 2011. On average, water supply services in CEE countries serve 9500 inhabitants.

The performance of utilities from CEE countries shows a great gap between the first wave of EU Member States and other countries. Whereas the former exhibit utilities with performance level reaching international best practices, the latter still lag behind. While non-revenue water amounts to 59 m^3/km/day in non-EU countries, the indicator drops to 14 m^3/km/day in CEE EU countries. Similarly, continuity of service reaches 24 hours per day in CEE EU countries compared to 17 hours per day on average in non-EU states. Metering level is high in CEE EU countries with 96%, while other countries reach 70% on average. Staff productivity, expressed in population served, is twice as high in CEE EU states compared to the other countries. The average rate of sewer blockage is four times higher in non-EU states than in CEE EU countries (World Bank 2015).

Affordability

As mentioned previously, tariffs have increased in the past 15 years in the region to fund WSS infrastructure improvements and upgrading. The increase over the past decade amounts to 5% to 10% per year. In the meantime, disposable income of domestic customers has also risen. As a result, water tariffs remain affordable for a large majority of the population in the region (Fig. 10.7). The average share of water expenditure in household total income remains below the threshold of 5%.[6] However, the share of households paying more than 5% for WSS provision amounts to 28.8% in Romania and 32.5% in Ukraine, thus showing important regional disparities.

When focusing on the bottom 40% of the population, affordability constraints appear in Moldova and Romania where, respectively, more than 80% and 100% of households spend more than 5% of their income

[6] Different donor institutions have applied different thresholds for assessing affordability constraints of utility services, including electricity, heating, water, and wastewater. An excellent overview of these thresholds is provided in Fankhauser and Tepic (2005, 5). For water and wastewater, 3% to 5% of total income is the typically applied benchmark to assess an affordability constraint.

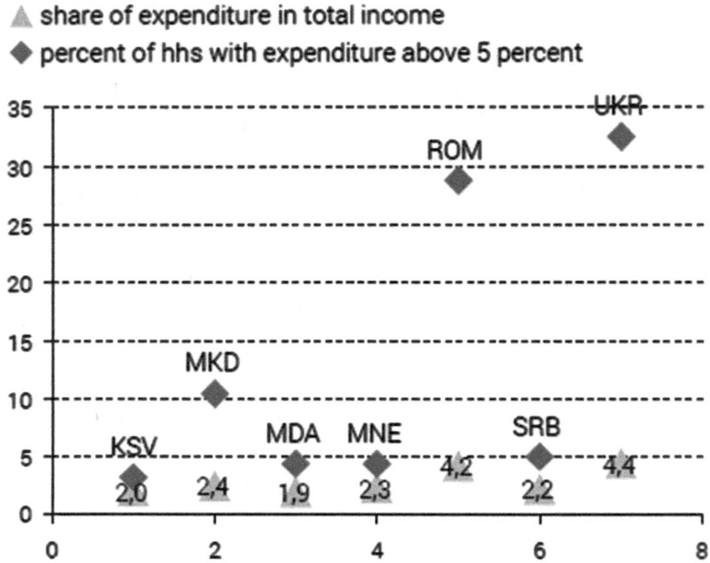

Fig. 10.7 Current average affordability for water and wastewater tariffs. Source: State of the Sector report (World Bank 2015)

on water expenditure. As stated in the report, "connecting the largely rural populations in Moldova and Romania to piped water and sewage systems would, at current tariff levels, not be affordable for large segments of the population." Bulgaria also displays potential affordability issues with close to 60% of its population spending more than 5% of their income on WSS services and close to 100% of the bottom 40% population (Fig. 10.8).

Taking into account this situation, several countries have set up formal subsidy schemes to support low-income water users. In Ukraine, subsidy schemes administered and funded at central level are available for households. In Croatia, cross-subsidies among customers are commonly applied, as well as a lower tariff for the first tariff block for low-income customers. Minimum consumption at subsidized rate is implemented for low-income customers at municipal level in FYR Macedonia.

Although the region has witnessed major change and improvements in the WSS in the past 15 years, some challenges are still to be addressed. As

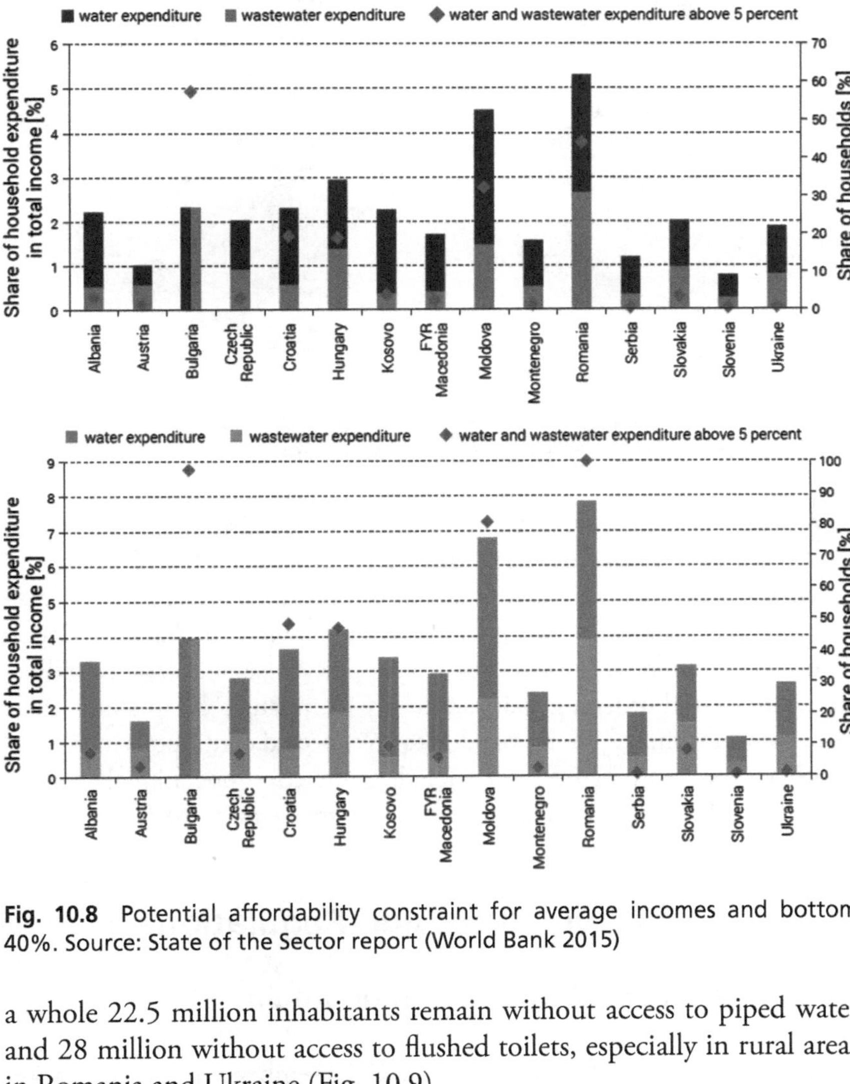

Fig. 10.8 Potential affordability constraint for average incomes and bottom 40%. Source: State of the Sector report (World Bank 2015)

a whole 22.5 million inhabitants remain without access to piped water and 28 million without access to flushed toilets, especially in rural areas in Romania and Ukraine (Fig. 10.9).

As stated in the report, "a significant number of people do not benefit from piped or public water services in the region, often because they lie outside of the services areas of utility companies. In the absence of better information on whom they receive service from, at what cost, and with what quality, and what would be the welfare and economic impact of

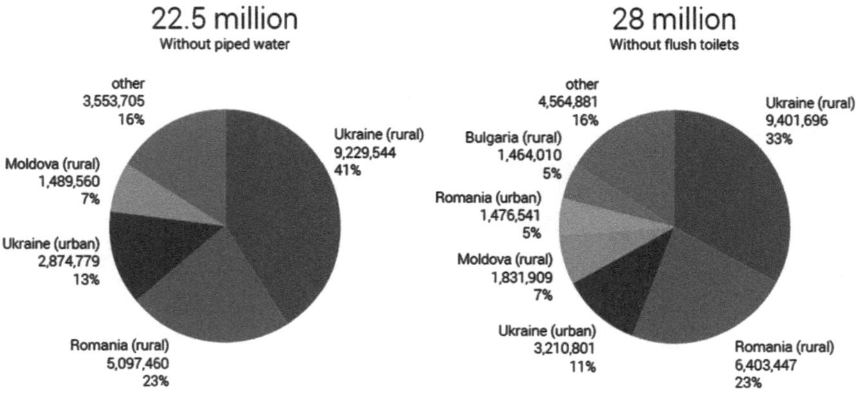

Fig. 10.9 Where are those without piped water or flush toilets in the Danube region? Source: State of the Sector report (World Bank 2015)

providing them with higher levels of service, it is challenging to determine how governments can ensure that their entire population benefits from sustainable services. More work should also be done to understand what least-cost or cost effective service provision technologies, models, or support mechanisms could be implemented to support those populations without necessarily overburdening existing utility companies by making them responsible for those." Moreover, due to an expected continuous increase of WSS tariffs to fund further asset upgrading and cover operational expenditure, affordability issues might become more stringent in the future.

3 Water Governance and Regulation

EU Accession and Acquis Communautaire

As outlined above, EU accession and EU membership have been key reform drivers in the water sector in CEE countries. Although the acquis communautaire would set quality requirements in the areas of water protection and wastewater treatment in the form of the WFD, the DWD, as well as the UWWTD, it was clear from the start of the EU's eastern enlargement that compliance with these quality targets would be costly. An initial estimate from the year 2000, quantifying the additional required investments for new Member States (CEE and Baltic countries) to reach compliance with the water directives of the EU, was as high as 16 billion euros (excluding WFD). In many cases this investment push constituted a clear paradigm shift compared to the chronic underinvestment during and before the 1990s. Perhaps it is of little surprise that the huge investment requirements in the water and even more so in the wastewater sector accounted for a large chunk of the overall funds the countries would receive through the EU structural funds. As Fig. 10.10 shows, water was at least 5% of total structural funds means during the period 2007 to 2013 for all CEE countries. In the case of Bulgaria, Slovenia, Romania, and Croatia, it accounted for more than 15% of the structural funds means (Fig. 10.11).

In absolute terms, large countries and those with significant investment needs like Poland and Romania received more than 3.5 billion euros through structural funds for water-related investments in the period 2007 to 2013. Although these are considerable investments in themselves, it is clear that total water investment would be a multiple of these amounts due to co-financing requirements and investment to simply maintain the current service levels. EU-related finance is nevertheless an important source of water-related investments in CEE countries, particularly in the process of upgrading the systems to achieve compliance with the EU directives.

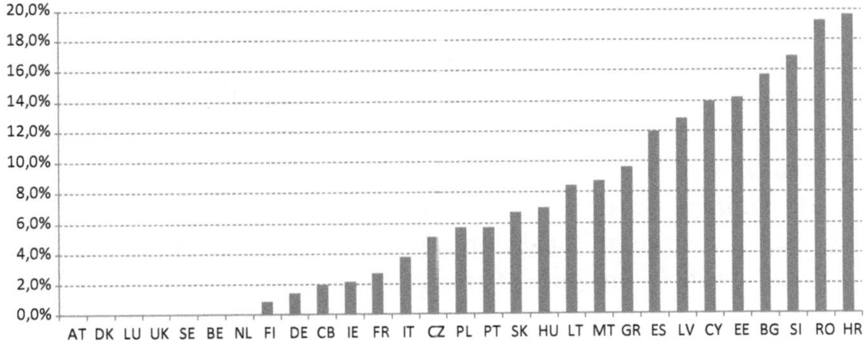

Fig. 10.10 Water-related funds (allocated) in % total EU funds (2007–2013). Source: European Structural and Investment Funds 2014–2020 (European Commission 2016)

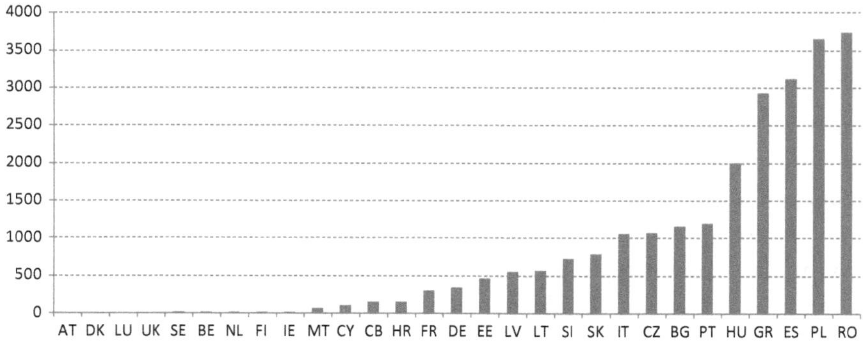

Fig. 10.11 Water-related funds (allocated) in Mio. Euro (2007–2013). Source: European Structural and Investment Funds 2014–2020 (European Commission 2016)

Another relevant aspect directly related to EU accession was the basic principle of cost recovery, prominently featured in the WFD. Although exceptions for social, environmental, and economic reasons are within the bounds of the directive (see WFD, Article 9), it also represents a marked difference to the prevailing situation in many CEE countries, where cost recovery was typically very low. Citizens were not used to pay for water services, particularly since in many cases service quality was very low. And despite improvements both in terms of quality and cost recovery, even two decades after the collapse of the Eastern Bloc, full cost recovery is still the exception rather than the rule in many regions in CEE countries (see World Bank 2015).

Overall, by enforcing higher service quality and cost recovery at the same time, EU accession has put significant (upward) pressure on prices. Although substantial EU funds were funneled into the WSS investment, a sustainable service would not be feasible without increasing the self-financing properties of the regional systems. Apart from simply increasing tariffs, countries in the region adopted two strategies to mitigate the political and social pressure:

- Independent regulatory bodies
- Utility aggregation

Both measures serve several purposes. Firstly, both aggregations and regulatory bodies are expected to increase efficiency and therefore should help to achieve higher performance without increasing costs. Hence, increasing sector efficiency in itself is a strategy to reduce the pressure to increase tariffs. Secondly, aggregation allows for a cross-subsidization within regions. Particularly geographically disadvantaged regions, very often in rural areas, would be beneficiaries of such a scheme. Having similar tariffs across jurisdictions/municipalities is also thought as to reduce dissatisfaction with "too high" tariffs. Thirdly, regulatory bodies not only enforce service standards and tariffs but can also assume responsibility for unpopular tariff increases. In a sense, having regulatory bodies enables local politicians and utilities to shift the blame for tariff increases to the regulatory agencies.[7]

To summarize, EU accession and the associated EU directives have had both direct and indirect effects on the water sectors in CEE countries. Apart from requiring higher quality standards and pricing policies, it also affected the governance of the sector by inducing aggregation reforms and the institution of regulatory bodies. These two crucial reform types are discussed more in detail in the two following subsections.

Regulatory Agencies

CEE countries exhibit different models of economic regulation in the water sector. Nine countries have set up a national independent regulatory agency, while seven other countries rely on self-regulation and contractual regulation. In countries with a central regulation by agency, the regulatory authorities have been set up during the past 15 years. Albania is the only country having installed its regulator during the 1990s (Fig. 10.12).

Two-thirds of these regulatory agencies are multi-sector regulators, while one-third is water sector specific (Fig. 10.13). This proportion matches the findings of the OECD Survey on the Governance of Water

[7] This role of regulatory bodies can be considered in the framework of strategic delegation, where politicians decide on certain "unpopular" tasks to be delegated to a bureaucrat (see Alesina and Tabellini 2008).

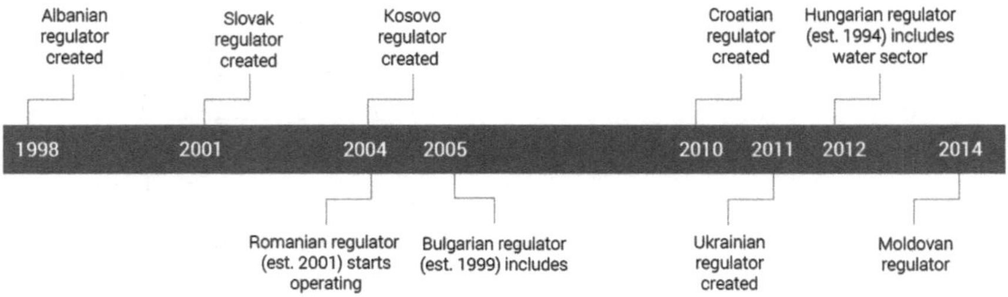

Fig. 10.12 Timeline of regulatory agencies established in CEE countries. Source: State of the Sector report (World Bank 2015)

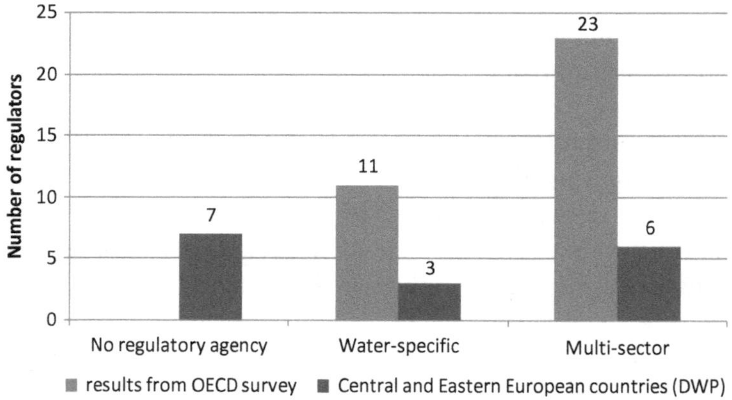

Fig. 10.13 Number of water-specific regulators and multi-sector regulators, results from OECD survey and Danube Water Program

Regulators (OECD 2015). For most multi-sector regulators, water has been bundled with energy (electricity and gas). Several arguments exist in favor of multi-sector regulators compared to sector-specific ones. For instance, such bundling can allow for transfer of regulatory expertise as well as synergies and economies of scale between sectors through joint administrative and support activities. As stated in OECD study (OECD 2015), some regulators also “report the sectors in their portfolio have different business cycles which allow technical staff to work transversally.” Moreover, multi-sector regulators are more likely to avoid sector capture and resist political pressure and interference. On the flip side, multi-sector regulators also report that water often takes a lesser importance than the electricity sector on the agenda sometimes resulting in postponed decisions.

Although every regulatory agency is established as independent since their creation, half of them depend on state budget for their financial resources. Direct comparison of staff across agencies is not easy as the number of regulated utilities varies from one country to another, and multi-sector regulatory agencies tend to be more staffed. As stated in the State of the Sector report from the Danube Water Program (World Bank 2015), "regulators that regulate mostly municipal utilities tend to have a ratio of one staff for three to four utilities, while agencies that regulate large regional operators (Hungary, Kosovo, Romania) tend to have around two staff for each utility" (Table 10.1).

All regulatory authorities have the competence of setting or approving water tariffs. Four regulators out of nine set tariffs, while others review and approve tariffs proposed by utilities. Hungary faces a specific situation as the tariff-setting formal decision rests within the line ministry, following the review and advice of the regulator. Tariff-setting methodology varies among CEE countries with two agencies using revenue cap, three agencies using price cap, and another three using rate of return.

In order to set or review prices, regulators use institutionalized and mandatory performance information systems that utilities have to fill in. Two regulators (Albania and Kosovo) implemented a regulatory benchmarking. All regulators, except the Bulgarian one, are in charge of issuing licenses to the operators, and two-thirds are competent to handle customer complaints. The time frame of the regulation commissioners mandate varies from four to seven years, with a possibility of renewal for half of the regulators. This finding is in line with the OECD survey (OECD 2015) where the most common mandate duration ranges from five to six years (Fig. 10.14).

Most regulators report to the parliament, while three of them are also accountable to the representatives of the executive power. In all CEE countries, water quality regulation is a competence of the public health ministry, while environmental regulation lies under the environmental ministry or agency.

Table 10.1 Main characteristics of regulatory agencies in CEE countries

	Albania	Bulgaria	Croatia	Hungary	Kosovo	Moldova	Romania	Slovakia	Ukraine
Regulates tariffs?	Yes	Yes	Yes	Yes	Yes	Yes	Yes	Yes	Yes
Licenses operators?	Yes	No	Yes	Yes	Yes	Yes	Yes	Yes	Yes
Handles customer complaints?	No	Yes	Yes	Yes	Yes	Yes	No	No	Yes
Annual budget	350,000 €	2,000,000 €	160,000 €	15,000,000 €	300,000 €	1,500,000 €			2,175,000 €
Sources of funds	Regulatory fee	Regulatory fee, fines through state budget	State budget	Regulatory fee, fines, other charges	Regulatory fee	Regulatory fee	State budget	State budget	State budget
Scope	WSS	Multi-sector	WSS	Multi-sector	WSS	Multi-sector	All local services	Multi-sector	Multi-sector
Water utilities regulated	58	64	157	41	7	40	42	14	147
Staff	5 commissioners	128 total (2 + 15 employees for WSS)	9 members (part-time) 1 technical	65	11 technical	60 people (7 for WSS)	96 people	6 people	Water 71, total 600
Appointment by?	Prime minister based on a short list	Parliament	Parliament	Prime minister	Parliament (on government proposal)	Parliament		President	President
Mandate	2 × 5-year	2 × 5-year	1 × 5-year	2 × 7-year	1 × 4-year	6-year		6-year	2 × 6-year
Reporting to?	Parliament, prime minister	Parliament	Parliament	Parliament	Parliament	Parliament	Ministry of Regional Dvpt	Parliament	President, parliament

Source: State of the Sector report (World Bank 2015)

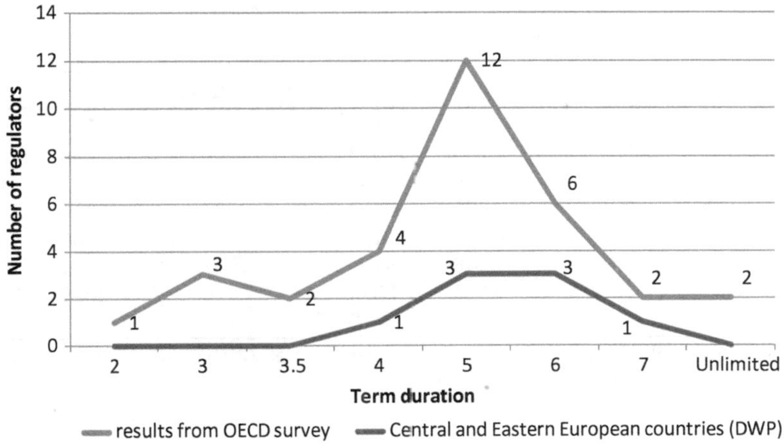

Fig. 10.14 Duration of the term of office of the agency or board members, results from OECD survey and Danube Water Program

Although half of the countries in the region have implemented policy reforms and set up national regulatory framework, "regulators often struggle to extend their regulatory reach over large numbers of local public service providers and achieve meaningful regulatory outcomes. Sector financing strategies have not been developed upon adoption of the cost recovery principle. Utility companies and management continue to be largely driven by local interests. In many cases, those reforms have not yet born fruit, and the analytical work done under the State of the Sector review shows that the long-term impact of such policies is still to emerge" (World Bank 2015).

Aggregation Reforms

One of the most important trends in the WSS in CEE is aggregation reforms. To understand this reform drive in the region, it is helpful to first consider the territorial fragmentation and its evolution since the breakdown of the Soviet Union. As Table 10.2 shows, the last three decades in the region were characterized by a number of intensive decentralization reforms. Particularly during the 1990s, in many countries the

Table 10.2 Number of municipalities in selected CEE countries

	Number of municipalities		
Country	1990	After decentralization	Currently
Bosnia and Herzegovina	109	143	142
Croatia	115	543	556
Czech Republic	4104	6230	6253
Hungary	1368	3133	3152
FYR Macedonia	30	123	80
Romania	2948	3190	3181
Slovakia	2694	2875	3029
Slovenia	62	210	210

Sources: Council of Europe (2009), De Ceuninck et al. (2010), Swianiewicz (2010), Carvalho et al. (2012), and World Bank (2017)

number of municipalities increased considerably. Similar to the argument of Foster (2005) for water sector reforms in Southern America, these territorial reforms appear to stem from a commitment to federalism rather than a step to address issues in water (or public service) governance. And although territorial fragmentation has come to be seen as a major obstacle for an effective functioning of the local government system and first reversal tendencies can be observed (see Swianiewicz (2010)), it is unlikely that the number of municipalities will reach the levels of 1990.

While territorial fragmentation appears stable, in the water sector, a number of countries have embarked on aggregation reforms in recent years to deal with fragmentation in the WSS. As a recent study on water utility aggregation shows, nowhere around the world are aggregation reforms as frequent as in this region. More than 50% of all aggregations worldwide took place in Europe and Central Asia, with Central and Eastern Europe contributing to the bulk of the aggregation cases (see World Bank 2017). In addition to a number of countries already having implemented reforms or with ongoing aggregations, a number of countries are at the stage of ex ante evaluations for potential aggregation reforms (see Table 10.3).

The specifics of the aggregations in CEE vary considerably. Although in most cases the process was mandatory, the speed and scale (level of aggregation) of the reforms varied considerably. To illustrate the various aggregation approaches in the region, in the following, the cases of Romania and Hungary are briefly described.

Table 10.3 Status of aggregation reforms in CEE countries

Country	Aggregation reform		
	Completed	Ongoing	Discussion
Albania			X
Bosnia and Herzegovina			
Bulgaria			X
Croatia		X	
Czech Republic			
Hungary	X		
Kosovo	X		
FYR Macedonia			
Moldova			X
Montenegro			
Romania		X	
Serbia			
Slovakia	X		
Slovenia			
Ukraine			

Source: Based on World Bank (2015).

Romania

Before 1990, water supply in Romania was heavily centralized at county level. Similar to most Eastern European countries, after the fall of the Soviet Empire, a process of decentralization was initiated, which also implied a shift of water provision to municipalities. Since then, water and wastewater are mainly carried out by local governments. As reported in Kruijf et al. (2009) and OECD (2009), a number of problematic issues prevail in the Romanian water sector. Firstly, there is a mismatch in responsibility because municipalities lack the financial resources to maintain the infrastructure and provide the service. In addition, financing is hampered because there are strong political reasons for local governments not to increase tariffs.

Given the chronic underinvestment and financing gap that translated into low connection degrees (only 50% of the population were connected in 2004), a large share of unaccounted-for water, as well as low cost recovery, Romania took several measures to improve the situation and finally meet EU standards. The core element of these reform plans is a regionalization initiative. Municipalities remain owners of the local infrastructure but delegate the service jointly (through a so-called Intercommunity Development Association (IDA)) to a Regional Operating Company (ROC). Hence, the relation between authorities and operators is changed in the sense that a whole regional service area is contracted to a regional operator. This requires that multiple local operators (more than 900) merge into one ROC (target 40–50), where municipalities share joint ownership (see OECD 2009). According to a study by Kruijf et al. (2009), the improved access to EU cohesion funds seems to have been an important driver, also giving local governments an incentive to support the regionalization plans. Most interviewees indicated that the establishment of the IDA was not such an important change. It was rather a formality that needed to be fulfilled in order to merge water services into one ROC and to get access to EU funds.

Hungary

Over the last 25 years, Hungary's water sector has experienced an important decentralization process which has been partially reversed over the last years. In the 1980s, there were 33 water and sewage companies in Hungary. They were state owned and centrally governed (see Regional Environmental Center 2009). Following the transition in Eastern Europe, in the early 1990s, assets of these companies were transferred to the municipal level. This decentralization pattern is consistent with the abovementioned regional fragmentation that was experienced by Romania and many CEE countries. However, governance of these companies with their large number of diverse municipal owners proved to be difficult due to the complicated and sometimes conflicting interests (see Regional Environmental Center 2009). As a reaction to these coordination problems, the companies were split up, resulting in over 300 utilities as of 2009. Many of them would serve a single municipality only.

More recently, the water sector in Hungary experienced a reversal of the previous disintegration trend. More and more small municipalities contract out their services to large water utilities. The main motives lie in the need to acquire technical expertise to meet increasing standards, which are also linked to EU accession, but also cost considerations.

Hence, today's situation is characterized by an increasing concentration of the sector, with small municipalities employing a kind of management contract to outsource service provision to large public utilities instead of providing the service themselves (see Regional Environmental Center 2009). A further step into this direction has been made by the Hungarian government with the 2011 Water Utility Act. The key pillars of this major reform project are centralized tariff setting, institution of a regulatory agency (Hungarian Energy Office), and the setting of the minimal utility size to serve at least 150,000 persons. Following the gradual implementation until 2016, the regulatory agency has started to give out operating permits, and it is expected that only 40 to 50 utilities will remain after 2016. Finally, to ensure comprehensive access to water services, utilities can also be obliged to serve more remote and disadvantaged areas.

As outlined above, an important goal of the aggregation reforms in CEE is to reduce the pressure on tariffs associated with an improved service quality and expanded access. Apart from efficiency gains through economies of scale, also the possibility of cross-subsidization between towns and regions plays a role. With respect to the economic motivations, they are largely based on the note that utilities serving larger towns tend to exhibit lower unit costs when providing WSS services. Although this relationship also appears present in the case of CEE, the evidence whether consolidations can reap these efficiency gains is mixed. Importantly, as shown in Klien and Michaud (2016), one-off cost increases and losses in service density can outweigh the economies of scale from larger operations. The loss in density is particularly pronounced for large utilities which initially served a single city. Depending on the setup, the results in Klien and Michaud (2016) suggest that unit costs of aggregated utilities can be higher even several years after the aggregation. Overall the design and process of aggregation seem important in determining whether aggregations lead to economies of scale. Despite these concerns, there is little evidence that the observed aggregation trend is about to come to a sudden stop.

4 The Role of the Private Sector for Water Services in CEE

After the fall of the Eastern Bloc, most CEE countries tried to restructure their economy toward a more market-based system by using privatization, albeit with large differences in intensity and design of the process (See Meyer 2003). Although the same is true for the water sector, a number of differences arise. Firstly, in contrast to the typical sale of state-owned enterprises in manufacturing, asset sales in the sense that the public infrastructure (pipe network, treatment facilities) was transferred to for-profit firms are rather rare. The typical approach for private sector participation in the WSS is through concession, lease, and management contracts or through types of PSP arrangements. Secondly, in a number of countries, the sale of public water- and sanitation-related assets is restricted. For example, despite the relative high level of PSP in Hungary, the public (usually the municipality) has to retain at least 51% of assets in mixed public-private companies. A first stocktaking of the presence of private sector participation can be found in World Bank (2003), which highlights the considerable country differences in PSP uptake.

As shown in Fig. 10.15, three countries (Czech Republic, Hungary, as well as Poland) along the former iron curtain account for the bulk of the PSP projects until 2003. Particularly the Czech Republic, with more than 25 projects until 2003, holds a particular position. Hungary with eight cases and Poland with five cases also distinguish themselves from the average. While the majority of countries in the region seem to have experimented with some sort of private sector participation, it was typically limited to one or a few singular cases. There appears to be a strong correlation between the level of development of a country (in a broad sense) and the take-up of PSPs. Among others, typical factors that appear to account for the large country differences are related to the institutional development, the state of the sector before the transition, the macroeconomic performance of a country, and household income (see World Bank 2003). Another common factor of the PSP experiences is that in a large number of cases major cities rather than small rural municipalities were involved. This focus on utilities with large customer bases is proba-

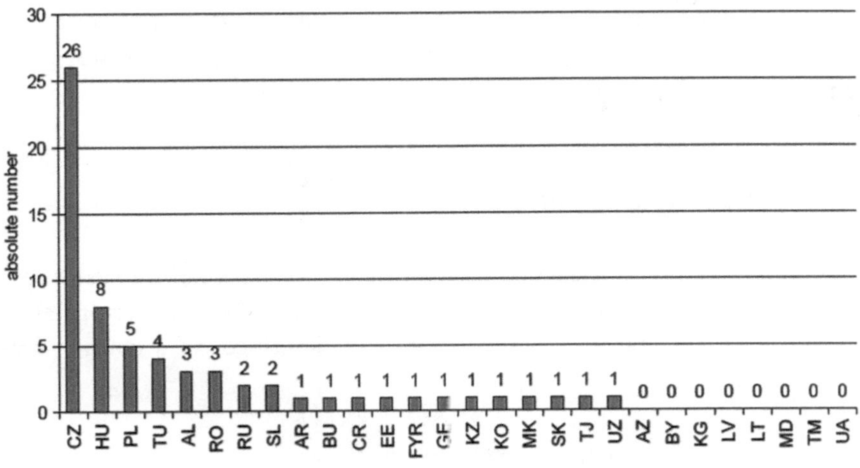

Fig. 10.15 Number of PSP projects in water in CEE countries until 2003. Source: World Bank (2003)

bly related to the considerable transaction costs related to setting up a PSP, requiring a certain size of contract to be financially interesting for private partners. Given this focus on large agglomerations, it also explains why the share of population served through these PSPs can be considerable despite the small number of PSP cases.

Although no recent and comprehensive study is available, auxiliary information from the IBNet database[8] suggests that there has been little systematic change in the use of PSP in CEE countries since the early 2000s. While the figures are not perfectly comparable to Fig. 10.13—partly due to varying samples—also in the restricted sample of IBNet utilities, the Czech Republic exhibits most PSP cases, and Hungary, Poland, along with Bosnia and Herzegovina (not covered in Fig. 10.14), and Romania exhibit a significant number of cases. Overall, the picture has changed little since 2003, with some countries adding PSP cases, whereas others reduced the number. The general stagnation is possibly

[8] The International Benchmarking Network for Water and Sanitation Utilities (IBNet) is an initiative to encourage water and sanitation utilities to compile and share a set of core cost and performance indicators and thus meet the needs of various stakeholders. It sets forth a common set of data definitions and a minimum set of core indicators and provides software to enable easy data collection and calculation of the indicators. It also provides resources to analyze data and present results.

related to the overall remunicipalization trend in water in Europe and worldwide, which has put a spotlight on local governments and their choice of service provision (Table 10.4).

5 Sector Performance

The most recent empirical evidence on the performance of the water sector in the CEE countries is World Bank (2015). Several performance dimensions are analyzed in the report and described on a country-by-country basis:

- **Service quality, customer satisfaction:**

- Water service is to a large extent continuous, and drinking water meets national quality standards. Service quality is higher in countries with higher standards of living, with Albania and Moldova experiencing the largest difficulties in ensuring continuous water supply. Service quality as measured by a composite water utility performance index (WUPI) is strongly positively correlated with measures of customer satisfaction (see Fig. 10.16). Although customer protection mechanisms are still rather rare in CEE countries, the independent regulatory agencies in many cases serve the role of a mediator.

At the same time, there is typically a large disparity between urban and rural regions. Similar to the abovementioned differences in access, service quality also tends to be considerably higher in urban regions or is not even assessed in rural areas.

Utility performance:

Both aggregate country level data and utility level data from IBNet suggest that utility performance in the CEE countries has improved over the last two decades. There is more widespread metering, staff productivity is improving, and commercial efficiency is on the rise. However, a few caveats remain. Firstly, while performance has most clearly improved, sometimes considerably, the case is less clear for efficiency in

Table 10.4 Cases of private sector participation in the sample of IBNet utilities in CEE countries

	Number of cases in IBNet										
	2003	2004	2005	2006	2007	2008	2009	2010	2011	2012	2013
Bosnia and Herzegovina	8	8									
Croatia	1	1									
Czech Republic	12	12	13	14	14	14	14	13	8	8	8
Hungary	4	4	7	3	3						
Macedonia, FYR						1	1	1	1		
Poland	3	3	3	3	3	3	3	3			
Romania	6	6	6	6	6	5	5	5			
Slovakia		1	2	3	3			1	1	1	1

Source: IBNet

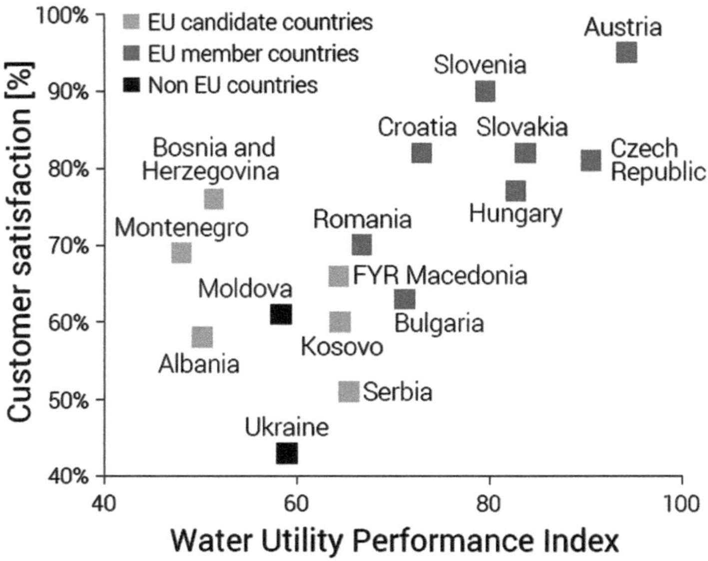

Fig. 10.16 Relationship between customer satisfaction and water utility performance index (WUPI). Source: World Bank (2015)

the sense of input-output ratios. For instance, WUPI scores have increased over time, and particularly badly performing utilities have made the biggest improvements. However, the performance improvements typically came at higher cost. The question if the utilities have moved closer to the efficiency frontier is an open issue. To some extent, the recent trends of utility aggregation and institution of regulatory agencies can be seen as reforms to raise efficiency. For instance, the findings in World Bank (2017) suggest that many aggregation reforms in countries with a sufficiently high level of service quality aim to reduce cost. This would suggest that the apparent increase in service quality is now followed by a stage that seeks to push utilities to the efficiency frontier

To summarize, the evidence on sector performance in CEE countries suggest that the service levels and service quality have been improving over time. Utilities and countries with low initial performance and efficiency improve the most—which is a sign of convergence in the sector. Nevertheless, the speed of adjustment is still varied, and also cases of performance decreases are not infrequent (see World Bank 2015). It therefore remains to be seen if the investment push through EU accession, which has been observed in many countries, has been enough to elevate the WSS to a high-level equilibrium, where high quality can be sustained through sufficient funding.

6 Conclusion

This chapter has tried to sketch the main features of water services in CEE countries. To understand the current situation, the chapter has analyzed the challenges of the sector—access, service quality, and affordability—which at least partly originate from the policies of centralization and chronic underinvestment until the fall of the Eastern Bloc.

Since then, the path to EU accession and particularly the acquis communautaire has been identified as a major game changer in water practices. Firstly, it has put in place significant and ambitious quality requirements for the water and wastewater sectors through various directives. Secondly, it has enforced the principle of cost recovery, which started from very low levels. This reform path has also been associated with two important trends in the CEE countries. The creation of independent regulatory bodies for water, on the one hand, and the frequent occurrence of aggregation reforms on the other. Somewhat in contrast, private sector participation was on the rise until the early 2000s but has largely stagnated since then and overall plays a minor role in the WSS. Finally, although the sector performance has improved consistently over the last two decades, the question is whether the utilities have been able to switch to a model of service delivery which can still be sustained in the absence of external funding. Efficiency-enhancing reforms, as potentially through aggregations, might be the next logical step in CEE countries.

References

Alesina, A., & Tabellini, G. (2008). Bureaucrats or Politicians? Part II: Multiple Policy Tasks. *Journal of Public Economics, 92*(3), 426–447.

Carvalho, P., Marques, R. C., & Berg, S. (2012). A Meta-regression Analysis of Benchmarking Studies on Water Utilities Market Structure. *Utilities Policy, 21*, 40–49.

Council of Europe. (2009). The Relationship Between Central and Local Authorities. Report of the European Committee on Local and Regional Democracy (CDLR). Retrieved June 4, 2014, from http://www.migm.gov.tr/AvrupaKonseyi/Ek41.pdf.

De Ceuninck, K., Reynaert, H., Steyvers, K., & Valcke, T. (2010). Municipal Amalgamations in the Low Countries: Same Problems, Different Solutions. *Local Government Studies, 36*(6), 803–822.

European Commission. (2016). Comparative Analysis of the Implementation of Cohesion Policy Programmes 2007–2013. Retrieved from http://ec.europa.eu/regional_policy/de/policy/evaluations/data-for-research/.

Foster, V. (2005). *Ten Years of Water Service Reform in Latin America: Toward an Anglo-French Model*. International Bank for Reconstruction and Development/The World Bank.

Klien, M., & Michaud, D. (2016). *Utility Consolidations: Review of Existing Evidence and Lessons from Central and Eastern Europe*. Washington, DC: World Bank.

Kruijf, J. V., Dinica, V., & Augustijn, D. (2009). Reorganization of Water and Waste Water Management in Romania: From Local to Regional Water Governance. *Environmental Engineering & Management Journal (EEMJ), 8*(5), 1061–1071.

Maslyukivska, O., Sohail, M., & Gentry, B. (2003). *Private Sector Participation in the Water Sector in the ECA Region: Emerging Lessons. Final Report*. World Bank.

Meyer, K. E. (2003). Privatisation and Corporate Governance in Eastern Europe: The Emergence of Stakeholder Capitalism. Keynote-address Chemnitz East Forum. March 19–22.

OECD. (2009). *Dealing with Post-Decentralisation Implications in the Water Sector: Based on Country Experience Cases*. Paris: OECD.

OECD. (2015). *The Governance of Water Regulators, OECD Studies on Water*. Paris: OECD Publishing. https://doi.org/10.1787/9789264231092-en.

Regional Environmental Center. (2009). *Strategies for Reform—A Manual for Water Utilities in South Eastern Europe*. Szentendre, Hungary: Regional Environmental Center.

Savedoff, W., & Spiller, P. (1999). *Spilled Water: Institutional Commitment in the Provision of Water Services*. Inter-American Development Bank.

Swianiewicz, P. (2010). If Territorial Fragmentation Is a Problem, Is Amalgamation a Solution? An East European Perspective. *Local Government Studies, 36*(2), 183–203.

World Bank. (2003). *Private Sector Participation in the Water Sector in the ECA Region: Emerging Lessons*. Washington, DC: World Bank.

World Bank. (2015). *Water and Wastewater Services in the Danube Region: A State of the Sector*. Regional Report, May.

World Bank. (2017). Joining Forces for Better Services? When, Why and How Water and Sanitation Utilities Can Benefit from Working Together. WSS GSG Utility Turnaround Series, June.

11

China
Water Governance Challenges in China

Olivia Jensen

1 Context

Water Resources

Water resources in China are limited and unevenly distributed across the country. China has 21% of the global population and about 6% of the world's freshwater resources. Total renewable water resources in China amount to 2018 m^3/capita/year, but availability is only 500 m^3/cap/year in the northern part of the country and 124 m^3/cap/year in the capital region of Beijing, compared to 25,000 m^3/capita in the south-western river basins. Northern China is undergoing continued urbanisation and industrial growth, putting further stress on the region's limited water resources. Resource shortages are exacerbated by high levels of water pollution throughout the country. In 2014, 8.9% of river monitoring points nationwide were classified as level 5 or below, a classification indicating

O. Jensen (✉)
Institute of Water Policy, Lee Kuan Yew School of Public Policy, National University of Singapore, Singapore, Singapore

S. Porcher, S. Saussier (eds.), *Facing the Challenges of Water Governance*, Palgrave Studies in Water Governance: Policy and Practice, https://doi.org/10.1007/978-3-319-98515-2_11

Table 11.1 Water resource availability and withdrawals

	2014 or latest		
Total renewable surface water (km^3/year)	2739		
Total renewable groundwater (km^3/year)	828.8		
Total renewable water resources (km^3/year)	2840	Total renewable water resources per capita (m^3/year)	2018
Fresh surface water withdrawal (km^3/year)	484.9	Surface water withdrawal as % of total renewable surface water	18%
Fresh groundwater withdrawal (km^3/year)	106.9	Groundwater withdrawal as % of total renewable groundwater	13%
Total freshwater withdrawal (km^3/year)	594.2	Total freshwater withdrawal as % of total renewable water resources	21%
Agricultural water withdrawal (km^3/year)	385.2	Agricultural water withdrawal as % of total water withdrawal	64.4%
Industrial water withdrawal (km^3/year)	133.5	Industrial water withdrawal as % of total water withdrawal	22.3%
Municipal water withdrawal (km^3/year)	79.4	Municipal water withdrawal as % of total withdrawal	13.3%

Source: FAO Aquastat Database (Downloaded 1 Dec 2017)

water which is not suitable for any productive purpose (Ministry of Environmental Protection China 2015) (Table 11.1).

Water Demand

Agriculture is the predominant source of demand for water, making up 65% of total withdrawals, compared to 23% for industry and 12% for domestic use (Table 11.1). Demand for water for agricultural use is highest in northern China which contains 65% of the country's agricultural land and produces half of the country's grains (FAO 2017). The share of consumption between sectors has been broadly stable over the last decade, as has the overall level of withdrawals (National Bureau of Statistics China 2017). Average per capita domestic consumption, meanwhile, has been on a declining trend since 2000 and stood at 175 litres per capita per day (lcd) in 2015 (Fig. 11.1).

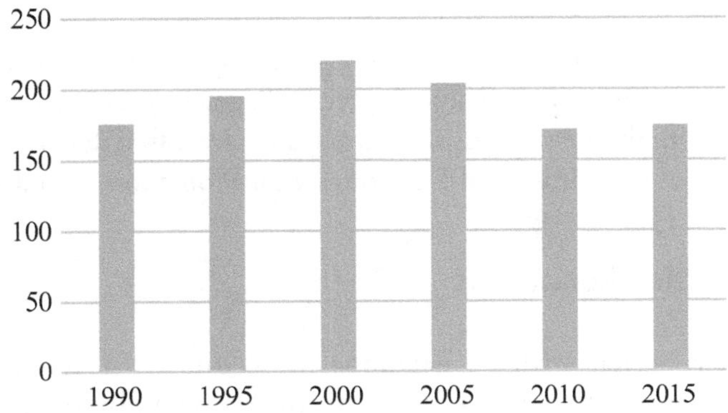

Fig. 11.1 Average household consumption (lcd). Source: MOHURD 2017

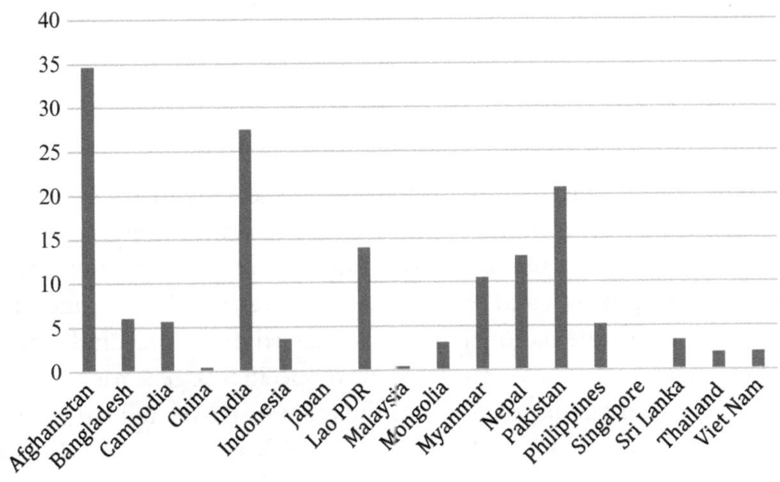

Fig. 11.2 Mortality rate attributed to exposure to unsafe WASH services (per 100,000 pop) (2012). Source: WHO Global Health Observatory (Downloaded 1 Dec 2017)

Public Health

Water-related mortality rates are considerably lower in China than in most other countries in the region. Figure 11.2 shows that in Asia only Japan and Singapore, both high-income countries, have a lower rate of water-related mortality than China.

However, there are very serious water-related health problems in specific locations in China usually associated with long-term exposure to environmental pollution. These "cancer villages" are characterised by a particularly high prevalence of cancers and are often located close to industrial plants (Ministry of Environmental Protection China 2011).

Cross-Border Resources

China shares rivers with 14 neighbouring countries, creating complex issues of transboundary water management relating to both quantity and quality of river flows (Su 2014). As the upstream riparian nation, the challenges for China lie in managing its obligations and liabilities to downstream nations.

2 Policy

The main policy objectives for the sector are set out in Five-Year Plans (FYPs) issued by the central government. These plans set broad goals and specific targets at the national level, indicate investment priorities and major national projects, and set out new initiatives, regulations, and incentive mechanisms. The national targets are passed down the administrative hierarchy where they are translated into specific targets for provinces, municipalities and counties. Some of the main water-related targets of the 12th (2011–2015) and 13th (2016–2020) FYPs are shown in Table 11.2.

Drinking Water Quality Regulation

The regulatory framework for piped water quality has been made consistently more stringent. An extended set of drinking water quality standards was introduced in 2012. These standards introduce stricter limits for some contaminants and increase the number of water quality indicators to 106 from the previous 35 indicators. Utilities were expected to meet these standards by the end of the 12th FYP, but most failed to do so, and targets were reiterated in the 13th FYP.

Table 11.2 Water targets in the 12th and 13th FYPs (2011–2020)

	Urban water supply	Rural water supply	Urban wastewater	Rural wastewater	Water use/ efficiency	Water supply
12th	Piped urban water supply of at least than 95%	Access to piped water supply from 64% to 80%	85% of the total wastewater generated in urban areas should be treated, and 20% of the treated wastewater should bc rcused	70% of wastewater generated in county-level towns and 30% in townships should be treated by 2015	Cap on total national water use 635 billion m^3/year by 2015; reduce industrial water use by 30% compared to 2010	Water supply capacity to increase by 40 km^3
13th	Upgrade water supply networks. Reduce leakage: upgrade networks in 100 cities. Implement pilot projects for performance-based leakage reduction contracts	80% tap water in rural areas Fully resolve the drinking water safety problem for poor areas	95% of wastewater generated in cities should receive centralised treatment Upgrade sewerage networks	85% of wastewater generated in county towns should receive centralised treatment	Cap on total water use of 670 billion m^3/year by 2020 Targets for water use per unit of industrial output	Increased use of alternative water sources (rainwater, stormwater, and reclaimed water) Rigid controls on groundwater use

Source: Author's elaboration

Environmental Regulation

Wastewater collection and treatment standards have been tightened progressively through industry-specific regulations. The enforcement of environmental standards has also been strengthened significantly. In 2014, the environmental protection law was revised to allow for criminal liability for causing environmental harm, civil cases against violators, and shutdown for persistent illegal discharges. The law establishes a role for environmental and other non-governmental organisations in monitoring enterprise environmental compliance and in bringing cases against violators in the courts (Ministry of Enviroment Protection Government of China 2014).

Large river basins and lakes are prioritised for action under the Water Pollution Prevention Plan on Key Basins (2011–2015). This plan supports coordinated actions between provinces and municipalities, but joint implementation of projects has proved difficult, and the central government has often had to intervene (Dai 2014). Monitoring of large river basins, lakes, and reservoirs is conducted at the national level. The national monitoring system is becoming increasingly sophisticated: the "Digital Yellow River System," for example, monitors water extraction and quality throughout the 5600 km river system.

Over the last 15 years, a number of market-based instruments to manage water and reduce water pollution have been piloted, including water rights trading (Bin 2014), discharge permit trading (Zhiping 2005), and payment for eco-services in catchment management (Dai 2014; Daniell et al. 2014), but their reach is so far limited.

Water Resource Regulation

The management of water resources has also been tightened progressively. Under the 13th FYP, China aims to achieve "the strictest water resource management regulations," making it more difficult to secure approvals for industrial projects in water-scarce areas. Water resource fees for abstractions from surface water bodies and groundwater were mandated by the central government in 2013. These specify a minimum fee for water resources by type of water and by province.

Industrial Policy

The government has put in place a set of industrial policy measures to support "environmental protection industries," a category which includes wastewater treatment and reuse technologies, sludge management, nutrient recovery, and combined energy generation, under its strategic plan for the development of environmental protection services (State Council, Government of China 2016). These include subsidies, tax relief, and other forms of preferential treatment for eligible companies and investments. National priority projects include catchment management, restoration of priority rivers and lakes, and urban river clean-up.

3 Governance

Public Administration

Public administration in China is divided both horizontally and vertically. Thus responsibility for different aspects of water management is split horizontally across ministries and other agencies of central government and vertically between levels of government.

The Water Law of 1998 and its amendments provide the over-arching legal framework for the sector. Under the Water Law, water resources are managed by the state on behalf of the nation. The central government defines plans which allocate water resources between provinces, develops and finances large-scale water infrastructure. (Pohlner 2016).

National ministries provide direction to the corresponding bureaus of lower levels of government—province (including five autonomous regions, four municipalities with provincial-level status [Beijing, Tianjin, Shanghai, and Chongqing], and two special administrative regions [Hong Kong and Macau]), municipal, county, and township. Thus, for example, the Ministry of Ecology and Environmental (MEE), formerly the Ministry of Environmental Protection (MEP), guides and sets targets for the provincial environmental bureaus (EPBs), which in turn guide and monitor city-level bureaus. In addition, provinces are also permitted to pass their own laws or set targets which can extend national regulations but must not come into conflict with them.

Some provincial governments have the same rank in the administrative hierarchy as certain ministries, making it impossible for those ministries to enforce policies in the face of opposing provincial interests. This has been particularly evident in the field of environmental regulation, where provinces have prioritised economic growth over the protection of the environment, frustrating the MEE's efforts to reduce pollution.

Some cross-provincial coordination on surface water management is provided by river basin commissions. These commissions have existed in China since before the establishment of the PRC. Today they cover the Yellow River, Changjiang (Yangtze), Huai River, Hai River, Pearl River, Songliao River, and Tai Lake. The best known is the Yellow River Conservancy Commission which manages abstraction from the river by nine provinces.

Provincial governments supervise utilities, approve tariffs, allocate financial transfers from the central government, monitor compliance with environmental (and other) standards, and approve the construction of water and wastewater infrastructure projects.

Water and wastewater service delivery is the responsibility of municipal governments. Water utilities fall under the municipal urban development or water affairs bureau. Most water and wastewater treatment assets and distribution networks are owned and operated by utilities under municipal government ownership. In larger towns, utilities are usually corporatised. In small cities and towns, local governments sometimes provide services directly, while in rural areas water systems are usually managed by village committees. Water utilities are responsible for the collection of water and wastewater fees from customers. The vast majority of water utilities are publicly owned, but the private sector plays a significant role in the development and operation of treatment assets, and there are also examples of private management of utilities under joint ventures (JVs) with municipalities.

The main central government bodies responsible for water issues are shown in Table 11.3.

Other ministries partially involved in water management are:

- Ministry of Agriculture (MOA) (agricultural water use)
- Ministry of Land and Resources (MLR) (planning, administration, protection, and use of groundwater)

Table 11.3 Central government institutions with water sector responsibilities

State Council	The highest administrative authority, responsible for the formulation and execution of administrative orders and regulations. Projects of critical national importance may be directly monitored by the State Council such as the Three Gorges Dam and the South-North Water Diversion Project. The State-owned Assets Supervision and Administration Commission (SASAC), a special commission directly under the State Council, is responsible for oversight of state-owned enterprises
National Development and Reform Commission (NDRC)	Macroeconomic management agency under the State Council that plays a role in investment planning and the allocation of funds
Ministry of Water Resources (MWR)	Executive government department responsible for managing water resources including flood control and drought relief, water resource protection and conservation, monitoring water quality in rivers and lakes, and issuing water extraction permits. River basin commissions fall under its supervision. Guides water resources for irrigation, drainage, and rural water supply
Ministry of Ecology and Environment (MEE)	Regulatory government department sets, supervises, and enforces water pollution laws, regulations, standards, and watershed/catchment protection and planning
Ministry of Housing and Urban-Rural Development (MOHURD)	Guides urban planning, urban infrastructure construction, and development projects including urban water supply and wastewater treatment assets and networks
Ministry of Finance (MOF)	Manages budget allocation, pollution levies, wastewater treatment charges, and water resource fees

- Ministry of Health (MOH) (drinking water standards)
- State Forestry Administration (SFA) (water and soil erosion in state forest management)
- State Oceanic Administration (SOA) (desalination projects)

Incentives for local officials to comply with central guidance are provided through the mayors' evaluation system for appointment and promotion. Mayors are evaluated on the implementation of city-level targets. In the late 1990s, access to water and wastewater services was included in the evaluations and was associated with strong improvements for these indicators. Since 2006, evaluations have also included quality of service and achievement of environmental targets (IFC private communication).

Monitoring and enforcement of provincial-level officials' environmental performance were carried out by MEE in about half of the provinces in 2016. The Ministry reported that more than 6000 people were held accountable for contravention of environmental regulations. In 2016, two provinces began implementation of monitoring and inspection of municipal-level officials, a reform that is expected to be rolled out across other provinces.

The fiscal system in China is decentralised: governments at the provincial level and below have their own revenue streams and are largely responsible for public expenditure on infrastructure. However, local governments face strict borrowing restrictions, and the municipal bond market is in its infancy. This has led municipalities to seek alternative ways to raise finance. A common strategy was for municipal governments to establish wholly owned companies or "local government finance vehicles" (LGFVs) and to inject municipal assets, including water treatment and distribution assets, into the companies. The companies were then able to raise money on private capital markets. This practice expanded rapidly with the 2008 economic stimulus plan.

However, the central government became increasingly concerned about the rapid increase in LGFV debt and introduced measures to reform municipal finance. A revised budget law passed in 2014 prohibits local governments from raising finance through LGFVs but allows them some flexibility to borrow without central government approval.

Public participation in policy-making is limited. The regulatory framework supports the involvement of the public in two aspects of policy: firstly, it supports the role of civil society organisations (CSOs) and individuals in monitoring the compliance of industry with environmental standards. CSOs have the power to bring civil cases against violators in the courts. Secondly, members of the public are able to participate in public hearings on tariff reviews. Further detail on this is provided in the section on tariffs. Public participation at the policy formulation stage, such as consultations on new regulations, is limited in both scope and impact.

Private Sector Participation

China opened the water sector (here the "water sector" is taken to include water treatment assets and networks, water reuse, desalination, wastewater collection and treatment) to private sector participation in 1995. Since then, it has become the world's largest market for water public-private partnerships (PPPs) in terms of the number of projects. Over 1500 contracts have been signed in total with a cumulative investment in that period in the range of $20 billion (RMB 126 billion) (Global Water Intelligence 2016). There is currently no comprehensive official national database of existing PPPs, although data on projects currently under planning and procurement is being collected by the China PPP Center under the Ministry of Finance (MOF). In the 1990s, private investment was focused in the most economically advanced coastal cities. Over time, the need for private sector finance and expertise to build and operate treatment facilities in these cities declined, and private participation spread to the next tier of coastal cities and into central provinces.

Estimates of the proportion of the population served (directly or indirectly) by a PPP using market data range from 10% (Global Water Intelligence 2017) up to 40% (IFC private communication). The wide variation between these estimates is due in part to the differing definitions of "private" and to the way in which population served is estimated for treatment plant PPPs.

Private investment in water and wastewater projects in China started cautiously in the mid-1990s. Initially, only foreign companies were eligible to enter the market, and a small number of contracts were signed which were mostly Build-Operate-Transfer (BOT)-type deals for water treatment plants (WTPs) (Chen and Messner 2005). In 2002, water treatment was opened to domestic investors, and the number and value of projects boomed. In the same year, 2002, water distribution and drainage systems were opened to foreign investment and sparked considerable interest from foreign companies (Lee 2010). However, municipal governments were concerned about the loss of control implied in these projects, and the model was not widely adopted.

Table 11.4 gives an indication of PPP contracts signed between 2001 and 2015. The number of contracts signed per year reached a peak in 2007 and then fell. Since 2013, the central government has provided renewed policy support to PPPs in water and other infrastructure sectors so the number of new projects may begin to rise again.

China does not have an over-arching law on PPPs. As of mid-2017, a draft law was under discussion. The legal framework for PPPs is set out in a large number of disparate regulations (denoted as circulars, opinions, measures, and notices) issued by several different branches of the central government. These regulations are not always totally consistent with one another. For example, the definition of PPP used by Ministry of Finance (MOF) differs from that employed by the State Council.

Overall the legal and policy framework is very supportive of private participation in municipal infrastructure, particularly in wastewater services, shown, for example, in a requirement introduced in 2016 that all new wastewater and domestic solid waste projects be considered for procurement via PPP (MOF 2016 Notice on the Promotion of Public-Private Partnership in Public Services).

Table 11.4 PPP contracts in China 2001–2015

	2001–2005	2006–2010	2011–2015
Total number of projects	320	637	370
Contract type			
BOT	68.4%	64.8%	73.0%
ROT	15.3%	10.8%	15.1%
TOT	1.6%	14.6%	8.4%
Other	14.1%	8.9%	3.0%
Developer			
Foreign private	19.1%	11.1%	5.7%
Local private	29.4%	28.9%	20.3%
SOE	48.1%	57.0%	72.4%
Region			
Coastal	48.1%	41.8%	38.1%
North-east	4.4%	8.2%	10.5%
Central	9.1%	33.6%	29.2%
West	5.3%	9.9%	18.1%

Source: Global Water Intelligence PSP Database 2016, Bloomberg
Note: *BOT* Build-Operate-Transfer, *ROT* Rehabilitate-Operate-Transfer, *TOT* Transfer-Operate-Transfer, *SOE* state-owned enterprise

Private participation in water resource infrastructure projects such as dams, reservoirs, and pipelines, often referred to under the collective term of "water conservancy," has been legally possible since 2004 but was not taken up by either subnational governments or private firms. In 2014, the central government reiterated its support for private investment in major water conservancy projects and announced measures to encourage PPPs covering construction and operation, and a first batch of projects for the development of reservoirs and water transfer pipelines are being developed. The first pilot PPPs for agricultural water management are also being pursued.

The involvement of foreign companies in water PPPs is subject to additional restrictions under foreign investment regulations. Since 2015, foreign investment has been restricted in systems in large cities with a population of over 500,000, and a Chinese JV party is required to hold majority equity ownership. On the other hand, foreign investment in environmental protection industries, which includes wastewater treatment, reuse, and sludge management, is encouraged.

The authority to award water and wastewater projects lies with municipal governments, and projects must be approved by the provincial government. Large projects require central government approval, as do projects under the national "PPP demonstration" scheme, which are eligible for central government subsidies.

All projects must be subject to competitive procurement. Procurement law specifies how the bidding process is managed and how bids are evaluated. Competitive bidding is preferred but competitive negotiation is allowed.

Contract Models

The legal framework provides for private ownership and operation of new and existing assets under BOT, Rehabilitate-Operate-Transfer (ROT), and Transfer-Operate-Transfer (TOT) structures as well as the granting of concession rights. Operations and maintenance (O&M) contracts are also legally possible, but neither public utilities nor private companies have shown interest in this model.

Most projects are structured as JVs between the private investor and the municipal government (Braadbaart et al. 2009). A new company is set up into which the municipality injects existing assets, while the private company invests cash. There is a wide range of ownership structures among these companies. It is common for the local government to hold a small majority of the equity shares, of 51–60%, but there are also examples of 50-50 ownership structures. Since 2014, projects under the central government-endorsed PPP programme must have majority private ownership. Private firms report that management control is genuinely shared in the majority of these JV companies.

Wholly owned private projects are rare in large cities. Chengdu No. 6 WTP, one of the earliest PPPs in the water sector in China, is an exception (Chen 2009). This project was structured as a BOT following international practice, with limited recourse financing and a detailed contract.

A contract model peculiar to the Chinese market is the "time-bound divestiture" in which the assets pass back to the public party at the end of the contract period of up to 50 years (Lee 2007). No contract of this kind has reached the end of its term, so it is not clear how the transfer of assets back to the public sector would work in practice.

Since 2015, a new project type has emerged which integrates assets and services for urban water management. These projects bundle together capital investment in treatment plants and networks along with services for the ecological and aesthetic improvement of highly degraded urban water environments and for stormwater management, referred to in PPP planning documents as "ecological restoration" or "sponge city" projects.

This new generation of PPP contracts is distinctive in several respects. Firstly, performance of the private party is measured according to multiple output indicators such as river quality. Secondly, the projects bundle together revenue-generating services—wastewater collection and treatment—with environmental improvement or stormwater management which does not have an associated revenue stream. Fees for the services will therefore need to be paid from the general municipal budget. The central government has sought to make the projects more attractive to local governments and developers by providing viability gap funding, which can be structured as a capital grant or as an operating subsidy.

Private Actors

Foreign companies were the first private entities to be involved in PPPs in China, but their role in the municipal sector has declined markedly over time. However, they continue to own equity stakes in some of the country's largest water utilities. French multinational company Veolia owns a 50% stake in the water utility serving Shanghai's Pudong district, while Suez Environnement holds a minority equity stake in Chongqing Water Group, a majority government-owned listed utility which serves almost nine million people (Global Water Intelligence 2017). However, in the last decade, almost all municipal PPPs have been awarded to Chinese companies.

This transition has occurred as a result of the transfer of technology and management skills, the development of local financial markets, and the desire of local governments to support local environmental services companies.

A distinctive feature of PPPs in China is the extensive involvement of state-owned enterprises (SOEs) as the "private" partner. Many of the local companies competing in the market are under majority state ownership. In China, these entities are classified as "social capital" which makes them eligible to bid as the "private" partner in a PPP. This definition contrasts with most other jurisdictions and with the definition employed by the World Bank in its PPI database, according to which these entities would be classified as public and the projects would therefore not qualify as PPPs.

Among these government-owned enterprises, three further types can be distinguished. Firstly, those in which the main shareholder is the central government, such as Everbright Water (EB Water) and General Water of China. A second category covers enterprises owned by provincial-level governments such as Beijing Enterprises Water Group, Beijing Capital, SIIC Environment (Shanghai), Tianjin Capital Environmental Protection, Chongqing Water, and Yunnan Water. Many of these entities are listed on Chinese and regional stock exchanges. In many respects, these two types of companies behave in a similar way to private companies. Managers are often hired from the private sector, and their remuneration is performance based. Their reported strategies are consistent with the objective of profit maximisation, and they are active in projects outside the administrative jurisdiction of their main shareholder. On the other hand, they enjoy certain advantages over pure private companies, notably access to subsidised capital and bureaucratic networks.

A third category is LGFVs under municipal ownership. According to State Council guidelines, LGFV projects can be counted as PPPs if the firm has undertaken market-oriented restructuring. Under MOF guidelines, LGFVs and SOEs owned by the municipal government in which the project is located would not count as PPPs. The MOF definition would be more consistent with the international understanding of PPP. These projects have not been included in the data presented in Table 11.4.

A few local private companies have played a leading role in certain segments of the market, such as Beijing Origin Water, which specialises in advanced wastewater treatment. At the same time, scores of small private companies are active in one or two projects in a particular locality but often sell equity to the large SOE groups when the project reaches the operational stage.

Economic Regulation and Tariff Setting

No economic regulators have been established for water services at any level of government in China. Municipal governments have the authority to set water tariffs following the guidance and procedures set out by the central government (PRC Price Law 1997, Regulations on City Water Price 1997, and subsequent guidance). These procedures are not "set in stone" (Li et al. 2017) but are generally followed (Zhong and Mol 2008). The process is shown in Fig. 11.3. First, the water utility prepares a tariff review case which can include changes to the level and the structure of the tariff (e.g., from linear volumetric tariffs to increasing block tariffs). This is submitted to the local price bureau and water bureau. The local price bureau considers the case and may organise public hearings to gather views on the adjustment. In principle, local delegates of the People's Congress, People's Consultative Conference, other government departments and water users should be invited to participate in the

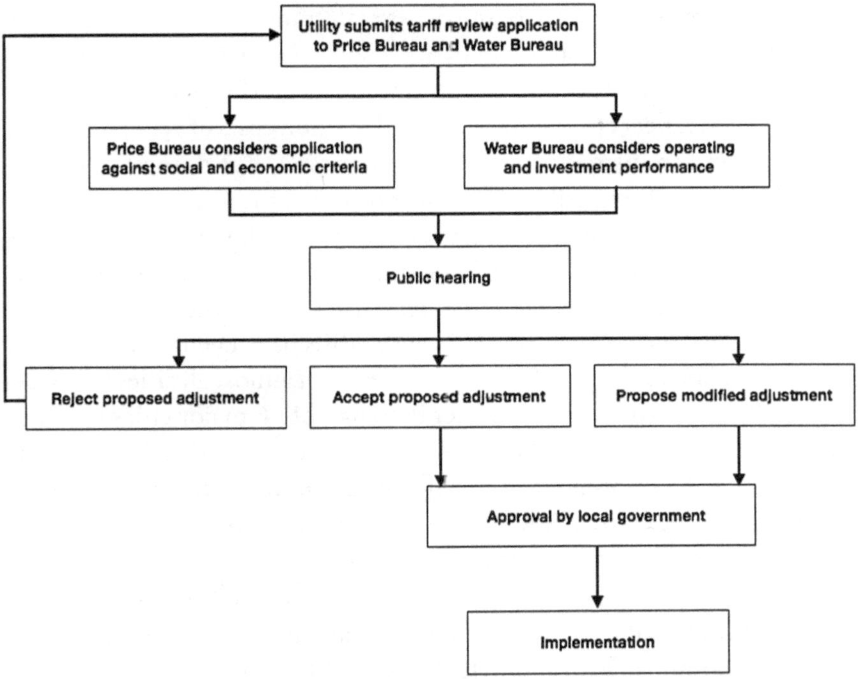

Fig. 11.3 Tariff review process. Source: Author's elaboration

hearings, but both the form and the identity of the participants are evolving (Zhong and Mol 2008). The price bureau then decides whether to accept the proposed tariff adjustment or to propose a modified adjustment. This is submitted to the local government which takes the final decision. If the tariff adjustment is not accepted, the utility must prepare a new plan and restart the process—there are no formal avenues of appeal or redress. If it is accepted, the local government submits the adjustment to the provincial government for approval. The new tariff should be announced to the public before implementation. In the small number of cases of water service PPPs in which the private party is directly remunerated by the end-user, tariff adjustment rules are often specified in the contract but must nevertheless undergo review by the price bureau and public consultation.

The customer bill is made up of several components: the tap water price (which covers costs of network construction and maintenance), wastewater treatment fee, and water resource fee. The water resource fee is intended to reflect environmental and depletion costs of water resources. Domestic and industrial users pay different rates, but domestic users all pay the same tariff regardless of income or type of residence.

According to central government guidance, local governments should set tariffs to recover full costs (operating and investment expenditure) and should employ an increasing block tariff (IBT) structure in order to incentivise water conservation. Although tariffs have been rising since the 1990s, they fall far short of cost recovery levels in almost all cities (Browder et al. 2007), Nitikin 2012 and fewer than half of 36 major cities in China had implemented IBT by the end of 2013 (Li et al. 2017). According to the National Development and Reform Committee, the average water treatment charge in the big cities is equivalent to two-thirds of the treatment cost, resulting in a shortage of investment in wastewater treatment (Li et al. 2017). The 13th FYP reiterates the goals of full cost recovery and IBT, but the central government does not have an effective mechanism to ensure the compliance of municipalities on this issue.

When setting tariffs, he price bureau primarily considers input costs and may also take into account profitability (for projects involving private participation). Although a comprehensive study on the application of tariff-setting principles by municipalities has not yet been carried out, anecdotal evidence suggests that price bureaus do not apply tests of efficiency and prudence of expenditure or engage in benchmarking utilities' performance.

4 Performance

Access

Access to improved water services has improved significantly in the last 15 years, reaching almost 97% in 2015 (UNICEF-WHO) from 78% in 2000. This is comparable to the rates of access achieved in Bangladesh, Malaysia, and Vietnam. Piped water coverage has increased rapidly in urban areas to 97% in 2015 from 48% in 1990, but around 240 million people, predominantly in rural areas, do not yet have access to piped water (Fig. 11.4).

Drinking water treatment capacity has also expanded, doubling between 1990 and 2015. Demand has increased more slowly, leading to a widening margin of capacity over demand (Fig. 11.5).

However, piped water is generally not potable either in urban or rural areas. Most of China's WTPs fail to meet the national drinking water standards (Tao and Xin 2014). Only a small number of WTPs in China use advanced technologies for water treatment.

Access to improved sanitation has also been extended rapidly since 2000, from 64% to almost 80%, a similar level to that of Indonesia and Vietnam (Fig. 11.6).

Investment

Public investment is channelled towards meeting government targets. China has invested heavily in the water sector for the last 15 years (Fig 11.7).

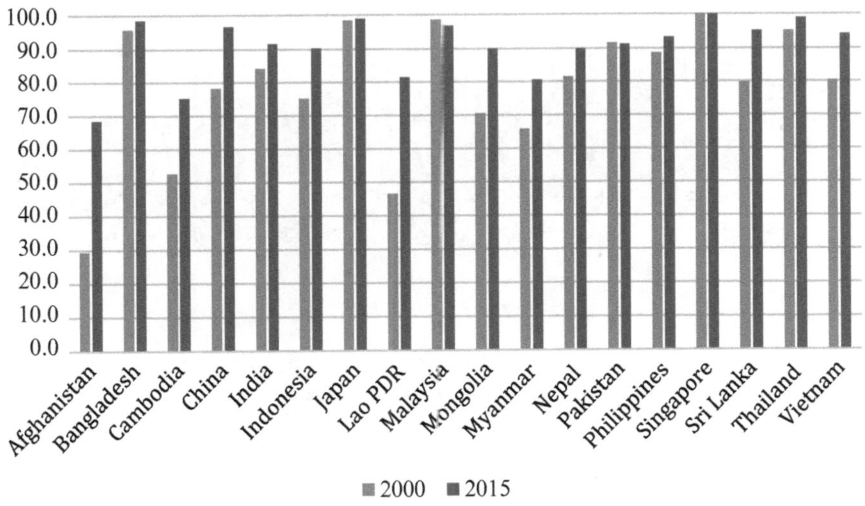

Fig. 11.4 Access to improved water services in Asian countries. Source: UNICEF-WHO Joint Monitoring Program (Downloaded 1 Dec 2017)

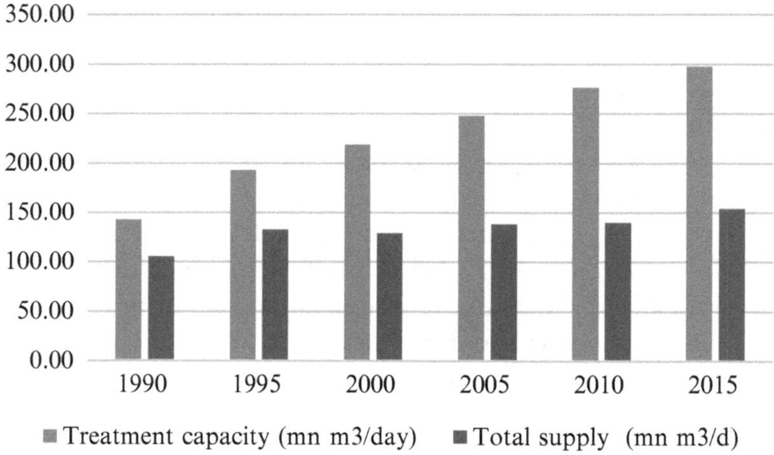

Fig. 11.5 Drinking water treatment capacity and supply. Source: MOHURD 2017

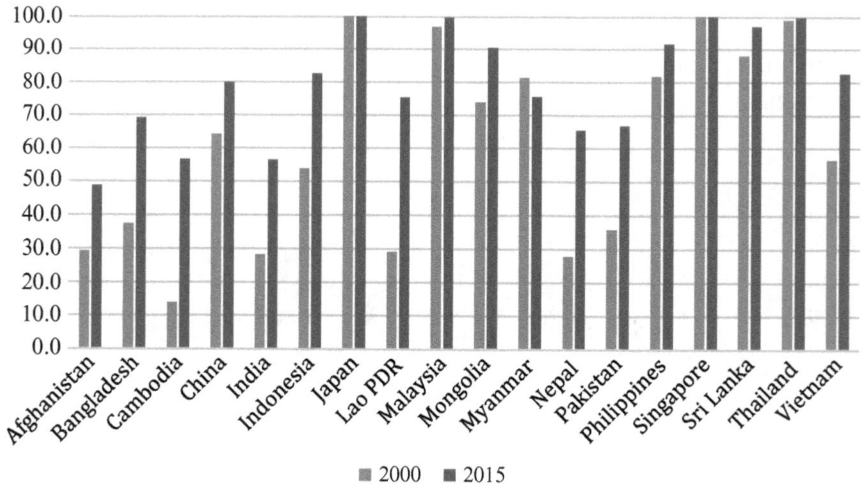

Fig. 11.6 Access to basic or safely managed sanitation in Asian countries. Source: UNICEF-WHO Joint Monitoring Program (Downloaded 1 Dec 2017)

During the 11th FYP (2006–2010), the total investment in fixed assets accounted for RMB 797 billion ($127 billion), and the annual investment growth on average remained above 30%. In terms of investment structure, most of the funding from 2006 to 2011 came from central and local governments and accounted for over 80% of the total investment (approximately 38% central and 45% local).

Investment from other sources has also been increasing, notably from the private sector. Investment in urban water-related projects during the 12th FYP (2011–2015) was estimated to be RMB 500–600 billion ($79.4–95.3 billion) (for all water-related projects including irrigation systems, the South-North Water Diversion Project, etc.). Out of the estimated RMB 500–600 billion ($79.4–95.3 billion) of urban water spending, municipal wastewater investment (including reuse) will account for RMB 380 billion ($60.3 billion) in 2011–2015. More than half of this is expected to come from private sources.

Efficiency

Data on utility performance are limited for China. Official data is collected at the provincial or municipal level rather than the utility level and is limited to a few indicators of treatment and supply. Provincial-level unaccounted-for water rates can be derived from the official data—these are presented in Fig. 11.8. They show a national average of 17% and a range of 8–31% suggesting that there may be considerable scope for improvements in operating efficiency. An industry association of water utilities has begun to collect data on a wider range of performance benchmarks from voluntarily participating utilities. In the future, these data may provide the basis for a full utility efficiency analysis (Fig. 11.8).

Local governments do not report separately on utilities' financial performance, but most water and wastewater utilities in China are thought not to be able to cover their operating costs.

Tariffs

Water tariffs in major cities increased up to 2007 but have levelled off in most places since then, leaving utilities dependent on transfers from local government to cover the gap (Global Water Intelligence 2016) (Fig. 11.9).

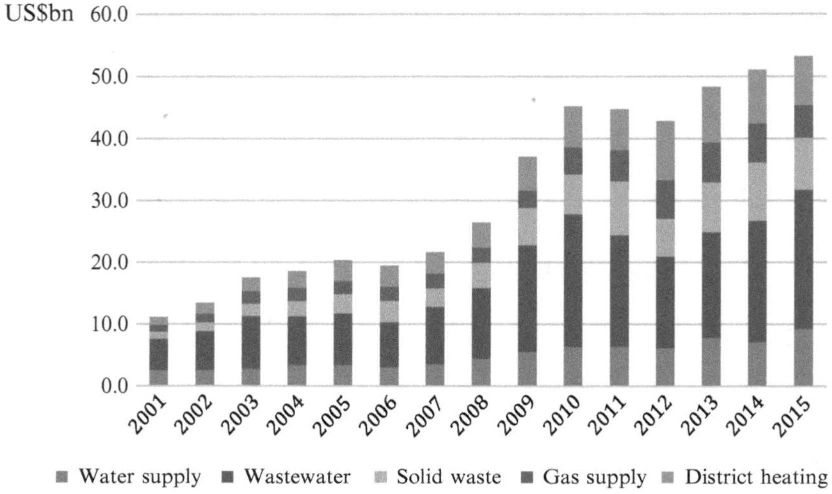

Fig. 11.7 National fixed asset investment in urban infrastructure. Source: MOHURD (Downloaded 31 August 2017)

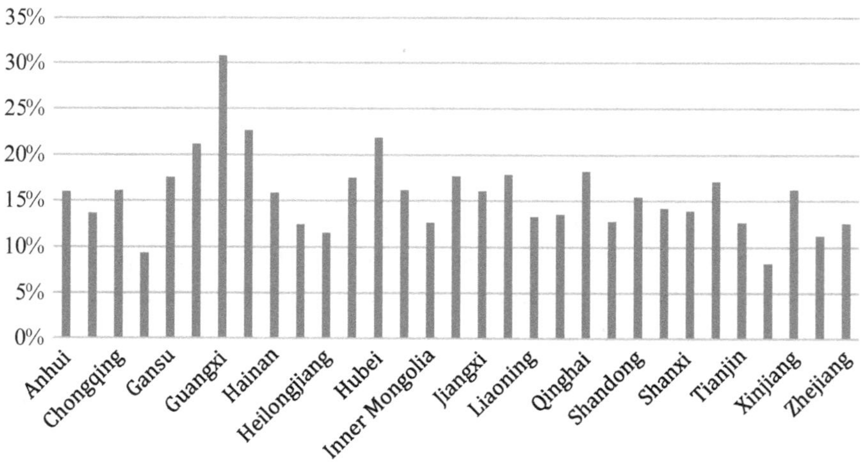

Fig. 11.8 Unaccounted-for water by province (2015). Source: MOHURD 2017

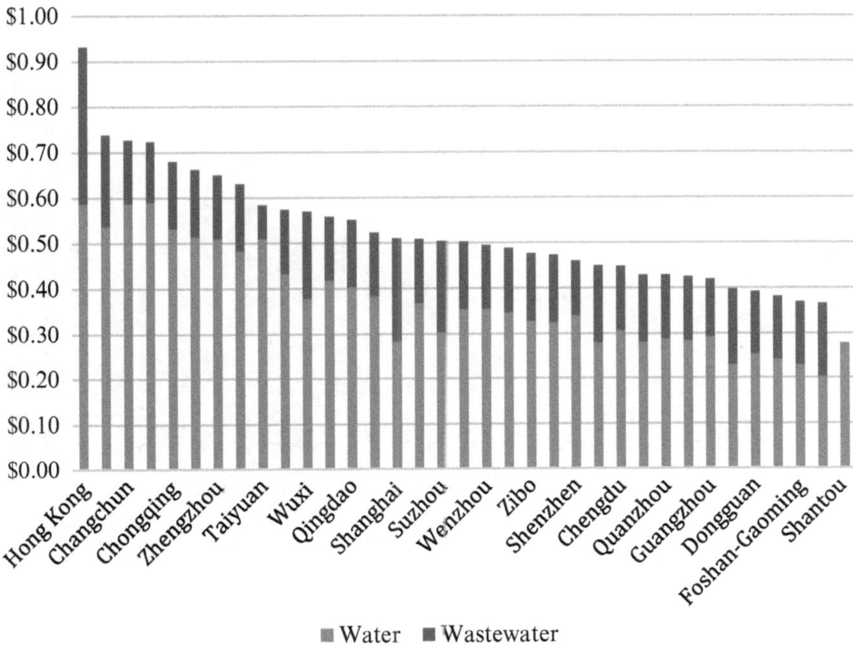

Fig. 11.9 Water and wastewater tariffs in selected cities (2017). Source: Global Water Intelligence (2017)

Environmental Performance

Wastewater collection and treatment have increased rapidly in the last decade. In 2015, there were 1944 municipal sewage treatment plants in China with a capacity to treat 140 million m^3/day and a treatment rate for municipal sewage of 91.9% (National Bureau of Statistics China 2017). This compares to a treatment rate of 82% in 2010 and 52% in 2005 (Fig. 11.10). Wastewater treatment in rural areas is much lower, around 60%, up from 14% in 2006. While this is a significant achievement, government authorities are concerned that much of the wastewater discharged by treatment plants does not meet the required standards. This suspicion is confirmed by persistently high levels of pollution in key river basins and lakes, although there has been a reduction in the number of nationally monitored water bodies in the lowest water quality category, as shown in Figs. 11.10 and 11.11.

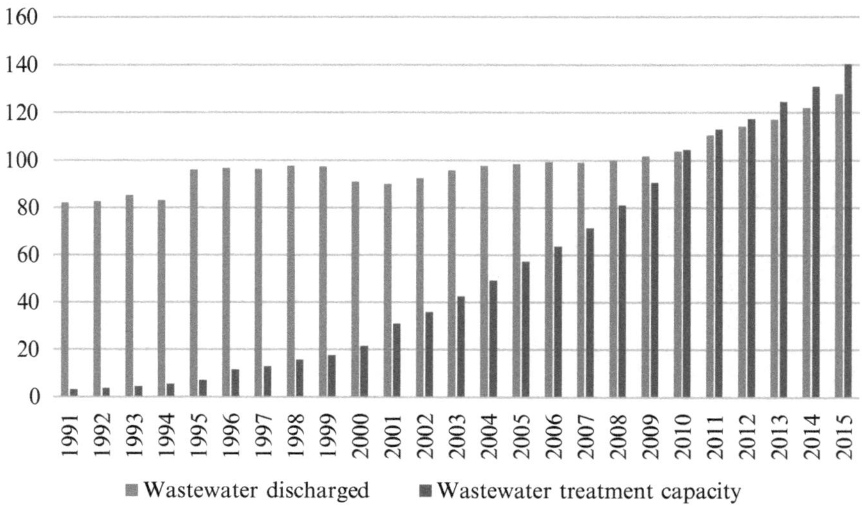

Fig. 11.10 Volume of wastewater discharged and treated (million m^3/day). Source: MOHURD (Downloaded 31 August 2017)

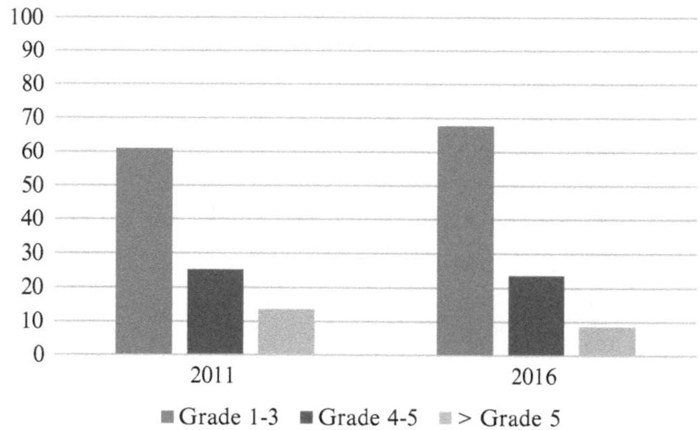

Fig. 11.11 Environmental water quality (National Surface Water Monitoring Program). Grade 1–3 water is suitable for drinking water after treatment, aquaculture, and recreation. Grade 4–5 water is suitable for industrial use, agriculture, and landscaping. Water below Grade 5 is not suitable for use. Source: MEP 2017 (Report on the State of the Environment in China 2016)

5 Discussion

China's achievement in expanding access to water services and raising wastewater treatment levels in the last 15 years has been remarkable. This has been achieved through a combination of public investment and active participation from the private sector. The central government has shown policy leadership, setting ambitious targets for lower tiers of government and allocating funds. It has used the performance assessment framework for mayors to provide strong incentives for compliance at the municipal level.

However, gaps remain in the governance framework. Firstly, little attention has been paid to efficiency incentives for utilities. There is no entity at any level of government in China that plays the role of an economic regulator in monitoring performance and providing efficiency incentives to utilities, and there are no incentives for utility managers to enhance efficiency (Li and Phillips 2017). The procedure employed by local price bureaus for the preparation of tariff recommendations does not require an assessment of efficiency or other economic analysis. Similarly, contracts with private providers usually contain a price adjustment clause based on the cost of inputs, with a guideline IRR of 8–12% set in national regulations, with no process or principles specified for the assessment of the efficiency of the operator's expenditure. However, the profit rates of listed investors in the sector are much higher.

The absence of economic regulation may account for the results found in empirical studies comparing PPPs with non-PPP water companies in China. Jiang and Zheng (2014) and Zheng and co-authors (2016) study the impact of private equity ownership on efficiency and performance outcome indicators for 200 urban utilities. They find a significant negative effect of PPP on employment size, managerial costs, and subsidies and a significant positive effect on profitability. They also find a significant effect on metering, but they are not able to find significant effects on operational efficiency measures like leakage rates or on coverage. Using a panel dataset of self-reported performance of a limited number of efficiency indicators, Li and Phillips (2017) find wide variation in efficiency levels between utilities and conclude that there is strong potential to raise utility efficiency in China through a regulatory framework that provides efficiency incentives to utility managers.

The power of local government to make the final decision on tariffs leaves further scope for arbitrary and politicised tariff adjustments. This suggests that water governance in China could be improved with clear guidance from the central government on the process and principles for efficiency-based tariff setting.

"Sunshine regulation" and benchmarking are impeded by the lack of comparable data for utilities. Benchmarking has great potential in China because of the large number of utilities operating under similar operating and governance conditions. Currently, however, utilities are not required to report on operating or financial performance indicators, and so comparison between them is not yet possible. The development of a national utility performance database would be of immediate value to local governments and could form the basis for nationwide benchmarking.

Incentives for investment efficiency have also been weak. Before 2014, local governments were not required to carry out economic viability assessments for proposed PPPs. Instead, they faced incentives to procure larger, gold-plated assets which contribute more to local investment figures. Public investment in wastewater collection networks lagged behind private investment in treatment plants. As a result, many plants are underutilised. Concerns about investment efficiency have been articulated by the national auditing body but an effective monitoring framework and incentive mechanism are yet to be established.

A second set of concerns relates to the growing role of SOEs in the sector. As noted above, the "private" partner in most PPPs in China is an SOE. This has an advantage: the lower cost of capital of SOEs compared to private firms should lower the total cost of building and operating infrastructure for local governments. However, it has potential disadvantages too. At the national level, SOE debt remains a contingent liability for government and therefore does not constitute a transfer of risk from the public to the private sector. Nor does it contribute to the government's stated objective of increasing the role of the market in the economy. At the level of the individual project, a concern is that SOEs belonging to higher tiers of government will be able to use their influence within the bureaucratic hierarchy to secure preferential treatment, raising their own profitability at the expense of local governments.

Thirdly, few mechanisms exist to encourage cooperation within the vertically and horizontally fragmented bureaucratic structure. River basin commissions have been partly successful, but their authority relies on the allocation of water resources to provinces set by the central government. Similarly, efforts to coordinate action across provinces to improve environmental quality in lakes and river basins have only been effective when the central government has intervened.

Finally, the level of transparency and scope for public participation in China's water sector is very limited. While public hearings are mandated as part of the tariff review process, these are not always held, the representatives selected to participate are restricted, and the level of information available to members of the public on performance is limited, making it impossible for the public to engage in the deliberations in a meaningful way. Strengthening the governance framework in these ways will help China to achieve the goals of greater water security and safe and sustainable access to water services.

References

Bin, L. (2014). Institutional Design Considerations for Water Rights Development in China. In D. Garrick, G. Anderson, D. Connell, & P. Bin (Eds.), *Federal Rivers: Managing Water in Multi-layered Political Systems*. Edward Elgar Publishing.

Braadbaart, O., Zhang, M., & Yang, W. (2009). Managing Urban Wastewater in China: A Survey of Build–Operate–Transfer Contracts. *Water and Environment Journal, 23*(1), 46–51.

Browder, G. J., Xie, S., Kim, Y., Gu, L., Fan, M., & Ehrhardt, D. (2007). *Stepping Up: Improving the Performance of China's Urban Water Utilities*. Washington, DC: World Bank.

Chen, C. (2009). Can the Pilot BOT Project Provide a Template for Future Projects? A Case Study of the Chengdu No. 6 Water Plant B Project. *International Journal of Project Management, 27*(6), 573–583.

Chen, C., & Messner, J. I. (2005). An Investigation of Chinese BOT Projects in Water Supply: A Comparative Perspective. *Construction Management and Economics, 23*(9), 913–925.

Dai, L. (2014). Exploring China's Approach to Implementing 'eco-Compensation' Schemes: The Lake Tai Watershed as Case Study Considered Through a Legal Lens. *Water International, 39*(5), 755–773.

Daniell, K. A., Coombes, P. J., & White, I. (2014). Politics of Innovation in Multi-level Water Governance Systems. *Journal of Hydrology, 519*, 2415–2435.

FAO (Food and Agriculture Organisation). (2017). Aquastat Database. Retrieved Februray 2, 2018, from www.fao.org/nr/water/aquastat/main/index.stm.

Global Water Intelligence. (2016). *Global Tariff Survey*. Retrieved from https://www.globalwaterintel.com/global-water-intelligence-magazine/tariff-survey/.

Global Water Intelligence. (2017). *Private Sector Participation in Water*. Oxford: Media Analytics.

Jiang, Y., & Zheng, X. (2014). Private Sector Participation and Performance of Urban Water Utilities in China. *Economic Development and Cultural Change, 63*(1), 155–189.

Lee, S. (2007). Public–Private Partnership in the Urban Water Sector of Shanghai. In F. Wu (Ed.), *China's Emerging Cities* (pp. 62–83). London: Routledge.

Lee, S. (2010). Development of Public Private Partnership (PPP) Projects in the Chinese Water Sector. *Water Resources Management, 24*(9), 1925–1945.

Li, F., & Phillips, M. A. (2017). The Influence of the Regulatory Environment on Chinese Urban Water Utilities. *Water Resources Management, 31*, 205. https://doi.org/10.1007/s11269-016-1519-z.

Li, L., Wu, X., & Tang, Y. (2017). Adoption of Increasing Block Tariffs (IBTs) Among Urban Water Utilities in Major Cities in China. *Urban Water Journal, 14*(7), 661–668. https://doi.org/10.1080/1573062X.2016.1236134.

Ministry of Enviroment Protection Government of China. (2011). Retrieved September 3, 2018, from http://english.mee.gov.cn/Resources/Plans/National_Fiveyear_Plan/201606/P020160601356854927248.pdf.

Ministry of Environment Protection, Government of China. (2014). Report on the State of the Envionment in China. Retrieved Februray 5, 2018, from english.sepa.gov.cn/Resources/Reports/soe/.

Ministry of Enviroment Protection Government of China. (2015). Report on the State of the Environment in China 2014. Retrieved February 5, 2018, from english.sepa.gov.cn/Resources/Reports/soe/.

National Bureau of Statistics China. (2017). *National Data*. Retrieved from data.stats.gov.cn.

Nitikin, D., Shen, C., Wang, Q., & Zou, H. (2012). Water Service Delivery Reform in China: Safeguarding the Interests of the Poor. *Annals of Economics and Finance, 13*(2), 463–487.
Pohlner, H. (2016). Institutional Change and the Political Economy of Water Megaprojects: China's South-North Water Transfer. *Global Environmental Change, 38*, 205–216.
State Council, Government of China. (2016). *Strategic Emerging Industry Plan and Guidance on the Development of Environmental Protection Services.*
Su, Y. (2014). Contemporary Legal Analysis of China's Transboundary Water Regimes: International Law in Practice. *Water International, 39*(5), 705–724.
Tao, T., & Xin, K. (2014). A Sustainable Plan for China's Drinking Water. *Nature, 511*(7511), 527.
Zheng, X., Jiang, Y., & Sugden, C. (2016). *People's Republic of China: Do Private Utilities Outperform State-Run Utilities?* Vol. 5. ADB East Asia Working Paper Series. ADB.
Zhong, L. J., & Mol, A. P. (2008). Participatory Environmental Governance in China: Public Hearings on Urban Water Tariff Setting. *Journal of Environmental Management, 88*(4), 899–913.
Zhiping, L. (2005). The Challenges of China's Discharge Permit System and Effective Solutions. *Temple Journal of Science, Technology & Environmental Law, 24*, 375.

12

US

From Excess to Entropy: Municipal Water Services Governance in the United States

Dan Zhao and A. Wren Montgomery

1 Introduction

In 2015 the World Economic Forum named "water crises" the largest risk to the global economy, ahead of disease and weapons of mass destruction. Water has also emerged into the global spotlight in recent years with businesses and industry associations recognizing the fundamental challenges growing pressures on water resources present (CDP and Deloitte 2011; KPMG 2012).

While long thought to be issues of the developing world, these crises have also manifested in the US in recent years, as several water crises and conflicts have made national and global headlines. First, there was the massive sell-off of municipal water systems following the Great Recession.

D. Zhao (✉)
Ross School of Business, University of Michigan, Ann Arbor, MI, USA
e-mail: danzhao@umich.edu

A. W. Montgomery
Ivey Business School, Western University, London, ON, Canada
e-mail: wmontgomery@ivey.ca

S. Porcher, S. Saussier (eds.), *Facing the Challenges of Water Governance*, Palgrave Studies in Water Governance: Policy and Practice,
https://doi.org/10.1007/978-3-319-98515-2_12

Then, there were the unprecedented droughts in California and Texas. More recently, the water shutoffs in Detroit and the water poisoning crisis in Toledo, Ohio, made the news. Finally, perhaps the most famous water calamity in decades, catastrophic lead poisoning crises in Flint, Michigan, and other US communities grabbed global headlines, placing water in the collective conscious like never before. These events speak to the urgency of better understanding the water governance landscape in the US, as well as examining emerging challenges including price, quality, funding, and environmental implications.

In this chapter, we begin in the second section with an overview of water services in the US including key issues at stake such as conflicts over governance forms, funding, price, and quality of water services. We then go on to present a summary of key water services regulatory mechanisms in the US including those that establish guidelines and standards for funding, quality, and price and follow this with a discussion of governance mechanisms and the emerging public versus private debate in the fourth section. Fifth, we provide a survey of the prominent extant empirical studies on water services in the US, including some of the latest developments in this literature. We conclude by discussing areas where future research may be needed and potential solutions for water services challenges in the US context.

2 Overview of US Water Services and Emerging Challenges

As with most OECD countries, public ownership of water services in the US has been entrenched in institutionalized systems of regulation for well over a century (Masten 2010; Pérard 2009). The US system developed as a "public health revolution" spread across developed economies from London's urban core, as local governments sought to address the outbreaks of waterborne disease and fire that were the result of fractured and unorganized water services (Solomon 2010, p. 255). Although a mix of ownership structures characterized nineteenth-century US cities, private ownership and control of local waterworks dominated. Yet, as systems were consolidated, publicly owned water services emerged in most US urban centers between the mid-1800s and early-1900s (Koeppel 2000) along with public delivery of healthcare and public education. Public water services soon became widely institutionalized and were underpinned by rules, regulatory structures, and normative expectations around the delivery of safe, clean, and low-cost water.

By 1896, all but 9 of 50 major American municipal water systems were publicly owned and operated (Jacobson and Tarr 1994). The fact that water and sewer systems became almost overwhelmingly publicly owned from the nineteenth century onwards, even in the face of large-scale privatization of other US public utilities, has been called an "aberrant" case (Priest 1993) and a "puzzle" (Masten 2010) in the utilities literature. The causes of this aberration have spawned a lively debate in regulatory economics and have been attributed to such factors as pricing disagreements on infrastructure extensions to less dense urban areas in private systems; the ability of private operators to hold cities "hostage" over rates; health and safety concerns with uncoordinated operations including the locations of water intakes and sewage outflows, as well as firefighting access; the relatively low costs and high efficiency of public provision; and the recognition of an important link between water security and economic growth (Beecher and Kalmbach 2013; Masten 2010).

This broad move to public ownership has remained relatively stable for the US, but private systems serving small communities and housing developments have become common in recent decades. Although the share of private water systems in the US increased from a low of 30% in the 1920s to roughly 50% by 2009, as illustrated in Fig. 12.1 (Pérard 2009), large municipal systems continued to be almost exclusively publicly owned throughout this period. Currently, only 10 to 20% of the US population is served by private systems (Pérard 2009), accounting for less than 10% of water sales (Beecher and Kalmbach 2013). However, this stable field appears to now be undergoing increased fluctuations and increased conflict. Food and Water Watch (FWW) estimates suggest that over 1400 US water systems have been privatized since 1991 (FWW 2010). As of 2010, as illustrated in Fig. 12.2, roughly 39 water systems serving over 11 million citizens were facing the possibility of privatization, with another 17 potential privatizations already suspended due to

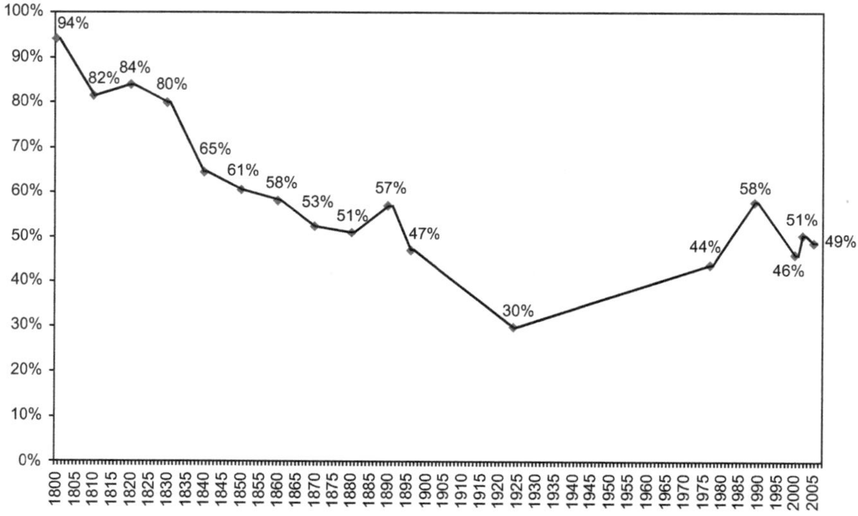

Fig. 12.1 Percentage of private waterworks in the US from 1800 (Pérard 2009)

public resistance (FWW 2010). In addition to increases in the aggregate number of privatized systems, these privatizations now include substantially larger municipal systems than ever before, with cities such as San Diego, Chicago, Indianapolis, Detroit, Miami, Tampa, St. Louis, New York, and Seattle all considering, or believed by activists to have considered, privatizing some or all of their water systems (FWW 2010; Qualitative Interviews).

In addition to these emerging conflicts and changes in governance and ownership of water services, the US has also begun to feel the effects of over a century of relative disregard for the health of both its water resources and infrastructure, both of which were long taken for granted, and, in many areas, their excess supported cheap industrial and agricultural production and rapid urban development and sprawl. However, years of neglect are now leading to entropy as US water quality becomes less guaranteed and infrastructure decays. For instance, the US Environmental Protection Agency (EPA) estimates an infrastructure funding gap for water and sewer services at $384 billion over the years to 2030 (EPA 2013). Indeed, it is widely understood that the US is facing a massive need for investment in a range of different types of infrastruc-

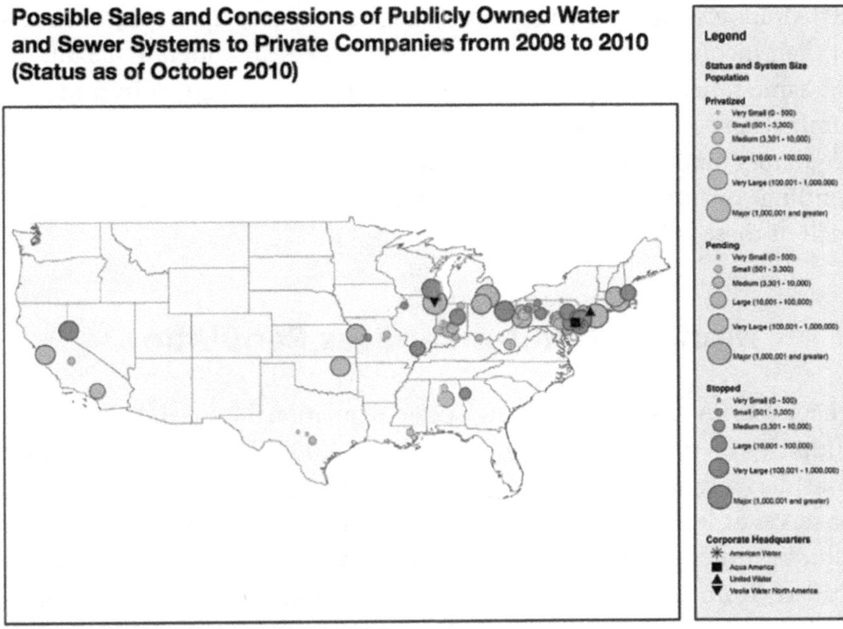

Fig. 12.2 Sales and concessions of US municipal waterworks 2008–2010 (FWW 2010)

ture. In 2017, the American Society of Civil Engineers gave US infrastructure a D+,[1] with drinking water systems receiving a D.[2] The American Water Works Association (AWWA 2011) estimates that maintaining US water service quality levels will cost an estimated $1 trillion over the next 25 years. A recent report prepared for the Department of Homeland Security notes of this funding gap, "The condition of our infrastructure seriously lags behind in an increasingly competitive global economy, but we have been unable to generate the overall public interest, support, and political will to reinvigorate it" (NIAC 2016, p. 3).

Aging infrastructure is causing water rates to soar with a recent study estimating a 41% increase in US rates since 2010 (Mack and Wrase

[1] ASCE. (n.d.). 2017 Infrastructure Report Card Website. Retrieved April 12, 2018, from https://www.infrastructurereportcard.org/.

[2] ASCE. (n.d.). 2017 Infrastructure Report Card. Retrieved April 12, 2018, from https://www.infrastructurereportcard.org/wp-content/uploads/2017/01/Drinking-Water-Final.pdf.

2017), increasingly driving poorer Americans to pay significant amounts of their income for water and wastewater services and often to default. At the same time, these aging lines are leading to increased quality and contamination concerns. These changes are beginning to challenge the extant regulatory and governance structures that long guaranteed adequate funding, quality water, and low-cost service in the US. We expand on each of these below.

3 Municipal Water Services Regulation

The US EPA oversees drinking water regulation in the US under the authority of the Safe Drinking Water Act (SDWA) of 1974. The EPA considers any water system that provides at least 15 service connections or serves at least 25 people for at least 60 days a year a public water system, regardless of its governance form. According to the data we downloaded from the EPA's Safe Drinking Water Information System (SDWIS), there are 152,060 such water systems nationwide in 2014. Based on its website, the EPA further classifies these 152,060 public water systems into three types.[3] Community water systems—that number 50,025—refer to public water systems that supply water to the same population year-round. Non-transient non-community water systems—that number 18,054—stand for public water systems that regularly supply water to at least 25 of the same people at least six months per year, like schools and hospitals. The remaining 83,981 are transient non-community water systems, that is, public water systems that provide water in a place such as a gas station or campground where people do not remain for long periods of time.

In this section, we focus on three critical aspects of the regulatory context for municipal water services in the US: funding structures, quality regulations, and price regulations. We also integrate case studies on crises in Detroit and Flint and data gathered in our research activities to illuminate our discussion and the growing challenges for municipal water service in the US context.

---

[3] U.S. EPA. (n.d.). Information about Public Water Systems. Retrieved March 1, 2018, from https://www.epa.gov/dwreginfo/information-about-public-water-systems.

National Water Infrastructure Funding

Based on the 1996 Amendments to the SDWA, the EPA sets up an EPA-states partnership called the Drinking Water State Revolving Fund (DWSRF) program to help water systems and states achieve the health protection objectives of the SDWA. The Congress provides funding for the DWSRF. EPA then awards capitalization grants to each state depending on feedback from the most recent Drinking Water Infrastructure Needs Survey and Assessment. After taking their set-asides, states combine their capitalization grant with the states' 20% match into a dedicated revolving loan fund to support water systems for eligible infrastructure projects. Through 2017, the state DWSRFs have provided more than $35.4 billion to water systems, of which over $19.1 billion comes from federal funding. Figure 12.3 illustrates the total national appropriation for the program from 1997 to 2017.[4] We see that there are two spikes in spending, in 1997 and 2010. Other than that, the annual spending has only grown very modestly over the years.

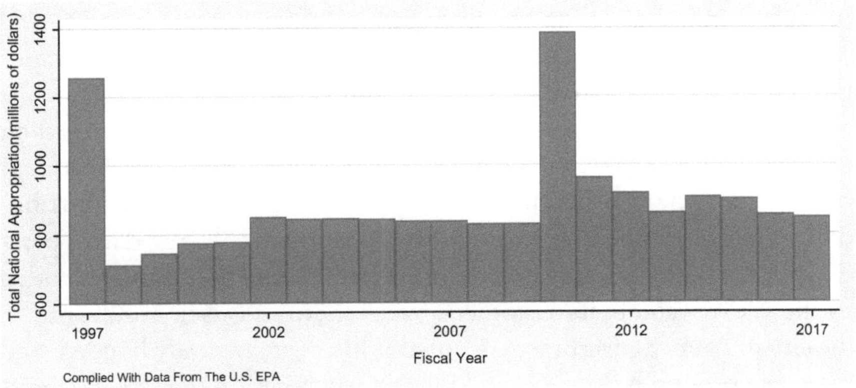

Fig. 12.3 Total national appropriation for Drinking Water State Revolving Fund 1997–2017

---

[4]U.S. EPA. (n.d.). Annual Allotment of Federal Funds for States, Tribes, and Territories. Retrieved May 2, 2018, from https://www.epa.gov/drinkingwatersrf/annual-allotment-federal-funds-states-tribes-and-territories.

In 2014, President Obama signed the Water Infrastructure Finance and Innovation Act (WIFIA), which established the WIFIA, a federal loan and guarantee program managed by EPA headquarters. This program has some interesting features like specifying a $20 million minimum project size for large communities. According to the Association of Metropolitan Water Agencies, the program provides a helpful supplement to the DWSRF as water infrastructure projects costing more than $20 million are generally too large to receive meaningful assistance through the DWSRF.[5] EPA invited 12 projects in 9 states to apply for over $2 billion in loans for WIFIA's inaugural round in 2017. In 2018, the EPA announced funding for $5.5 billion in loans to leverage over $11 billion through the WIFIA program. The program also received $63 million from the Consolidated Appropriations Act signed by President Trump in the same year,[6] which is a relatively insignificant number compared with other federal funds granted to water infrastructure.

Local Water Infrastructure Funding

As public information on the financing of water systems is scarce and varies drastically across jurisdictions, we base our discussion of local funding models on qualitative data gathered in two research projects in the state of Michigan, specifically in Detroit and Ann Arbor. In Ann Arbor, the potable water system, which includes the drinking water treatment plant, remote pumps stations and storage tanks, and the water distribution system, is funded by water rates. More specifically, the daily operations of the utility are operated by an enterprise fund that is solely funded by the sale of water to its customers, which means it gets no funding from the city or county government. Capital projects are typically bonded, and the rates cover both principal and interest payments on the capital investment. The water utility, however, is a part of the City of Ann Arbor gov-

[5] Association of Metropolitan Water Agencies. (n.d.). Water Infrastructure Investment & Financing. Retrieved January 21, 2018, from https://www.amwa.net/water-infrastructure-investment-financing.

[6] U.S. EPA. (2018, April 4). EPA Announces New Funding for Water Infrastructure Projects. Retrieved April 4, 2018, from https://www.epa.gov/newsreleases/epa-announces-new-funding-water-infrastructure-projects-0.

ernment structure, but again its funding source is separate from the rest of the city. State laws are in place to govern how water systems finance their operations in a very prescriptive fashion. The manager of the water system confirms that the model of funding for the Ann Arbor water system is typical among other publicly owned water systems.

The funding structure becomes more complicated when there is a change of governance form. One such case is the Detroit Water and Sewerage Department (DWSD). DWSD serves over 200,000 Detroit residential and commercial accounts that include a city population of nearly 680,000. In 2016, the City of Detroit leased regional water and sewer infrastructure, together with water and wastewater treatment facilities, to the Great Lakes Water Authority (GLWA) which serves 126 municipalities in 7 counties in southeast Michigan. The agreement between DWSD and GLWA includes a $50 million annual lease payment for 40 years to replace and restore DWSD's aging water and sewer system. The ownership of these assets still resides with DWSD.[7]

Similar to the Ann Arbor water systems, the DWSD is funded by rates and bond issuance. According to the 2016 Water Supply System Capital Improvement Program of DWSD, its capital financing strategy is designed to (a) recover the costs of capital investment over the useful lives of the capital assets, (b) minimize the impact of the capital programs on water and sewage rates, and (c) protect and enhance the department's financial position. This implies that water rates cover some of the capital investments. According to DWSD's 2012 Debt Management Policy, DWSD annually develops and adopts a five-year Capital Improvement Program for both the water system and the sewer system. The policy also states DWSD will aim to fund a portion of routine capital needs with revenue financed capital and that it will use its best efforts to utilize subsidized debt financing (i.e., State Revolving Fund loans as mentioned in the previous section) to the greatest extent feasible. This funding model can be confirmed by checking the financial statements of City of Detroit Water Fund (an enterprise fund), which shows the system is either funded by revenue or bond issuance.

[7] City of Detroit. (n.d.). Capital Improvement Program. Retrieved August 15, 2017, from http://www.detroitmi.gov/DWSDimproves.

Drinking Water Quality Regulation

Drinking water quality regulation in the US is based on the SDWA, the central federal law to ensure the quality of drinking water. Initially passed by Congress in 1974, the law was later amended in 1986 and 1996. It aims to protect public health by regulating the nation's public drinking water supply. The SDWA stipulates many required actions to protect drinking water and its sources: rivers, lakes, reservoirs, springs, and groundwater wells. States also publish their own SDWAs based on the federal SDWA. The SDWA requires the US EPA to set drinking water standards that public water systems must meet. Ninety contaminants now have official EPA standards. Under the SDWA, states that meet specific requirements, including setting regulations that are at least as stringent as US EPA's, can apply for primary enforcement authority, or primacy. The EPA also stipulates on a very detailed level the treatment, monitoring, and reporting rules required to meet the drinking water standards for systems using various types of filtration systems.

According to EPA's latest 2016 National Public Water Systems Compliance Snapshot, there are 150,438 public water systems in the country, among which 51,573 have at least one violation. Among these systems with violations, 11,938 have health-based issues, while 39,580 monitoring and reporting issues. However, only 4470 are serious violators, with small systems constituting the majority of these. Correspondingly, 30,478 enforcement actions were taken by the regulators.[8]

According to the EPA, the monitoring logistics are as follows: individual water systems collect samples of their water and submit them for laboratory testing to verify that the water they provide to the public meets all federal and state standards. How often and in what manner samples for what contaminants are taken vary according to numerous conditions and are detailed in EPA regulations. Monitoring schedules are contingent upon the type of contaminant, the type of source water used to produce drinking water, and the population served by the public water system, among other

[8] U.S. EPA. (n.d.). Providing Safe Drinking Water in America: National Public Water Systems Compliance Report. Retrieved March 23, 2018, from https://www.epa.gov/compliance/providing-safe-drinking-water-america-national-public-water-systems-compliance-report.

things. The laboratories that conduct the analyses are certified by states or the EPA. The EPA regulations specify the methods that must be used to analyze drinking water samples. On top of these routine, relatively high-frequency sample collection procedures, the EPA and state governments also conduct on-site visits to the water systems to make sure they are complying with all the regulations. Some on-site visits are triggered by previous violations of SDWA, while others are regularly scheduled. Sanitary surveys are a proactive type of on-site visits conducted by primacy agencies to provide a comprehensive overview of the water systems' capabilities to supply safe drinking water and educate them if necessary. Sanitary surveys are undertaken relatively infrequently, though, ranging from once every three years to five years.

Once violations are detected in any of these procedures, legal actions or compliance orders may ensue. Violations may be remedied by compliance/enforcement actions, such as improved filtration techniques or changes in procedures. Some examples of violations include maximum contaminant level (MCL) violations, treatment technique (TT) violations, failure to replace lead service lines, monitoring and reporting violations, and procedural violations. Among these, MCL and TT violations are the two major types of violations that deserve particular notice. MCL denotes the highest level of a contaminant that EPA allows in drinking water without posing either a short-term or long-term health risk. TT refers to a required process meant to reduce the level of a contaminant in drinking water. To ensure these violations, or the lack thereof, are communicated with the consumers served by the water systems, the EPA requires community water systems to deliver to the customers a Consumer Confidence Report, a form of annual drinking water quality report. If a severe violation is detected, the regulators may issue public notification requests to make sure the consumers are aware of water quality emergencies in a speedy manner. Consumers can also acquire information about public water systems and their violations of EPA's drinking water regulations through the SDWIS and Enforcement and Compliance History Online (ECHO) portal provided by the EPA. Despite the rigid regulations listed above, the SDWA specifies that states and the EPA may have the authority to grant "variances" that allow public water systems to use less costly technology and exemptions to allow public water systems more time to comply with a new drinking water regulation.

Regardless of the sophisticated water quality regulations in place, in the past several years, there have been multiple public health emergencies caused by lead poisoning in drinking water. The most famous one of these is the lead poisoning crisis in the water system of Flint, Michigan, that poisoned over 100,000 residents and led former President Obama to sign an emergency declaration, outlined in greater detail below.[9] In latest developments, large-scale lead poisoning has been found in hundreds of households in Chicago.[10] Lead has not been the only issue. The water crisis in Toledo, Ohio, in 2014 caused widespread panic,[11] and the mayor even compared it to 9/11. The cause was algae growth fed by phosphorous, which came mainly from agriculture and the excessive amount of fertilizer and animal manure that many farmers apply to their fields. Water quality problems can also be very costly financially. Toledo spent about $4 million in 2013 on maintaining water quality but could not prevent the 2014 crisis, and another $300 million in plant renovations is planned.[12]

Emerging Water Quality Challenges: The Case of Flint.[13] Flint, Michigan, the birthplace of General Motors (GM), was once a prosperous city. Yet despite decades of prosperity by the 1980s and 1990s, GM began shutting down plants in Flint. Unemployment rates began to increase, and Flint's population fell to 98,310 by 2015 down from almost 200,000 in the 1960s, as Flint was named one of the poorest cities in America with 40.1% of residents living in poverty.[14] On April 16, 2013, bankrupt,

[9] Fieldstadt E. (2016, January 16). Flint, Michigan, Water Crisis: Obama Signs Emergency Declaration. Retrieved January 17, 2016, from http://www.nbcnews.com/news/us-news/flint-michigan-water-crisis-obama-signs-emergency-declaration-n498001.

[10] Hawthorne M. & Reyes C. (2018, April 12). Brain-damaging lead found in tap water in hundreds of homes tested across Chicago, results show. Retrieved April 17, 2018, from http://www.chicagotribune.com/news/watchdog/ct-chicago-water-lead-contamination-20180411-htmlstory.html.

[11] CBS/AP. (2014, August 2). Toxic tap water causes state of emergency in Toledo. Retrieved August 2, 2014, from http://www.cbsnews.com/news/toledo-ohio-officials-warn-against-drinking-toxic-tap-water/.

[12] Troy T. (2014, August 19). Mayor says water crisis is similar to 9/11. Retrieved August 19, 2014, from http://www.toledoblade.com/Politics/2014/08/19/Mayor-says-water-crisis-is-similar-to-9-11.html.

[13] Adapted from Aquino, N. and Montgomery, A.W. (2018). "Water, Water Everywhere? The Flint Water Crisis" SAGE Business Cases.

[14] United States Census Bureau. (n.d.). Flint, Michigan Quick Facts. Retrieved September 1, 2016, from http://www.census.gov/quickfacts/table/PST045215/2629000.

under state-appointed emergency manager, and trying to save money, Flint decided it would stop buying water from the City of Detroit which had supplied Flint's municipal water services since 1967.[15] Upon hearing that Flint would be building its own pipeline, the DWSD informed Flint on April 17, 2013, that it would terminate its water supply services in one year.

As Flint's contract with the DWSD approached expiration, the city opted to use its local river water, the Flint River. To some local leaders, this was a historic moment for the City of Flint; however, many residents were worried. On April 25, 2014, Flint officially made the switch to the Flint River. It would later come to light that an employee of the Michigan Department of Environmental Quality (MDEQ) warned other MDEQ officials about serious health problems linked to Flint's river water prior to the switch.[16] Less than a month after the switch, residents started to complain about the water, including its color and odor. Residents were advised by authorities to boil their water.

One resident, LeeAnne Walters, reached out to Marc Edwards, a professor at Virginia Polytechnic Institute, in August 2015. Professor Edwards and his graduate students came to Flint just two days after speaking with Walters. Their results indicated high lead levels in Flint's water with 40.1% of the samples over 5 ppb and 16.7% of the samples over 15 ppb. Some samples even came back above 100 ppb, and one sample, collected after 45 seconds of flushing the hydrants, came back at 1000 ppb. Edwards issued a warning to Flint residents urging them to not drink the water. Yet even after these warnings, city officials insisted the water was safe to drink. Further warnings of lead contamination came from Dr. Mona Hanna-Attisha, a Flint pediatrician who had been seeing an increasing number of sick children. Dr. Hanna-Attisha began to compare patients' blood lead levels before and after the municipal water switch. The results were alarming and indicated that the number of chil-

[15] Carmody, S. (2013, April 16). Flint ending Detroit water contract, DWSD looking at its options. Retrieved September 1, 2016, from http://michiganradio.org/post/flint-ending-detroit-water-contract-dwsd-looking-its-options#stream/0.

[16] Ridley, G. (2016, February 12). Flint water switch was approved despite DEQ health warnings a year earlier. Retrieved September 1, 2016 from http://www.mlive.com/news/flint/index.ssf/2016/02/deq_official_warned_of_flint_r.html.

dren with lead poisoning in the city had almost doubled.[17] Dr. Hanna-Attisha urged city officials to stop using the Flint River water, as city officials responded that her numbers were incorrect. After approximately 18 months, officials finally admitted to the dangers of the Flint River water.

Emerging evidence now suggests that this crisis could have been easily averted. In February 2015, an EPA expert from Region 5 discovered the lead problems in Flint and asked the MDEQ if they were using phosphates for corrosion control, a roughly $200 per day treatment. An MDEQ official, one of the three officials charged for their involvement in the crisis, responded with the following statement: "The City of Flint … has an Optimized Corrosion Control Program [and] conducts quarterly Water Quality Parameter monitoring at 25 sites and has not had any unusual results." This was later found to be an outright lie.[18] Today, Flint has switched back to Detroit water and is still attempting to replace lead-lined pipes as well as treat the many young citizens suffering from the life-long effects of lead poisoning.

Drinking Water Price Regulation

Unlike the case for water quality, there isn't a central agency in the US that regulates and monitors the pricing of water systems. The EPA does issue guidelines to assist water and wastewater systems when designing rate structures. Specifically, EPA guidelines suggest that water and wastewater services combined should not exceed 4.5% of household income. The guidelines propose a number of areas of consideration when determining water rates. Affordability, especially for low-income households, is an essential point of concern. Nevertheless, the EPA encourages full cost pricing, which factors all costs into prices, including past and future, operations, maintenance, and capital costs. In case these costs do not suf-

[17] Gupta, S., Tinker, B., & Hume, T. (2016, January 22). Flint doctor's fight to expose lead poisoning. Retrieved August 25, 2016, from http://www.cnn.com/2016/01/21/health/flint-water-mona-hanna-attish/.

[18] Roy, S. (2015, September 30). COMMENTARY: MDEQ Mistakes and Deception Created the Flint Water Crisis. Retrieved August 26, 2016, from http://flintwaterstudy.org/2015/09/commentary-mdeq-mistakes-deception-flint-water-crisis/.

ficiently take into consideration the externalities generated by water use, it also emphasizes pricing structures that encourage conservation.[19]

There are, however, some private agencies that monitor and record water pricing. For example, non-profit water news service Circle of Blue gathers single-family residential water rate information from the websites of each of 30 major US cities' water utility or through phone calls and emails to the utilities. According to Circle of Blue's annual water pricing survey, the average monthly cost of water for a family of four using 100 gallons per person per day has gone from $44.39 in 2010 up to $68.14 in 2017.[20] AWWA has been producing a biennial survey of water and wastewater charges for US utilities since 2002.[21] This provides the most comprehensive survey of US water and wastewater rates available. The 2016 survey contains data for 264 utilities providing water service and 182 wastewater utilities from 42 states as well as the District of Columbia, serving from 1000 to 5 million customers. A recent study employing this data estimates that while 11.9% of US households currently find their water rates unaffordable (as per EPA guidelines), this number is likely to rise to 35.6% of the population if conservative estimates of rate increases over the next five years come to pass (Mack and Wrase 2017).

Emerging Water Access and Affordability Challenges: The Case of Detroit. Until 2015, Detroit's water and sewer services were owned and operated by the publicly held DWSD, one of the oldest and largest water utilities in the US. The DWSD provided water and sewer services to over 127 local communities across Wayne County and nearby counties and despite population declines remains the third largest system in the country serving almost 4 million people (DWSD 2015). While the DWSD is responsible for residential services and billing within the City of Detroit proper, where the shutoffs occurred in 2014, it provides only bulk water delivery and sewage disposal to the neighboring counties and communities who manage their own residential billing and local infrastructure and distribution systems.

---

[19] U.S. EPA. (n.d.). Pricing and Affordability of Water Services. Retrieved February 3, 2018, from https://www.epa.gov/sustainable-water-infrastructure/pricing-and-affordability-water-services.

[20] Circle of blue. (n.d.). The Price of Water. Retrieved April 7, 2018, from http://www.circleofblue.org/waterpricing/.

[21] AWWA. (n.d.). Water & Wastewater Rates. Retrieved April 7, 2018, from https://www.awwa.org/resources-tools/water-and-wastewater-utility-management/water-wastewater-rates.aspx.

In 2014, in the midst of Detroit's bankruptcy and facing financial pressures and past debts, the DWSD announced they would reinstate and begin to escalate water shutoffs in the city, sending tens of thousands of past-due and shutoff notices to residents. Although these shutoffs were not new for Detroiters, their scope and scale were. An outside contractor was hired to resume and escalate water shutoffs with the DWSD's own estimates indicating that over 130,000 households, or almost half of its residential customers, had accounts over $150 or 60 days' overdue and therefore subject to disconnection (DWSD 2014). By the end of May 2014, over 7000 households had been disconnected, a number which reached over 33,000 by the end of the year (DWSD 2015). With a population that averaged over a 40% poverty rate, well above the US average of 15.9%,[22] and water rates that at an average of $75 per month for a family of four were double the national average according to EPA data,[23] many of the city's residents faced frequent challenges paying their water and utility bills. However, the DWSD's services were the only ones targeted specifically in this push.[24]

Social justice, environmental, and labor groups had been protesting and speaking out on water affordability issues, water quality, and potential water privatization in Detroit for almost a decade individually as well as under a coalition termed the Detroit People's Water Board. With monthly meetings and the occasional rally, this board had had some early success on passing a water affordability plan through city council almost a decade earlier, although this had never been implemented. But, in the spring of 2014, with Detroit under the governance of an emergency manager as the largest city in the US ever to declare bankruptcy, several events and circumstances coincided to make the shutoffs, and the DWSD itself, a news headline around the world for the weeks and months to come. The shutoffs hit national and international news making headlines

[22] U.S. Census Bureau. (2015). State and County Quick Facts: Detroit, Michigan. Retrieved September 4, 2015, from http://quickfacts.census.gov/qfd/states/26/2622000.html.

[23] Clark, A. (2014, July 3). Going without water in Detroit. Retrieved July 4, 2014, from http://www.nytimes.com/2014/07/04/opinion/going-without-water-in-detroit.html?_r=0.

[24] The regional and privately owned DTE Energy Co. that delivered both gas and electric services to the area had more comprehensive bill monitoring and assistance programs in place and so did not have large numbers of overdue accounts.

from *The New York Times* to *The Economist* to *Al-Jazeera* and from Canada to China, as the United Nations declared Detroit to be in violation of its 2010 Human Right to Water and Sanitation.

4 Water Services Governance

Many of the discussions on the issues we have touched upon so far take place as part of a broader debate over the merits of different types of water governance forms. In light of the existence of a range of drinking water governance forms that entail both public and private participation, it would be tempting to view water governance forms in terms of a strict dichotomy between public and private. The reality, however, resembles a spectrum with purely public on one end and purely private on the other.

These drinking water governance forms are characterized broadly into three categories by Bakker (2003), all of which are present in the US. First is the "public utility" municipal model that was the dominant form of water provision over the past century. These systems are owned and primarily operated by governments, are generally non-profit, and often have subsidized rates. Second is the private sector "commercial" model under which systems are managed and sometimes owned by private companies. One common type in this category is private sector participation (also termed public-private partnerships). This mode, which has been gaining popularity, uses private companies to manage infrastructure owned by municipal governments. Due to the stigmatization of private governance form in certain contexts, publicly owned systems may use management contracts instead of direct privatization in order to avoid being seen as private. Private systems are typically for profit and apply a market-oriented pricing scheme. The third model, lying somewhere between the aforementioned two, is the community "cooperative" model which is an enterprise owned and controlled by the users of the services provided. This model often finds its existence in rural areas.

Smith's (2003) report on privatization of small water systems argues there are a number of other models involving public-private partnership. First, most publicly owned systems outsource certain parts of their services to private parties, such as engineering work to private engineering firms. A second model is Design/Build/Operate (DBO), in which the owner of a municipal system hires a private contractor to design, build, and operate a facility for a set time. Taking this a step further is Design/Build/Own/Operate/Finance. While in the DBO model the municipal government always owns and pays for the construction of the facility, this new model is applied when the owner (municipal government) doesn't have sufficient funds to fund the project and the operator funds the project from the revenue obtained. Finally, the asset lease model involves a public system owner leasing the system to a private party to operate it.

Even though a multitude of models have been discussed above, they are still not comprehensive. As the title implies, the models form a continuum rather than be discrete, and there may be special arrangements for every model that fit into particular circumstances at the time. From a broader perspective, both public and private parties play a significant yet different role in US water provision, and their cooperation has been extensive and creative.

The Public Versus Private Debate

As discussed in other chapters of this book, the appropriate, most efficient, and most politically acceptable governance form for water systems is a global issue. There was a notable trend toward privatization of water systems around the world, especially in developing countries, in the late twentieth century (Bakker 2010). In the US the commodity versus commons and public versus private arguments are becoming more salient as a variety of stakeholders debate the usage of increasingly scarce water resources ranging from the Ogallala Aquifer to the Colorado River. Related to this is the growing debate over water privatization as corporations buy land and their attached water rights or invest in previously publicly owned water systems, such as in Atlanta, Indianapolis, or Detroit. Several activist groups are actively tracking and organizing to protest such sales.

Following the Great Recession of 2009, this debate became much more prominent as many municipal governments, especially those in the US Rust Belt, considered privatizing their water systems to cope with fiscal challenges (FWW 2010). The financial conundrum of these municipal governments and the subsequent sell-off of their water systems shed light on the debate over the choice of governance forms. Advocates of privatization often argue it will drive down costs and increase efficiency, but social activist groups take a very different view. FWW (2010) claims that privatization amounts to a very expensive loan and estimates that privatization added 64%, or $153 per annum, to the water bill of a typical New Jersey household.

As privatization has gained attention, municipalization of water systems has also become more common (FWW 2012). In fact, according to the data collected from the SDWIS, the number of people served by private systems dropped by 16% between 2007 and 2011, while those served by publicly owned systems rose by 8%. Underlying these numbers is the high frequency with which local governments have been acquiring private systems. For example, from 1998 to 2010, 379 privately owned and operated water systems were consolidated with government-owned water systems in Georgia. Activists contend that one of the three primary reasons why communities take public control of water and sewer systems is to improve water quality and service (FWW 2012).

5 Extant Literature: US Water Services

Despite attracting growing attention in the public arena, water supply is understudied in the business literature, which is lagging behind in studying the business implications of a water-constrained world and the stakeholder and institutional forces that businesses face around water (Kurland and Zell 2010). Moreover, there is little research exploring how alternative approaches to managing water, for example, public versus private ownership, perform. As for the economics literature, although there is a substantial discussion on privatization, it has focused on cost, pricing, and efficiency (Bel and Warner 2008). Scant attention has been paid to the issue of quality of service and in particular compliance with quality standards among water systems of different governance forms. In this section we review the relatively nascent literature on the governance of US water services.

Governance Form and Efficiency

As outlined above, Masten (2010) points out that water and sewage are the only public utilities that are predominantly publicly owned in the US. He lays out several possible causes of the historical municipalization of water and assesses them using data from the turn of the nineteenth century. He finds that what Wilcox (1911) calls the "administrative simplicity" of early water supply systems made them less vulnerable to poor management by government bureaucrats and that the contentiousness of any attempts to raise water prices made private ownership less attractive. Troesken and Geddes (2003) use a dataset on municipalization of private water companies in the US from 1897 to 1915 to support a transaction cost theory for public acquisition. They conclude that since governments cannot credibly commit to eschew opportunistic expropriation from private companies once they have made investments, private companies have weak incentives to make investments, and governments accordingly may need to acquire private water systems in order to induce appropriate investments.

As for the consequences of water privatization, Bel and Warner (2008) conduct a review of all published econometric studies of water and waste production since 1970. They find little support for a connection between privatization and cost savings. For example, no evidence of cost savings exists in water delivery, and savings are not systematic in waste.

In general, studies on the efficiency of different water governance forms are beleaguered by the inadequate data on drinking water pricing. Since there is no federal government agency that collects water rates data across the nation, water rates often need to be acquired by either contacting the water systems and communities directly or using data on a very limited number of water systems collected by Circle of Blue and AWWA mentioned above. Unfortunately, it is not clear if this situation is going to improve in the immediate future.

Governance Form and Quality

The studies on the quality effect of privatization or municipalization of water systems seem to reach substantially different conclusions. In a rare study of the impact of privatization on service quality, Troesken (2001) finds African Americans to be a major beneficiary of water municipalization, as it decreases the spread of waterborne disease in these communities. However the effect is much smaller on whites suggesting that public water companies provided blacks with better service than did private water companies.

Konisky and Teodoro (2015) develop a political theory of the compliance behavior of US public versus private water systems, arguing that regulators may be politically unable to impose penalties on publicly owned systems that are as high as those imposed on private systems. However, their theory ignores the fact that private systems have high-powered incentives to cut costs that can undermine quality. They test their theory using water system data in the US on violations from 2010 to 2013 and claim that publicly owned systems incur more violations than private ones.

Wallsten and Kosec (2008) use a system-level panel from 1997 to 2003 to test the effects of ownership and benchmark competition on violations of the SDWA in the US. They conclude that compliance is not affected by system ownership, although the detailed findings indicate a much more complex pattern. Most of the literature discussed above, even though adding to our understanding of water governance on multiple fronts, are limited in their ability to convincingly prove causality due to the lack of control for endogeneity of ownership forms. Granted, ownership forms are intrinsically endogenous and can be ranked as one of the most endogenous variables in the industrial organization field of economics. Many of the previous papers also suffer from relatively short time spans in the data used. A short time span means that the results could be driven by idiosyncratic events happening during the observed period. It also means that we may not observe long-term trends in the performance of different governance forms.

In recent developments, Lyon et al. (2018) present the first large-scale empirical study of the ownership-service quality relationship that examines both privatization and municipalization in the US drinking water supply. Using a panel of 65,025 water systems from 2006 to 2014, and combining matching and difference-in-differences methods to construct comparable counterfactual control groups, they find privatization consistently produces quality improvements, but municipalization does not. Leveraging the different financing structures of public versus private water systems, the authors also find that privatization is more beneficial in communities with more government financing challenges as represented by higher government borrowing costs.

Information Disclosure and Quality

Finally, another stream of literature involves the impact of information on the provision of water quality in the US. Bennear and Olmstead (2008) study the 1996 Amendments to the SDWA, which stipulates that water systems issue annual Consumer Confidence Reports to their customers and that systems serving over 10,000 people mail the reports directly to consumers. They conclude that the requirement to mail the reports directly to households decreased violations by water suppliers in Massachusetts by about one-fifth of a violation per year. To the contrary, Johnson (2003) uses an experiment in which US customers read different versions of a water quality report (Consumer Confidence Report) to show that, in contrast to what many expected, the content of the reports per se probably makes little difference to consumers. This finding suggests that consumers of water are not always paying attention to water quality, even when information is provided directly to them. Montgomery et al. (2018) observe some heterogeneity in the role attention plays in drinking water quality. Using local educational attainment as a proxy for stakeholder attention, they find that publicly owned systems' compliance with SDWA regulations improves more rapidly with stakeholder attention. Together, these papers show the importance of consumer attention to water quality information.

6 Concluding Remarks

As we summarize above, attitudes and approaches to water resources and municipal water services in the US have changed markedly over the past decade. From seeing water as a cheap and virtually infinite resource to be exploited and overused to mounting challenges ranging from infrastructure decay to quality and affordability crises in once-affluent cities, this landscape has changed markedly.

Although several past and new studies attempt to quantify these issues, data limitations make these tasks both challenging and complex areas for future research. Yet, these are also areas where research insight is timely and important. The existing literature on US water services suggests that governance form, and changes to governance form, of water systems may have substantial impacts on costs, productivity, prices, and quality. These impacts, however, may depend in subtle ways on details of the empirical setting, such as the scale of operation and demographic factors such as race or consumer attention to service quality. In examining studies across nations, we may also appreciate the important role institutional backgrounds play in shaping the effectiveness of a governance form. Future research that can better understand these factors, and which are key to ensuring effective water governance, is urgently needed. In addition, one major area where future studies may make a substantial contribution to the literature and practice is on water rates. As we outline, data on water rates is largely insufficient and incomplete in the US. As a result, a large-scale empirical study that collects nationwide water rates would benefit our understanding of the efficiency implications of governance form, regulatory regime, and tracking and reacting to growing affordability challenges. Practitioner partners might also participate in these efforts through centralized tracking and monitoring of rates data, perhaps through existing SDWA and EPA structures.

Thoroughly understanding the merits of both publicly and privately owned water governance in the US context, as well as the varied impacts of the unique regulatory and institutional environment, is extremely important in improving the future performance of water services. Such quality and affordability improvements are essential for American communities that are becoming increasingly vulnerable due to socioeconomic factors, while also protecting, preserving, and restoring the US's wealth of natural water resources.

References

AWWA. (2011). *Buried No Longer: Confronting America's Water Infrastructure Challenge*. Denver, CO: American Water Works Association.

Bakker, K. (2003). *Water: Commons or Commodity?* University of British Columbia, Alternatives. Spring Issue.

Bakker, K. (2010). *Privatizing Water: Governance Failure and the World's Urban Water Crisis*. Cornell University Press.

Beecher, J. A., & Kalmbach, J. A. (2013). Structure, Regulation, and Pricing of Water in the United States: A Study of the Great Lakes Region. *Utilities Policy, 24*, 32–47.

Bel, G., & Warner, M. (2008). Does Privatization of Solid Waste and Water Services Reduce Costs? A Review of Empirical Studies. *Resources, Conservation and Recycling, 52*(12), 1337–1348.

Bennear, L. S., & Olmstead, S. M. (2008). The Impacts of the "Right to Know": Information Disclosure and the Violation of Drinking Water Standards. *Journal of Environmental Economics and Management, 56*(2), 117–130.

CDP and Deloitte. (2011). *CDP Water Disclosure Global Report 2011*. New York: Carbon Disclosure Project.

DWSD. (2014). *Finance Committee Binder*. Detroit, MI: Detroit Water and Sewerage Department.

DWSD. (2015). *Finance Committee Binder*. Detroit, MI: Detroit Water and Sewerage Department.

EPA. (2013). *Providing Safe Drinking Water in America: 2013 National Public Water Systems Compliance Report*. Washington, DC: U.S. Environmental Protection Agency.

FWW. (2010). *Trends in Water Privatization*. Washington, DC: Food and Water Watch.

FWW. (2012). *Municipalization Guide*. Washington, DC: Food and Water Watch.

Jacobson, C. D., & Tarr, J. A. (1994). The Development of Water Works in the United States. *Rassegna: Themes in Architecture, 57*, 37–41.

Johnson, B. B. (2003). Do Reports on Drinking Water Quality Affect Customers' Concerns? Experiments in Report Content. *Risk Analysis, 23*(5), 985–998.

Koeppel, G. T. (2000). *Water for Gotham: A History*. Princeton, NJ: Princeton University Press.

Konisky, D. M., & Teodoro, M. P. (2015). When Governments Regulate Governments. *American Journal of Political Science, 60*(3), 559–574.

KPMG. (2012). *Expect the Unexpected: Building Business Value in a Changing World*. Switzerland: KPMG International.

Kurland, N. B., & Zell, D. (2010). Water and Business: A Taxonomy and Review of the Research. *Organization and Environment, 23*(3), 316–353.

Lyon, T. P., Montgomery, A. W., & Zhao, D. (2018). *Do For-Profit Firms Degrade Quality?: Evidence from Drinking Water Systems in the U.S.* University of Michigan Working Paper.

Mack, E. A., & Wrase, S. (2017). A Burgeoning Crisis? A Nationwide Assessment of the Geography of Water Affordability in the United States. *PLoS One, 12*(1), e0169488.

Masten, S. E. (2010). Public Utility Ownership in 19th-Century America: The "Aberrant" Case of Water. *The Journal of Law, Economics, and Organization, 27*(3), 604–654.

Montgomery, A. W., Lyon, T. P., & Zhao, D. (2018). Not a Drop to Drink? Drinking Water Quality, System Ownership, and Stakeholder Attention. *Research in the Sociology of Organizations, 56*(7), 207–245.

NIAC. (2016). *Water Sector Resilience, June 2016 Draft*. Washington, DC: National Infrastructure Advisory Council.

Pérard, E. (2009). Water Supply: Public or Private? An Approach Based on Cost of Funds, Transaction Costs, Efficiency and Political Costs. *Policy and Society, 27*(3), 193–219.

Priest, G. L. (1993). The Origins of Utility Regulation and the "Theories of Regulation" Debate. *The Journal of Law and Economics, 36*(1, Part 2), 289–323.

Smith, H. J. (2003). *Privatization of Small Water Systems*. National Rural Water Association.

Solomon, S. (2010). *Water: The Epic Struggle for Wealth, Power, and Civilization*. New York: Harper Perennial.

Troesken, W. (2001). Race, Disease, and the Provision of Water in American Cities, 1889–1921. *The Journal of Economic History, 61*(3), 750–776.

Troesken, W., & Geddes, R. (2003). Municipalizing American Waterworks, 1897–1915. *Journal of Law, Economics, and Organization, 19*(2), 373–400.

Wallsten, S., & Kosec, K. (2008). The Effects of Ownership and Benchmark Competition: An Empirical Analysis of U.S. Water Systems. *International Journal of Industrial Organization, 26*, 186–205.

Wilcox, D. F. (1911). *Municipal Franchises: A Description of the Terms and Conditions upon Which Private Corporations Enjoy Special Privileges in the Streets of American Cities* (Vol. 2). Gervaise Press.

World Economic Forum. (2015). *The Global Risks Report 2015* (11th ed.). Geneva: World Economic Forum.

13

Japan
Recent Policy Changes in the Japanese Water Supply Industries

Takuya Urakami

1 Introduction

Cleanliness and the abundant supply of inexpensive water are the philosophies behind the water supply business in Japan and clearly stated in Article 1 of the Waterworks Act. In fact, the Japanese water industries have succeeded in reaching their goals by supplying high-quality water services to the entire population at a low cost. However, the industry is facing a transitional period reflecting changes in populations, natural disasters, the ageing water pipelines and facilities, and difficulties associated with inheriting legacy technologies. The Ministry of Health, Labour and Welfare (MHLW), which is the regulatory agency for the water supply businesses in Japan, is aiming to revise the Waterworks Act to strengthen the foundations of the water utility management. In September 2015, the MHLW convened a Study Meeting Strengthening the Foundations of the Water Supply Businesses (Suidojigyo Kiban Kyoka

T. Urakami (✉)
Faculty of Business Administration, Kindai University, Higashi-osaka, Japan
e-mail: urakami@bus.kindai.ac.jp

S. Porcher, S. Saussier (eds.), *Facing the Challenges of Water Governance*, Palgrave Studies in Water Governance: Policy and Practice,
https://doi.org/10.1007/978-3-319-98515-2_13

Hosaku Kentokai), discussing in detail methods to strengthen the management base of water supply business, along with a report due towards the end of January 2016. There were two key conclusions from the Study Meeting which described in a final report that (a) further wide-area consolidation should be promoted within relatively smaller water utilities and (b) public-private partnership (PPP) could be an appropriate tool for sustainable water supply business. In a response to the final report submitted by the Study Meeting, the Committee on Living Environment and Water Supply Health Science Council (Kosei Kagaku Shingikai Seikatsu Kankyo Suido Bukai) of the MHLW was held in March 2016 which agreed to establish a Special Committee on the Maintenance and the Improvement of Water Supply Businesses (Suidojigyo No Iji Kojo Ni Kansuru Senmon Iinkai). Furthermore, since the meeting, the Special Committee have discussed matters to be included in the revised Waterworks Act and compiled a final report entitled 'Measures to be Taken to Strengthen the Foundation of the Water Supply Business to Support the People's Living' (Kokumin Seikatsu Wo Sasaeru Suidojigyo No Kiban Kyoka Tou Ni Mukete Kouzubeki Sesaku Ni Tsuite). Finally, the government submitted a bill to the House of Representatives to amend part of the Waterworks Act in March 2017. Unfortunately, the bill did not go through in the ordinary Diet session of spring in 2017 and 2018 but will be submitted again after the extraordinary Diet session to be held in autumn 2018.

The objectives of this chapter are to provide an overview of the past and current Japanese water industries and to examine the course of reform of the water supply businesses in Japan.

The chapter is structured as follows. Firstly, a historical perspective in the development of the Japanese water supply businesses is provided to clarify the regulatory framework and its market structure. This will provide a basis for discussions focused on the financing and pricing systems, allowing for a clearer picture of the direction of regulatory reform of water industry to be determined. Finally, the chapter will summarize the main findings.

2 History of the Japanese Water Industry

The first modern water supply system (WSS) was introduced in Yokohama in 1887 with a role of preventing waterborne infectious diseases and fire prevention. It was for these reasons that initial WSSs were constructed near trade ports towards the end of the nineteenth century. For example, the WSS in Hakodate was constructed in 1889, Nagasaki in 1891, Osaka in 1895, Tokyo in 1898, Hiroshima in 1899, and Kobe in 1900. In the twentieth century, the numbers of WSSs increased dramatically and especially during the period of economic growth (the numbers of WSSs totalled 383 in 1950 and 1896 in 1980). Therefore, it is said that Japan realized its goal of supplying water services to its entire population. Figure 13.1 shows the growth in the population of Japan as well as water-supplied population from the 1930s to the current day along with the calculated coverage ratio. Japan is made up of many islands the majority of which are covered by forests. Therefore, it can be said that 97.7%[1] of

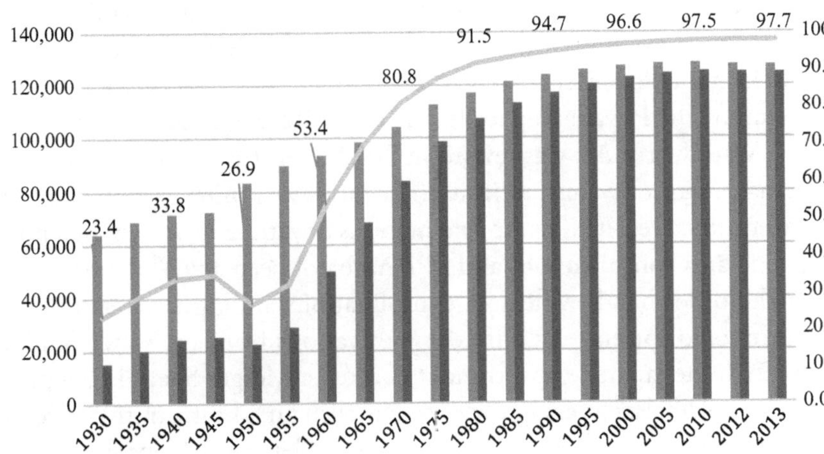

Fig. 13.1 Total population, water-supplied population, and coverage ratio (in thousands). Source: JWWA

[1] The Statistics of Waterworks, Japan Waterworks Association.

water supply penetration ratio represents a high level of water customers being serviced.

Along with the spread of WSS has come the development of legal systems. The first law pertaining to the water supply business was the Water Supply Ordinance (Suido Jorei) enacted in 1890. Under this law, the management of the water supply business was limited to municipalities such as cities, towns, and villages. After the end of World War II, a new Waterworks Act was enacted in 1957 and the idea of ownership was changed slightly. WSSs were in principle still managed by municipalities, but their ownership had expanded to include prefectures and partial affairs associations, and if municipalities admitted it, private entry had become possible.

During the period of economic and population growth, the need of broadening WSS was at the forefront of discussions because of the proliferation of water shortages, increases in water charges, and associated construction costs resulting in a shortfall of sourcing quality water. The Japanese government undertook various initiatives to tackle these issues. A national subsidy system for wide-area water facilities was established in 1967, and the provision of the wide-area water supply improvement plan (Koikiteki Suido Seibi Keikaku) was added to the Waterworks Act in 1977. Furthermore, the Private Finance Initiative (PFI) law was enacted in 1999, and the Waterworks Act was revised in 2002 to entrust the operation and the management of water purification facilities to a third party including private entities. Despite the implementations of various governmental policies, wide-area consolidations and PPP have not been actively pursued.

On the other hand, wide-area consolidations amongst water utilities were promoted for entirely different motivations from the water policy. Since 1999, the Japanese government enacted a Merger Special Law promoting municipal mergers in order to establish an administrative base of the basic municipalities (referred to as the Great Heisei-Era Consolidation). As a result, many WSSs have been consolidated due to the municipal mergers.

At the beginning of the twenty-first century, Japan's water supply business faced a major turning point. As depicted in Fig. 13.1, the total population of Japan peaked in 2008 but has been decreasing thereafter. In 2004, the MHLW announced the Water Supply Vision aimed to provide comprehensive and concrete measures to realize goals, such as safety, stability, sustainability, and environmentally friendly water supply system, but decided to revise this because of major social structural changes, formulating a New Water Supply Vision in 2013. There are three concepts underlying this vision—safety, sustainability, and resilience. Safety relates to supplying everyone with sufficient water of the highest quality anytime and anywhere. Sustainability relates to maintaining a sound and stable financial management system despite declines in population and water demand. Resilience relates to minimizing damages caused by natural disasters and restoring water utilities as soon as possible. The ambitions of the MHLW are for these goals to be delivered over the next 50–100 years and jointly accepted by all the stakeholders including governments, water utilities, residents, private companies, and academicians/researchers.

3 Regulatory and Market Structure

The MHLW is the regulatory authority of the water supply business in Japan. The roles of the MHLW are:

- Governance of the Waterworks Act
- Issuing business licences
- Establishment and revision of water standards for facilities, drinking water, and water service installations
- Supervisory guidance and site inspections
- Financial assistance
- Research and development of technologies

The Waterworks Act specifies drinking water quality standards for facilities, the application for a water utility licence and its approval, the operation of a water utility, supervision of water utilities, and the government's financial assistance along with other rules. The details of the Waterworks Act are highlighted in Appendix. Before a water supply business or bulk water supply business is established, water utilities have to obtain a business licence from the MHLW or prefecture governor. The MHLW approves business licences for water utilities supplying populations greater than 50,000,[2] whereas the prefecture governor approves business licences for other smaller water utilities.

Under the Waterworks Act, WSSs are categorized into four types, namely, large water suppliers (LWSs), small water suppliers (SWSs), small private water suppliers (SPWs), and bulk water suppliers (BWSs). LWSs are defined as utilities supplying water to a population of over 5001, and SWSs are defined as supplying water to populations of between 101 and 5000. Typically, SPWs are found in buildings equipped with holding water tanks with a capacity of more than 10 m^3 and can receive potable water from large or small water supply systems. By contrast, BWS supplies potable water to large or small water supply systems but not to end users. BWS plays an important role in developing water resources and supplying purified water to both LWSs and SWSs where their own water resources are insufficient to meet demand. Table 13.1 shows the number of municipality, water utility, and licence.

As already mentioned, water utilities are basically owned and operated by municipalities. It is also worth noting that water utilities sometimes have multiple licences due to mergers between municipalities which have been promoted since 1999 (Great Heisei-Era Consolidation), but such licences were not consolidated. The reasons behind this may be due to the existence of a service level gap and/or price level gap among merged utilities. Therefore, the number of licences granted exceeds that of the number of water utilities. Table 13.2 shows the number of LWSs.

Of particular note is that more than 80% of water utilities are operators supplying populations of less than 100,000. Moreover, the total number of LWS has decreased during the sample period, especially between 2005 and 2006, due to the Great Heisei-Era Consolidation of municipalities. The rate of decrease in the number of LWSs is particularly noticeable in relatively smaller LWSs. Although municipal mergers have resulted in slight increases in the size of water supply businesses, many small businesses still exist.

[2] Strictly speaking, the MHLW approves only water supply business that uses surface water as the main water source. For example, Kumamoto City, which depends 100% of water source on underground water, has a population exceeding 600,000 but has been approved by Kumamoto Prefecture.

Table 13.1 The number of municipality, utility, and licence

	Municipality			Utility			Licence				
								LWS		SWS	
	City	Town	Village	BWS	LWS	SWS	BWS	Public	Private	Public	Others
2001	675	1981	562	86	1905	1648	111	1946	10	7487	1303
2002	677	1957	552	87	1898	1644	111	1946	10	7415	1184
2003	695	1872	533	86	1870	1587	109	1926	10	7267	1093
2004	739	1317	339	85	1651	1232	107	1802	9	7029	1039
2005	779	844	197	79	1346	909	102	1593	9	6802	992
2006	782	823	195	79	1330	891	102	1563	9	6653	977
2007	783	811	193	79	1326	872	102	1547	9	6483	930
2008	783	799	191	78	1317	849	101	1510	9	6276	876
2009	786	757	184	77	1289	809	101	1456	9	6045	841
2010	786	750	184	74	1284	794	98	1434	9	5874	813
2011	787	748	184	73	1281	780	95	1420	9	5672	783
2012	789	746	184	73	1281	769	95	1405	9	5494	763
2013	790	746	183	73	1279	760	95	1392	9	5356	749
2014	790	745	183	72	1276	749	94	1379	9	5166	724

Source: The Ministry of Internal Affairs and Communications, JWWA

Table 13.2 Number of LWS with respect to water supplied population size

Water supplied population (in thousands)	2001	2002	2003	2004	2005	2006	2007	2008	2009	2010	2011	2012	2013	2014
>1000	13	13	14	14	14	14	14	14	14	14	14	15	15	14
500~1000	9	9	9	9	10	10	10	10	11	11	11	11	11	12
250~500	52	52	51	53	57	57	57	57	56	56	56	56	56	56
100~250	131	132	128	133	133	134	134	136	140	143	145	144	143	144
50~100	188	193	200	204	208	206	209	211	206	213	208	209	208	206
10~50	924	917	896	826	715	709	701	677	655	630	629	622	616	611
5~10	490	494	491	429	347	333	323	307	277	273	267	261	252	243
<5	136	135	139	132	107	100	99	100	101	97	96	93	96	98
Under construction	13	11	8	11	11	9	9	7	5	3	3	3	4	4
Total	1956	1956	1936	1811	1602	1572	1556	1519	1465	1440	1429	1414	1401	1388

Source: JWWA

Table 13.3 Annual amount of delivered water (10^6 m^3)

	Delivered water volume				
		Effective water volume			Ineffective water volume
			Revenue water volume	Non-revenue water volume	
LWS	14,567	13,441	13,042	399	1076
SWS	608	463	463	NA	145
PWS	29	20	20	NA	9
Total	15,154	13,925	13,526	399	1229

Source: JWWA

Table 13.3 shows an annual amount of water delivered with respect to WSS categories. LWS supplies more than 96% of the total water delivered to the consumers.[3] Surprisingly, the water loss ratio, defined as inefficiencies of water volume per delivered water volume, is about 7.4%, which is the lowest in the developed world, for example, the water loss ratio exceeds 20% in the UK, France, Spain, and Italy.

4 Financing and Pricing

Financial Situation

The total operating revenue of the LWS in Japan in FY 2014 was 2745 billion JPY. Figure 13.2 shows that 82% of the revenues comes from water supplies and 9% comes from operating subsidies. In principle, subsidizing the operational costs of the water supply business is not permitted; however, subsidies from general accounting are being introduced due to local governmental welfare policies and/or to lower the water prices in places where it is deemed to be high.

The operating expenditure of the LWS over the same time period was 2601 billion JPY leaving a surplus of 144 billion JPY. At first glance, the water supply business seems profitable, where preferential policies such as subsidies for construction costs and exemption from taxes and duties may

[3] All empirical analysis using data of the water service business in Japan adopts only LWS samples. This is because financial and operational data are available only for LWS not for SWS and PWS.

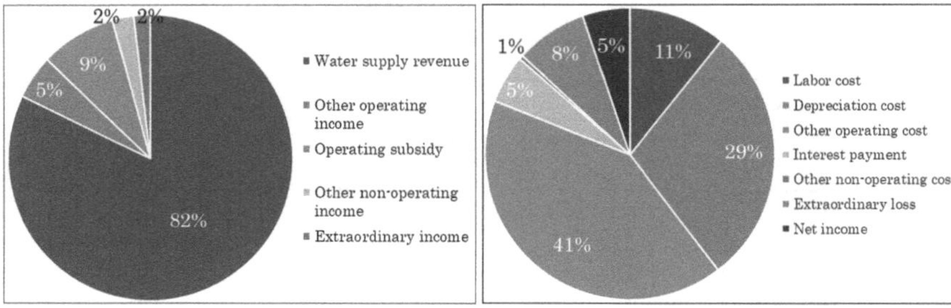

Fig. 13.2 Operating revenue and expenditure (FY 2014). Source: JWWA

influence the underlying results. However, it is of concern that the necessary capital investment (e.g., renewal investment of aged pipeline) has yet to be undertaken. The MHLW (2016) reveals that the rate of pipeline renewals has decreased year-on-year (from 1.54% in 2001 to 0.76% in 2014), despite the ageing pipeline infrastructure. A possible reason for this lack of infrastructure renewal is the fact that water utilities are publicly owned, and therefore, investments have been suppressed with the aim of keeping water tariffs lower.

Returning to Fig. 13.2, there is a large proportion of other operating costs. Purified water has to be purchased from BWS if water utilities do not have sufficient water sources to meet demand. Urakami and Parker (2011) stated that purchased water costs are a major factor in the high cost structure of water supply businesses.

Figure 13.3 shows capital income and capital expenditure. The total capital expenditure of LWS was 1403 billion JPY in FY 2014 with a total capital income of 435 billion JPY. This equates to a negative total balance of 968 billion JPY, with the shortfall being met with cash from internal profit and loss accounts such as accumulated depreciation and retained earnings.

According to Fig. 13.3, 57% of the capital income is corporate bonds and 21% is subsidies. The Japan Water Works Association (JWWA), a public interest incorporated association, whose aim is the spread of water supply and development of water supply technologies in Japan, reported that subsidies have been decreasing—the national subsidy was 191 billion JPY in 1995 but has fallen to 48 billion JPY in 2016. Furthermore,

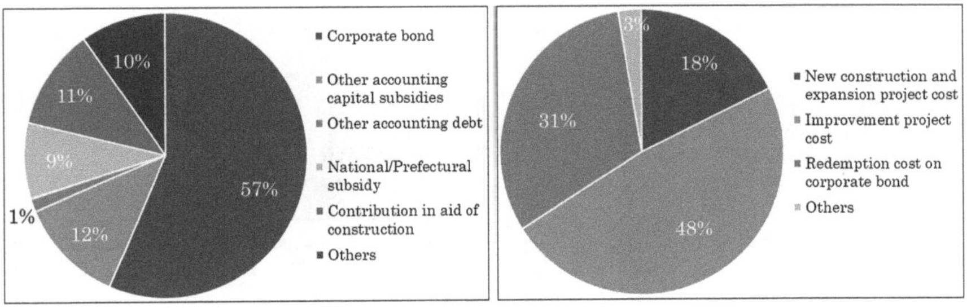

Fig. 13.3 Capital income and expenditure (FY 2014). Source: JWWA

the government and municipalities will face further financial pressures in the future because of potential declines in the population, and so it is expected that the number of subsidies will also fall.

Turning to the capital expenditure, 48% of capital expenditure are investment costs for equipment renewal, and 31% are expenditure expenses for corporate bond redemption. If income from subsidies will decrease in the future, the proportion of corporate bonds in capital income and the proportion of corporate bond redemption in capital expenditure will increase in the future. To make matters worse, ageing facilities are a serious concern facing the water supply businesses, and it is expected that investment costs to renew such infrastructure will increase further in the future.

Pricing System

The Local Public Enterprises Law and the Waterworks Act stipulate that water tariffs should be set based on fair cost incurred as a result of efficient management. Tariff levels and structures, however, are not defined by these laws, but guidelines for calculating tariffs are implied by the JWWA. The full cost pricing is adopted and calculated based on the formulas shown in Fig. 13.4.

Water tariff levels should be set to a level that equates to the full costs, being defined as the sum of the operating and capital costs. The capital costs are also defined as the sum of interest payments and asset maintenance costs. The JWWA proposes that asset maintenance costs are calcu-

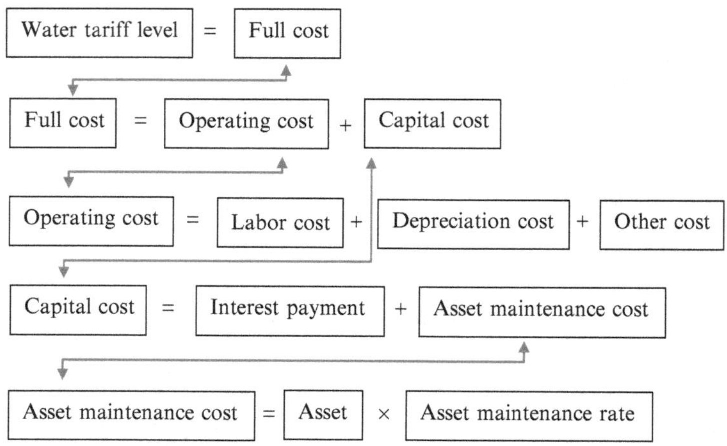

Fig. 13.4 Full cost pricing. Source: JWWA

lated by multiplying the assets by a certain rate, referred to here as an asset maintenance rate. The JWWA proposes that the asset maintenance rate should be 3% after conducting detailed simulations. Although regional council approval is required to alter water tariff levels and pricing structures, water utilities can themselves in principle exclusively set them. Due to this, the asset maintenance rate is also set arbitrarily, and sometimes there is either no rate or the rate is set lower than the standard. The JWWA conducted a questionnaire survey about pricing system for LWSs and got responses from 201 LWSs (JWWA 2015). A total of 67 LWSs answered that they included asset maintenance costs in their total costs. Furthermore, 44 LWSs set the asset retention rate of which 13 were those with an asset maintenance rate of 3% or higher and 30 were those with asset maintenance rate less than 3%. In response to this result, the MHLW believes that water utilities do not properly calculate the cost of setting water tariffs ordering them to include asset maintenance costs appropriately.

Table 13.4 shows an average cost and average price of LWS. The table shows that relatively smaller LWSs have higher average costs and they are unable to recover such costs by water revenue alone. This situation is expected to continue in the long term, and the MHLW is considering promoting wide-area consolidation and PPP as measures to strengthen the management base of small to medium LWSs.

Table 13.4 Average cost and average price

Water-supplied population (in thousands)	Average cost (JPY/m^3)	Average price (JPY/m^3)	Average price/ average cost
>1000	177.7	182.7	1.028
500~1000	142.2	152.6	1.073
250~500	149.5	165.7	1.108
100~250	154.8	164.3	1.061
50~100	164.0	173.0	1.055
30~50	172.3	173.2	1.005
20~30	170.6	173.0	1.014
10~20	173.6	172.9	0.996
5~10	206.1	192.2	0.933
<0.5	262.5	155.8	0.594
Total	165.1	172.3	1.044

Source: JWWA

5 Wide-Area Cooperation and PPP

Wide-Area Cooperation in Water Supply Business

Historically, wide-area cooperation in the water supply business began in 1919 when Edogawa Water Supply Authority (Edogawa Josui Choson Kumiai) was established in Tokyo. The first prefectural water utility was founded in Kanagawa Prefecture in 1936. After that, a national subsidy system was established in 1967 in order to address a shortage of water sources and to improve the vulnerability in the management of the small-scale water supply business. In 1977, the Waterworks Act was revised, and the Wide-Area Water Supply Development Plan (Koikiteki Suido Seibi Keikaku) was established. Water utilities established for the purpose of wide-area cooperation numbered 194 in 2001 (Development Bank of Japan 2015).

The MHLW announced the Water Supply Vision in 2004 and decided to promote wide-area cooperation in order to strengthen the management and the technology base of water utilities. In 2008, the MHLW set out 'Guidelines for Investigation of Wide-Area Cooperation of Water Supply Business' (Suido Koikika No Tebiki) and extended the definition of wide-area cooperation of water supply business as shown in Fig. 13.5. Consolidation (Jigyo Togo) refers to many water supply businesses that merge to become a single operator. Integration of management (Keiei No Ittaika) relates to a

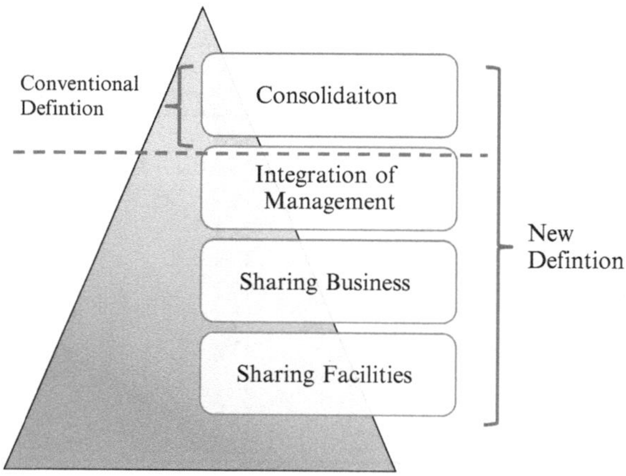

Fig. 13.5 New concept of wide-area cooperation. Source: JWWA

single operator having multiple licences. Sharing business (Kanri No Ittaika) relates to a collaborative implementation of maintenance work. Sharing facilities (Shisetsu No Kyodoka) means sharing facilities with other water utilities. Traditionally, wide-area cooperation in the water industry meant that numerous water supply businesses merged to become single entities. However, the new definition of this took into consideration partial or full integration of the management of many water supply businesses.

The numbers of water supply businesses implementing wide-area cooperation between 1989 and 2013 are shown in Table 13.5. There are 17 examples of integration of which 6 are vertical and 11 are horizontal. There is a single case of integration of management and 10 are cases of sharing facilities. Twenty-four cases are under consideration for wide-area cooperation, but it is expected that wide-area cooperation itself is still not promoted between the utilities.

Public-Private Partnership

PPP in water supply business meant the outsourcing of supplementary tasks such as meter reading, tariff collection, and the operational management of water purification plants during the nights and over holidays. The definition of PPP was stipulated by the PFI law enacted in 1999 and

Table 13.5 Number of incrementations of wide-area cooperation

	Completed	Under consideration
Consolidation		
Vertical	6	6
Horizontal	11	11
Vertical and horizontal	–	1
Integration of management	1	2
Sharing business	–	1
Sharing facilities	10	–
Other	–	3
Total	28	24

Source: MHLW

led to the revision of the Waterworks Act in 2002, entrusting the operation and management of water purification facilities to a third party. A designated administrator system was introduced in 2003 where concessions were introduced by a revision of the PFI law in 2011. The MHLW classifies PPP depending on the level of involvement of private business operators (Fig. 13.6).

Eleven cases involving PFI have been introduced in the water business since 2014. As for concessions, the Cabinet Office of Japan established 'PPP/PFI Promotion Action Plan' in 2013 (revised in 2017) and decided to focus on promoting the use of concessions. In 2017, Veolia Japan was selected as a concession operator of a sewage treatment plant based in Hamamatsu City. Although there are limited cases available where concessions have been introduced, the possibilities applicable to many fields are currently being studied: The Cabinet Office of Japan has announced that the concession project is reviewing seven cases at the airport, two cases at the water, four cases at the sewerage, and one case on the road (the Cabinet Office web site).

6 Direction of the Reform

As stated earlier, the MHLW is revising the Waterworks Act in order to strengthen the management foundations of water utilities to address the issues facing the industry such as population declines, frequently occur-

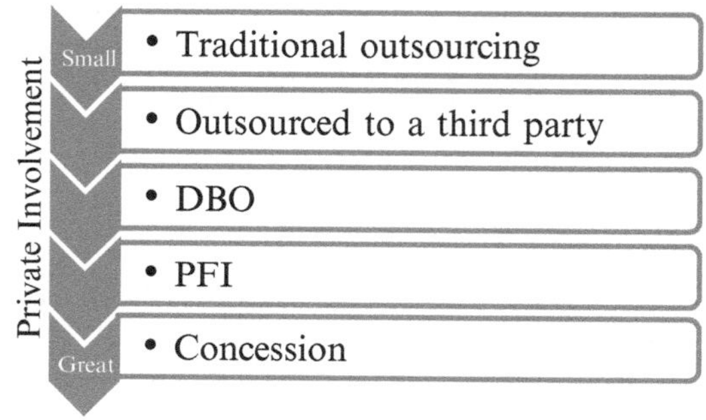

Fig. 13.6 Public-private partnerships. Source: MHLW

ring natural disasters (such as earthquakes, typhoons, and heavy rains[4]), the ageing water pipelines and facilities, and the difficulties associated with inheriting legacy technologies all of which are recognized since the arrival of the twenty-first century. Therefore, the MHLW established a Special Committee on the Maintenance and the Improvement of Water Supply Business in March 2016 and compiled a final report. There are several issues regarding the direction of the reform claimed in the report.

The Era of Maintenance and Management

The Waterworks Act was enacted in the era of new construction and the expansion of water facilities. Recently, the drinking water coverage ratio exceeds 97% so that maintenance and renewal investment become more important than the construction of new facilities. Due to declines in the population and subsequent reduced water demand in the long term, water facilities must be appropriately downsized in the future. Therefore, the relevant provisions of the Waterworks Act must be revised to address the current era of maintenance and management.

[4] Sudden and heavy downpour events are termed in Japan as 'Guerrilla heavy rainfall'.

Clarification of Roles of National and Prefectural Governments

As alluded to in this chapter, the water supply business is facing some difficult aspects, especially so in the situations where there are numerous small water supply businesses. Here, the situation has become severe due to several factors which include higher cost structures and higher water charges, the lack of engineers, and the inability to deal with future risks. The MHLW anticipates that prefectures will play a role of coordinating small water supply businesses and promoting wide-area cooperation in addition to PPP between small water supply businesses and private sectors. Therefore, the MHLW is considering establishing 'Basic Policy (Kihon Hoshin)' in the revised Waterworks Act to address the issues facing the industry and to establish 'a Council (Kyogikai)' in prefectures to discuss wide-area cooperation and PPP.

Creating Fixed Assets Register

Although water supply businesses expanded rapidly during the period of high economic growth in Japan, it is evident that the fixed assets register is not sufficiently updated by some water utilities. The fixed assets register is indispensable for planning facility renewals and also when downsizing facilities in the future. Therefore, the MHLW is contemplating whether to enforce all water utilities to ensure the fixed assets register is updated.

Appropriate Water Tariff

Water tariffs should be determined on proper costs and applied under efficient management as stipulated in the Waterworks Act. However, many water utilities have set water tariffs as low as possible since cleanliness and the abundant supply of inexpensive water are clearly stated in Article 1 of the Act as the original mission of the water supply business. As already shown in Table 13.4, water utilities are no longer able to cover costs from water revenues alone. To make matters worse, expenses for future renewal investment are not accounted for in the current water charging structures. While national subsidies and payments from general accounts are expected to decrease in the future, the MHLW proposes an appropriate cost calculation to be applied to water tariffs, paying attention to fairness and the burdening of rising costs to future generations.

Concession

PPP is not merely a means of cost reduction but is considered as one of the important strategies for ensuring the water supply businesses are fit for the future. The Cabinet Office of Japan established 'PPP/PFI Promotion Action Plan' and decided to focus on promoting the use of concessions. The MHLW plans to include concessions as part of the revisions of the Waterworks Act.

7 Summary

The water supply business in Japan has overcome geographical and topographical conditions realizing drinking water coverage ratios exceeding 97%, as well as meeting two original goals of securing public health and preventing fire. However, since the arrival of the twenty-first century, the water supply business has faced various challenges. The population of Japan peaked in 2008 and has continued to decline thereafter and is expected to fall below 90 million in 2060. The decline in population will lead to a decline in water demand resulting in reduced water revenues for water businesses. Furthermore, water supply facilities developed during the period of high economic growth period now need to be renewed. The facility infrastructure is ageing and renewal investments will increase rapidly. Moreover, the frequency of earthquakes as well as heavy rains has intensified in recent years. Therefore, costs for earthquake resistance and preparation for natural disasters are expected to increase in the future.

In principle, the water utilities in Japan are operated by municipalities leading to many small entities, with more than 80% of water utility operators supplying areas with populations of less than 100,000. It will be challenging for such small water utilities to independently tackle the various problems mentioned above. Therefore, the MHLW, the regulator of water industry in Japan, is considering amending the Waterworks Act to strengthen the business foundation of the water supply business.

No one knows what kind of future the water supply business will face due to changes in the social structure. Water supplies are a social infrastructure, indispensable to our daily lives. Not only government but also all citizens must understand and consider ways to overcome difficulties.

Acknowledgement This work was supported by JSPS Grant-in-Aid for Scientific Research(C) Grant Number 24530284.

Appendix: Waterworks Act

Waterworks Act was promulgated on June 15, 1957, and came into effect on December 14, 1957. After that, the Act has been amended several times. The items of the Act are as follows (the MHLW web site):

Chapter 1 General Provisions (Article 1 to 5)

Article 1 Purpose of This Act
Article 2 Responsibility
Article 3 Definitions of the Terms
Article 4 Water Quality Standards
Article 5 Facility Standards

Chapter 1.2 Wide-Area Water Supply Reinforcement Program (Article 5.2)

Chapter 2 Water Supply Business

Section 1 Business License, etc. (Article 6 to 13)
Section 2 Services (Article 14 to 25)
Section 3 Designated Water Service Installation Works Contractor (Article 25.2 to Article 25.11)
Section 4 Designated Examining Body (Article 25.12 to Article 25.27)

Chapter 3 Bulk Water Supply Business (Article 26 to 31)
Chapter 4 Private Water Supply System (Article 32 to 34)
Chapter 4.2 Private Water Supply Facility (Article 34.2 to Article 34.4)
Chapter 5 Supervision (Article 35 to 39)
Chapter 6 Miscellaneous provisions (Article 40 to 50.3)
Chapter 7 Penal Provisions (Article 51 to 57)
Supplementary Provisions

References

Development Bank of Japan. (2015). Current Situation and Issues of Water Supply Business in Japan (Wagakuni Suidojigyosya No Genjo To Kadai). (in Japanese).

Japan Waterworks Association. (2015). Survey Results on Water Tariff System in Japan (Suido Ryokin Seido Ni Kansuru Chosakekka Ni Tsuite). (in Japanese).

The Cabinet Office, Government of Japan, Concession Business (Concession Jigyo). Retrieved August 12, 2017, from http://www8.cao.go.jp/pfi/concession/concession_index.html (in Japanese).

The Ministry of Health, Labour and Welfare. (2016). Measures to Be Taken to Strengthen the Foundation of the Water Supply Business to Support the People's Living' (Kokumin Seikatsu Wo Sasaeru Suidojigyo No Kiban Kyoka Tou Ni Mukete Kouzubeki Sesaku Ni Tsuite). (in Japanese).

The Ministry of Health, Labour and Welfare, Water Supply in Japan. Retrieved June 22, 2017, from http://www.mhlw.go.jp/english/policy/health/water_supply/menu.html.

Urakami, T., & Parker, D. (2011). The Effects of Consolidation Amongst Japanese Water Utilities: A Hedonic Cost Function Analysis. *Urban Studies, 48*(13), 2807–2827.

FSC
www.fsc.org
MIX
Papier | Fördert gute Waldnutzung
FSC® C083411

Kolibri 360 GmbH
Ferdinand-Jühlke-Straße 7
99095 Erfurt, Deutschland
produktsicherheit@kolibri360.de